WAR OF THE WOLF

TEXAS' MEMORIAL SUBMARINE:
WORLD WAR II'S FAMOUS *USS SEAWOLF*

STEPHEN L. MOORE

© Stephen L. Moore 2008

Manufactured in the United States of America
All rights reserved

10 9 8 7 6 5 4 3 2 1

No part of this book may be reproduced or utilized in any form or by any means, electronic, or mechanical, including photocopying, recording, or by any information storage and retrieval system without permission in writing. All events, persons, locations, personal accounts and organizations are based on true accounts relayed to the author by the individuals who experienced them. Any resemblance to any other experience is unintended, and entirely coincidental. Inquiries or requests for permission to reproduce material from this work should be sent to:

Atriad Press LLC
13820 Methuen Green
Dallas, Texas 75240

The paper used in this book meets the minimum requirements of the American National Standard for Permanence of Paper for Printed Library Materials, Z39.48.1984.

Library of Congress Cataloging-in-Publication Data

Moore, Stephen L.
War of the Wolf: Texas' memorial submarine: World War II's famous USS Seawolf/ Stephen L. Moore.
p. cm.
Includes bibliographical references and index.
ISBN-13: 978-1-933177-11-3 (pbk. alk. paper)
ISBN-10: 1-933177-11-X
1. World War, 1939—Naval operations—Submarine. 2. World War, 1939—1945—Naval operations, American. I. Title

Library of Congress Control Number: 2008925913

Book layout and cover design by Stephen L. Moore. Cover photos and adjacent title page photos are official U.S. Navy photos, courtesy of Robert N. Hanson, Henry H. Thomson, Peter Lober and Mike McCoy.

WAR
OF THE
WOLF

STEPHEN L. MOORE

Other Books by Stephen L. Moore

Presumed Lost. The Incredible Ordeal of America's Submarine Veteran POWs of World War II. Annapolis, MD: Naval Institute Press, 2009.

Savage Frontier: Rangers, Riflemen, and Indian Wars in Texas. Volume III: 1840–1841. Denton, TX: University of North Texas Press, 2007.

Spadefish: On Patrol With a Top-Scoring World War II Submarine. Dallas, TX: Atriad Press, 2006.

Savage Frontier: Rangers, Riflemen, and Indian Wars in Texas. Volume II: 1838–1839. Denton, TX: University of North Texas Press, 2006.

Eighteen Minutes: The Battle of San Jacinto and the Texas Independence Campaign. Plano, TX: Republic of Texas Press, 2004.

Savage Frontier: Rangers, Riflemen, and Indian Wars in Texas. Volume 1: 1835–1837. Plano, TX: Republic of Texas Press, 2002.

Taming Texas. Captain William T. Sadler's Lone Star Service. Austin, TX: State House Press, 2000.

With William J. Shinneman and Robert W. Gruebel. *The Buzzard Brigade: Torpedo Squadron Ten at War.* Missoula, MT: Pictorial Histories Publishing, 1996.

For more information, visit www.stephenlmoore.com

Contents

Prologue vii
USS *Seawolf* World War II Roster 1
USS *Seawolf* Final Roster 8

First War Patrol
1. "We're at War" 11
2. Faulty Fish 43

Second–Third War Patrols
3. Special Missions 67

Fourth War Patrol
4. "Into a Hornet's Nest" 93
5. "Fearless Freddie" 113

Fifth War Patrol
6. The Jinx 137

Sixth War Patrol
7. "A Marvelous Spectacle" 163

Seventh War Patrol
8. *Sagami Maru* 181
9. "Circular Run!" 197

Eighth War Patrol
10. "Hope This Old Man Knows His Stuff" 213

Ninth War Patrol
11. "The Next One's Gonna Get Us" 251

Tenth War Patrol
12. "Looks Like a Long Chase" 281

Eleventh War Patrol
13. All Torpedoes Expended 307

Twelfth War Patrol
14. "A Masterful Performance" 323

Thirteenth War Patrol
15. Lifeguard League 349
Fourteenth War Patrol
16. Guerrilla Warfare 371
Fifteenth War Patrol
17. On Eternal Patrol 389

18. Epilogue 401

APPENDICES:
A. Top U.S. Submarines by Tonnage Sunk 417
B. Top U.S. Submarines by Ships Sunk 418
C. *Seawolf* Sinkings and Special Missions Summary 419
D. Comparison of *Seawolf* Sinkings by Source 421
E. Awards Given to *Seawolf* and Crew 422

Chapter Notes 427
Bibliography 439

Prologue

Looking out over Galveston Harbor is one of a small number of World War II submarines that has survived both the great war itself and the postwar scrapyards. She is landlocked to prevent her from rusting in the nearby saltwater, although renovations are an ongoing necessity. More than 60 years after being launched, the proud submarine in Seawolf Park is still open daily as part of an interactive museum.

Visitors stride past Mark 16 torpedoes and commemorative plaques to climb up onto her teakwood-covered main deck. Moving forward, they pass her conning tower and descend into her hull through the forward torpedo room hatch. Inside, they are able to experience the confined spaces and 1940s-era equipment that made up a World War II fleet submarine of the United States Navy. With a little imagination, youngsters can envision the adrenaline rush of torpedo attacks and the anxieties of depth charge counterattacks that a crew of 80 once routinely endured.

While visitors to Seawolf Park often leave with the belief that they have toured *Seawolf*, it is *Cavalla* (SS-244) that resides on Pelican Island—a small island that creates the Galveston Ship Channel with Galveston Island. *Cavalla* became a museum thanks to the work of U.S. Submarine Veterans of World War II, whose state chapters began organizing memorials to the men and boats lost during the war. The Texas unit had been chartered as the "Submarine *Seawolf* Commission" and in May 1967, they dedicated a block granite marker capped by a torpedo and bronze plaque to commemorate *Seawolf*. It was located near the battleship *Texas*, but the Submarine *Seawolf* Commission still hoped to add a World War II submarine to their efforts. Delays in preparing the park site caused their first secured submarine, *Cabrilla*, to deteriorate. The Texas SubVets then managed to secure newly decommissioned *Cavalla* in 1971.

Cavalla was opened to the public for tours on 11 April 1971—on the 71st anniversary of the U.S. Submarine Service. Seawolf Park, opened in 1974 as Texas' tribute to the lost submariners of World War II, also features the destroyer escort *Stewart* (DE-238). *Seawolf* once visited Galveston Island in the spring of 1940, while on her shakedown cruise, and passed right by the spot where *Cavalla* now stands in her memory.

SS-244 is equally representative of the fighting spirit of her lost predecessor, SS-197. She earned a Presidential Unit Citation, four total battle stars, and destroyed more than 34,000 tons of shipping. One former *Seawolf* officer was aboard *Cavalla* when she sank the mighty Japanese aircraft carrier *Shokaku*—one of the carriers responsible for the "day of infamy" at Pearl Harbor on 7 December 1941.

The namesake submarine of Texas' Seawolf Park went to war just hours after the Japanese planes departed from the Hawaiian Islands. At the time Pearl Harbor was being attacked, *Seawolf* was laid up in Manila Harbor in the Philippines awaiting an engine overhaul. That same evening—8 December in Manila—she was out to sea and in search of Japanese shipping targets.

Seawolf became famous during the war for her exploits. Her first skipper, Lt. Cdr. Frederick Warder, earned the nickname "Fearless Freddie" for boldly entering shallow enemy harbors, for making three attacks against heavily armed Japanese cruisers, and for even taking on Navy brass when necessary. Warder's sinking totals would have been much higher had he been working with good torpedoes. He was among the first skippers to rat out the Navy's torpedo scandal. Long after the war had started, testing proved that the magnetic exploders on the Mark 14 torpedoes were faulty and the torpedoes generally ran at least ten feet deeper than their settings—therefore passing harmlessly under shipping targets. While improvements were made, the torpedo issue still allowed Japanese ships to escape well past 1942. For example, Japanese documents show that *Seawolf* hit one tanker with no less than five dud torpedoes—fired in multiple attacks—on 9 November 1943.

Seawolf's other skippers—Roy "Googy" Gross, Richard "Ozzie" Lynch, and Al Bontier—were equally valiant. Lt. Cdr. Gross would end up as eighth best U.S. sub skipper in terms of enemy shipping sunk. Under the command of Lynch and Bontier, *Seawolf* ran special operations missions in 1944. She performed a special photographic reconnaissance probe of the Palau Islands, rescued aviators downed in the Battle of the Philippine Sea, and landed supplies and men to support the commandos in the Philippines.

The *Wolf* was delivering other special ops forces—an elite Army unit known as the Alamo Scouts—when she was lost with all hands. Although she was officially listed as "overdue and presumed loss," the cause of her loss was soon painfully apparent. A Japanese submarine torpedoed a U.S. destroyer near Morotai Island, just northwest of New Guinea. Carrier planes spotted *Seawolf* diving and marked the area with dye. The destroyer escort *Richard M. Rowell* soon arrived and established contact with the unknown submarine.

Seawolf's radiomen attempted to warn the destroyer that they were a friendly sub, but their signals were not understood. *Rowell* delivered a deadly series of hedgehog attacks, and her depth charges brought up air bubbles and debris. She moved on, oblivious to the fact that she had just destroyed the proud *USS Seawolf*.

Ninety-nine officers, enlisted men, and Alamo Scouts perished when the *Wolf* was fatally damaged. Her remains lie some two miles below the surface, where these brave souls remain on eternal patrol. Seawolf Park in Texas specifically honors SS-197, but is part of a national effort by the Submarine Veterans of World War II to properly honor the 3,500-plus men who perished in the Silent Service. U.S. submarines destroyed more than 1,300 Japanese ships during the war for a 5.3 million gross ton total. Although the submarine service only accounted for 1.5 per cent of the U.S. Navy, it accounted for 55 per cent of all Japanese vessels sunk in World War II.

Against this brilliant accomplishment was the loss rate of the submarine service. Fifty-two boats were lost to all causes. Roughly 20 percent of the men who volunteered to fight from below the waves never returned. No other branch of the military suffered such a loss rate in World War II.

During the war, *Seawolf* was credited with sinking 20 ships for 109,600 tons and for damaging another 13 ships for 66,500 tons. The Navy officially credited *Seawolf* with 18 sinkings of ships over 500 tons, plus shared credit for another ship that she pursued, pounded with gunfire, and herded to a sister sub to kill. Her gun crews additionally sank four smaller vessels and helped finish off one torpedoed cripple. More recent studies of all sinkings by U.S. submarines shows that *Seawolf* destroyed 26 total ships for 95,060 tons. She was awarded two Navy Unit Commendations and 13 battle stars for her patrols. Some of the photos taken through her periscope have been published in *Life* magazine, wartime newspapers, and numerous postwar books about the submarine war.

In addition to sinking ships, *Seawolf* was equally called on to perform rescues and special missions. During the fall of the Philippines, she evacuated Navy personnel, submarine command staff, Navy and Army pilots, and even a British Secret Intelligence Service agent. She also hauled in ammunition to help support the troops on Corregidor. Later in the war, the *Wolf* landed commandos and brought in supplies for the guerrillas who were ashore in the Philippines to undermine the Japanese war efforts. She also rescued two Avenger torpedo bomber flyboys who were forced to ditch during the battle of the Philippine Sea.

Muster rolls, patrol reports, deck logs, commendation citations, and other official Navy reports helped create the backbone of the *Seawolf*'s history. Adding the human element of life aboard the *Wolf*, however, would not have been possible without the help of those who served aboard SS-197. These men openly shared their photographs, personal papers, and memories to help add color to the official reports. Those serving aboard her before her loss who contributed are: John H. Bilkey, Allen B. Bingham, Francis "Jim" Cashero, Robert R. Curtin, Ferdinand V. Cucchi, Lewis S. Donche, Olin R. Fogle, Robert N. Hanson, William R. Harlow, Norman Kisver, George D. Leffingwell, Robert W. Lents, Peter N. Lober, Delbert Mar, Rex L. Mickey, Edward L. Milas, Lucien T. Rajotte, Joseph H. Strong, Henry H. Thomson, Charles C. Woodard, and Paul Zimmerman.

In addition, families of those now deceased or who were lost with *Seawolf* also contributed stories, photos, and papers pertaining to their loved ones. Among these were Keith Bjerke, Marion Bjerke, Nancy V. Dombroski, Carl H. Enslin Jr., Joan George, Earl G. Hill Jr., Timothy Holden, Ruth Johnson, Cora Enslin Kobesky, Karl J. Kramer, Susan Leffingwell, Bobbie Mallough, Margarite Mills, Becky Palazolo, Hope Syverson, Pauline Whitman, and Amy Wiegenstein. For the rescue of the two *Wasp* Avenger aviators, I am indebted to John C. "Jack" Bramer, who shared his story and photos of the rescue. Charles Houston, author of *Flying With Iron Angels*, put me in contact with Jack and helped provide information on Bramer's pilot. Special thanks are also due to retired World War II submarine veterans Kevin D. Harty and Charles H. Ver Valin for reading over this manuscript for factual errors. I regret anyone whom I have neglected to mention here or in the bibliography.

Mike McCoy, nephew of one of the men lost with *Seawolf*, shared a wealth of information from his family's collection. Walter Glen "Bud" McCoy was aboard *Seawolf* throughout the war and was the senior enlisted man aboard at the time of her loss. Bud kept a diary during some of this time, the contents of which have been in the McCoy family until this time. His surviving brothers, Claude and Harold McCoy, graciously allowed Mike to share this diary with me. Mike also kindly shared a number of photos from his *Seawolf* collection and helped look over rough drafts of the text for errors.

In his diary, Bud McCoy described the ship's attacks and counterattacks, the motormacs' struggles to keep the engines running at times, and the anticipation of R&R after a successful patrol run. McCoy's reflections of life on the *Wolf* are especially poignant since he did not survive the ship's final patrol.

To him and the other 98 men whose final resting place is within the crushed hull of *USS Seawolf*, this book is dedicated. Texas' *Seawolf* Park is a lasting tribute to their sacrifice, and the sacrifice of thousands of other U.S. submarine veterans who did not survive World War II. The efforts of the Silent Service certainly helped expedite the ending of World War II. The sacrifice of such brave men—who volunteered to work in the most dangerous occupation of the war—is part of the price of our precious freedom.

USS Seawolf World War II Roster
First through Fourteenth Patrols

Note: The first rank or rate for officers and crew indicates highest held while aboard *Seawolf*. Previous ranks and rates held are also shown.

1c	First Class	Lt.(jg)	Lieutenant, Junior Grade
2c	Second Class	MoMM	Motor Machinist's Mate
3c	Third Class	MM	Machinist's Mate
Bkr	Baker	OS	Officers' Steward
BM	Boatswain's Mate	PhM	Pharmacist's Mate
C	Chief Petty Officer (CPO)	QM	Quartermaster
CCS	Chief Commissary Steward	RM	Radioman
Cdr.	Commander	RT	Radio Technician
Ck	Cook	S	Seaman
EM	Electrician's Mate	SC	Ship's Cook
Ens.	Ensign	SM	Signalman
F	Fireman	ST	Steward
FCS	Fire Controlman, Surface Weapons	StM	Steward's Mate
		TM	Torpedoman's Mate
GM	Gunner's Mate	WM	Warrant Machinist
Lt.	Lieutenant	Y	Yeoman

OFFICERS

Name:	Rank:	Patrols:
Gross, Royce Lawrence	Cdr.	8-12
Warder, Frederick Burdett	Cdr./Lt. Cdr.	1-7
Garnett, Philip Weaver	Lt. Cdr.	10
Hess, John Bordon	Lt. Cdr.	13
Lynch, Richard Barr	Lt. Cdr.	13
Risser, Robert Dunlap	Lt. Cdr.	9-10
Deragon, William Nolin	Lt. Cdr./Lt.	1-8
Stephan, Edward Clark	Lt. Cdr.	5
Holden, Richard (n)	Lt./Lt.(jg)	1-7
John, Clary Leonard	Lt.	9-13
Smith, William Lee	Lt.	11-13
Syverson, Douglas Neil	Lt./Lt.(jg)/Ens.	1-12
Mercer, James (n)	Lt./Lt.(jg)/Ens.	1-10

Whitman, William Alexander	Lt./Lt.(jg)	2-11
Casler, James Burr Jr.	Lt.(jg)/Ens./CQM	1-9
Kennelly, John J.	Lt.(jg)/Ens.	8-9
Robinson, Dougald "G"	Lt./Lt.(jg)	10-12
Butler, David (n)	Ens./WM/CMoMM	1-7
Clevenger, Raymond Eugene	Ens.	11
Crane, Lawrence William Jr.	Ens./WE/EM1c	1-7
Gluski, Hubert Eugene	Lt.(jg)/Ens.	8
Goudy, David Ernest	Ens./CRM	10-11
Leffingwell, George Darrell	Ens/CFC/FC1c	1-9, 11-12
Lober, Peter Nicholas	Ens./WM/CMM/MM1c	1-7
Sullivan, John Edward	Ens./CY/Y1c	1-7
Bilkey, John Howard	WE/CEM/EM1c	1-9, 11-14

ENLISTED MEN:
(* indicates service aboard only between war patrols.)

Name:	Rank:	Patrols:
Adams, Glenn Tommins	CMoMM	7
Anderson, James Joseph	S2c	5-6
Anderson, Owen Stanley	EM2c	12-14
Armbruster, Chester Leroy	GM3c/S1c	11-12
Arpia, Eugene Furtado	MoMM2c	13-14
Baker, Auston	SC1c	1-4
Barboni, Albert	SC3c	8-9
Bateman, Roy Walker	CMoMM/1c/MM1c	1-9, 11-12
Beatley, John Wesley	CQM	7-9
Bennett, John Nelson	CGM/GM1c	1-10
Bingham, Allen B.	Y2c	14
Bjerk, Keith Allen	TM2c/3c	1-7
Brengelman, Henry Bernard	CEM/EM1c	1-8
Bugawisan, Mariano	OS1c/2c/3c	1-7
Burruss, John Martin	Y2c/3c	8-9
Caillier, Leonard Joseph	CMoMM	11-12
Cani, John (n)	SC2c/3c	10-12
Capece, Edmund Currie	CEM	1-5
Cashero, Francis James	TM2c/3c/S1c	3-9
Chubbuck, Wilbur Harver	TM2c	1-7
Coyne, Edward J.	S2c	6
Cronk, Kenneth Eric	MoMM1c/MM1c/2c	3-8
Cross, Orval Clyde	MoMM2c/MM1c/2c	1-7
Cucchi, Ferdinand Victor	Bkr3c	*
Curtin, Robert Roy	MoMM2c	10-12

USS *Seawolf* Wartime Muster Roll

Name	Rate	Patrols
Delnigro, Albert (n)	MoMM2c/MM2c/F1c	1-7
Denemore, Howard McLeod	MoMM2c	11-12
Dishman, Otis Charles	CMoMM/1c/MM1c	1-8
Dobbel, Carlton Edgar Jr.	TM2c	10-12
Donche, Lewis Sigmund	MoMM2c	10-12
Eckberg, Joseph Melvin	CRM/RM1c	1-5, 7
Edmonds, Leroy (n)	SC1c	12
Elliott, William L.	S2c	6
Enslin, Carl Henry	CMoMM/CMM	1-9
Faber, Lee M.	S2c	6
Faciane, Frank L.	S2c	6
Fenwick, David James	CCS/SC1c	9-11
Ferguson, Joe Carlton	RM2c	1-5
Ferreira, John Frutado Jr.	S2c	8-9
Fogle, Olin Richard	EM3c	9-12
Fox, John Joseph	Y2c	13
Franz, Frank (n) Jr.	SM1c	1-7
Fuller, Vernie Marion	S2c	2
Galvan, Basilio	OS3c	*
Garrett, George Booth	CMoMM/1c	11-14
Gervais, Joseph Rudolph Oliver	TM1c/2c/3c	1-8
Gibson, John Stafford Jr.	CTM/TM1c	1-8
Glimsdale, Carlson Eldred	MoMM1c/MM2c	8-11
Gordon, Charles Walter	S1c/2c	12
Handley, John Jr.	MoMM2c/F1c/2c	8-11, 13
Hanson, Henry Howard Jr.	MoMM2c/F2c/1c/2c/S1c/GM3c/S1c	1-9
Hanson, Robert Norman	MoMM2c/F1c/2c	1-9
Harlow, William Robert	F1c/2c	11-12
Harrington, Earl Lee	MM2c	7
Hershey, Albert Earl	CMoMM/CMM/MM1c	1-8
Hickerson, Peter Thomas	F2c	*
Hicks, James Clarence	FC(S)2c/FC2c	8-11, 13
Hinson, Edward Elie	RM1c/2c	6, 8-9
Hughes, Jimmie Ross	EM1c/2c/3c	7-9
Hutchison, Robert Lester	EM1c	1-5, 7
Irvin, James Willard	CEM	13
Jancik, Calvin George	SC2c	13-14
Jenkins, Maurice (n)	CMoMM/CMM/MM1c	1-7
Jennings, Harry Augustus Jr.	F3c	12
Jobe, Clinton (n)	CEM	1-4, 6
Jobe, Jesse (n)	EM2c/3c/S2c	1-10

Johnson, Charles Alfred	MoMM2c/F1c/2c/S1c/2c	1-9
Jones, Scott Weldon	QM3c/S1c	9-10, 12
Kasloski, Robert John	FC3c	*
Kellerer, John Randolph	EM1c	11
Kennedy, Denver Guy Jr.	EM2c	7
Kibbons, Clarence Vernon	TM1c/2c	1-9
Kincaid, Wilford Charles	CEM	12
Koehler, Robert Henry	EM3c/S2c	1-8
Klimes, Joseph Jr.	MoMM1c/2c	10-12
Kraght, Henry G.	F1c	4-5
Kisver, Norman (n)	QM3c/S1c	1-7
Kropp, Dale Henry	GM2c	12
Lamberson, Arthur Earl	TM1c/2c	1-8
Langford, Robert (n)	CTM	1-4
Larson, James Edward	GM2c/3c	2-7
Laumann, George Arthur	EM2c/3c	10-12
Lear, Earl William	CMM	7
Lewis, John Miles	EM2c	8-9
Loaiza, Frank (n)	CPhM/PhM1c	1-8
Magnuson, William H.	S2c	1-4
Maley, Paul Leroy	RM1c/2c	1-9
Mallory, William Cleaver	SC1c/2c	1-9
Mallough, Kenneth Geiser	MoMM1c/MM1c	4-9
Maples, John Alonzo Jr.	MoMM2c/F1c	9-10
Mar, Delbert	FC3c/S1c/2c	8-10, 12
Mathews, Curtis Arden	TM3c	*
McGee, William Timothy	S2c	*
McMullen, Clarence R.	S1c	*
McNabb, John Joseph	F2c	*
McTavish, John Francis	S1c	1-6
Mehner, Waldemar Richard	SC3c	*
Metz, William T.	CMM	1-4
Mickey, Rex LaVere	GM2c/3c	8-11, 13-14
Milas, Edward Lawrence	EM2c/3c	5-7
Miller, Allen Thomas	QM3c	10
Mills, Wilson (n)	S1c/2c	8-11
Mocarsky, Alexander Peter	CEM	1-7
Moye, George (n)	TM3c	*
Munger, Alva Arnold	F1c/2c/S2c	6-8
Myers, Walter Donald	MoMM1c/2c/F1c/2c/S2c	6-10, 12
Neil, John Spence Jr.	TM2c/3c/S1c	1-10
Newman, Kenneth Eugene	GM2c/3c	10-11

USS *Seawolf* Wartime Muster Roll 5

Newton, Joseph Lincoln	MoMM2c/F2c/3c	8-12
Noble, Jack Howard	EM3c	8-10
O'Brien, Roger Allen	Y1c/2c	10-12
Ostrander, Alfred George	TM2c	9-12
Ottaway, Samuel Adrain	RM1c/2c	9-11
Parden, Lee Bob	EM1c/2c	1-7
Patterson, Donald Richard	F2c/3c	10-11
Perry, Victor Irving	Sp(P)1c	13
Pierce, Arthur Calhoun	PhoM1c	13
Poe, James Cavanaugh Jr.	EM2c	13-14
Poole, Mason (n)	CEM/EM1c/2c	1-13
Pulsifer, Charles Lee	F1c/2c/S1c	11-12
Rajotte, Lucien Thomas	CMoMM/1c/MM1c	1-10
Randazzo, Salvatore (n)	CMoMM/1c/MM1c	1-8
Roberts, Lawrence (n) Jr.	TM3c	9-11
Romito, Frank Deminic	SM1c/2c	10-12
Russ, Lamar George	F1c	*
Russell, Benjamin Earl	MoMM2c	*
Sandridge, Lloyd W.	CMM	1-4
Schultz, Paul Joseph	SC3c	10-11
Scott, Theron	CRM	*
Sherman, John J.	EM1c	7
Short, William Edward	TM2c	7
Snyder, James Earl	MM2c	1-5
Souza, Edwin Enos	CTM	1-7
Stark, Robert Merrill	TM2c/3c	10-12
Steward, Harry Carlyle	MoMM2c	9
Street, John Edwin	MoMM1c/MM1c	1-7
Strong, Joseph Hale	RM2c	8
Tamayo, Brigido (n)	OS1c/2c/3c	1-9
Taylor, Wendell Broadus	TM2c	7
Thomson, Henry Hanford	CQM/1c/2c	1-9, 11-12
Times, Charles Henry	F2c/S2c	6-7
Tranquilly, Robert (n)	Y2c	8
Vineyard, John Leslie	MoMM3c/F1c	9-11
Vitello, Donato (n)	Bkr1c	12
Weade, Claiborne Hoyt	TM1c	4-5
Wilkerson, Robert Goley	EM1c/2c	9-11
Wilton, Henry Peter	SC1c/2c	5-7
Woodard, Charles Clark	RM1c/2c	6-7
Wright, Harold Lankford	SC1c/2c	1-8
Yersick, Paul Anthony	MoMM2c/F1c	12-14

Zimmerman, Paul (n)	MoMM1c/2c/MM2c/F1c	1-9
Zirkel, Frederick Andrew	CMM/MM1c	1-5

PASSENGERS:

U.S. Navy:	Rate/Rank:	Patrol:
Benedict, C. P.	ACMM	3
Borowski, A. A.	Y1c	2
Bramer, John Conrad Jr.	ARM2c	13
Bryant, Eliot Hinman	Cdr.	2
Carlisle, Lawrence L.	EM2c	10
Carrico, Harold	TM1c	10
Clark, J. W.	AMM1c	3
Combs, H. V. Jr.	Lt.(jg)	2
Connolly, Joseph Anthony	Cdr.	2
Cook, R. F.	AMM1c	3
Entire, R. K.	Lt.(jg)	2
Fife, James Jr.	Cdr.	2
Holeman, Victor Rollo	MoMM2c	10
Irish, Donald W.	RM1c	2
Kelly, D. M.	ACMM	3
Llanes, A.	OS2c	2
Mangigian, Paul	TM3c	10
Mumma, Morton Claire Jr.	Lt. Cdr.	2
Payne, Earl D.	ACMM	3
Pew, L. A.	Lt.(jg)	3
Pollock, Thomas F.	Lt.(jg)	3
Seward, N. E.	Y1c	3
Swenson, H. R.	Lt.(jg)	3
Thomas, Marion K.	MoMM1c	10
Utter, Harmon T.	Lt.	3
Vanzant, Ralston B.	Lt. Cdr.	2
Walraven, Albert Tavel	Lt.(jg)	13
Williamson, L. H.	Ens.	3
Woodard, Charles Clark	RM2c	2

U.S. Army:	Rate/Rank:	Patrol:
Bender, Frank Peter	1st Lieut.	3
Carpenter, John W.	1st Lieut.	3
Croxton, Warner W.	1st Lieut.	3
Eubank, William E. Jr.	Capt.	3
Hoevet, Dean C.	1st Lieut.	3

USS *Seawolf* Wartime Muster Roll

Marrocco, William A.	Capt. (Marines)	3
McAfee, James B.	1st Lieut.	3
Pease, Harlan Jr.	1st Lieut.	3
Shedd, Morris H.	Capt.	3
Stafford, Robert F.	2nd Lieut.	3
Stephenson, Glenwood G.	Capt.	3
Vance, Reginald F. C.	Major	3

British Army:	Rate:	Patrol:
Wilkinson, Gerald	Major	3

Filipino Guerillas:	Rate:	Patrol:
Cabais, Eutiquio B.	Mst. Sgt.	14
Cuteran, J.	Sgt.	14
Daelto, Marcuano R.	Sgt.	14
Dagandan	T/5	14
Goloyugo, V. C.	Sgt.	14
Marquina, L.	T/4	14
Teo, Konglan	Lt.	14
Vergara, T. E.	T/4	14
four others unknown		

USS *Seawolf* Final Roster
Fifteenth Patrol

OFFICERS:

Name:	Rank:	Patrols:
Bontier, Albert Marion	Lt. Cdr.	14-15
Cox, Robert Leon	Lt.	12-15
Doane, Paul (n)	Lt.	14-15
O'Brien, Edward Francis Jr.	Lt.	15
Asa, Marion Lee	Lt.(jg)/Ens.	12-15
Miller, Ralph Van Dorn	Lt.(jg)	14-15
Szendrey, Edward John	Lt.(jg)/Ens.	12-15
Reiland, William Frederick Jr.	Ens./CTM/TM1c/2c	1-11, 13-15
Van Andel, John	Ens.	13-15

ENLISTED MEN:

Name:	Rank:	Patrols:
Astarita, John Michael	S1c/2c	13-15
Balch, Lloyd Richard	EM1c	13-15
Ballard, Francis Arden	GM2c/3c	11-15
Bannister, Jack (n)	Bkr3c	15
Bargenquast, Arnold Frank	MoMM1c/2c	12-15
Beck, William Barndt	F1c	15
Bekke, Gerald Edgar	CRM/RM1c	11-15
Bennett, Robert Jordan	S1c	13-15
Bergevin, Patrick Kenneth	S1c	13-15
Bolon, Dallas Victor	F1c/2c	13-15
Call, James Burdell	RM2c/3c	10-15
Carithers, James Purcell	F1c	15
Carnegie, Robert Jack	EM3c	15
Cash, Wilfred Leslie	MoMM3c/F1c	13-15
Chapman, Edward (n)	MoMM1c/2c/ MM2c/F1c	1-10, 12-15
Coon, Norman "D"	RT2c/RM2c/3c	11-15
Copas, Chester Mayo	Y1c	15
Cotton, Wayne Houston	SC3c	13-15
Cunnally, James Patrick	MoMM1c/2c	11-15

Eternal Patrol: *Seawolf's* Final Roster   9

Devitt, Robert Floyd	MoMM2c/3c/F1c/2c	1-13, 15
Ewing, John Louis	QM3c	15
Fixler, Robert Nelson	S1c/2c	13-15
Flynn, Kenneth Judd	EM1c/2c/3c	10-15
Franco, Peter (n)	MoMM2c/3c/F1c/2c	10-15
George, Lloyd (n)	EM3c/S1c/S2c	12-15
Grimes, James (n)	QM2c/3c/F2c	8-15
Hadley, William Thomas	CPhM/1c	9-15
Harris, John Gordon	F1c	15
Howard, Alfred Herman	TM2c/3c	10-15
Huff, Roy Edward	MoMM2c/3c/F1c/2c/3c/S2c	8-15
Johnson, James Everard	RM2c/3c/S1c/2c	8-10, 12-15
Jurnic, Michael (n)	SC1c/2c	12-15
Kenney, Jack Edward	S1c	12-15
Krempa, Charles Stanley	MoMM1c	11, 13-15
Kuehn, Alfred Eric	QM1c/2c/3c	8-10, 13-15
Lawson, Chester Gelean	TM2c/3c/S1c	10-15
Leeman, Merlin Hibbard Jr.	S1c/2c	13-15
Likert, Gilbert Roland	BM1c	13-15
Lynch, Carl Dean	EM2c/3c	12-15
Malone, Dallas Leroy	TM2c/3c/S1c/2c	8-11, 13-15
Marston, George Franklin	TM2c/3c/S1c	11-15
Maus, Charles Robert	SM2c/3c/S1c/2c	8-15
McCoy, Walter Glen	CMoMM/MoMM1c/MM2c	1-10, 12-15
Michael, Forrest Samuel	EM3c	13-15
Miller, Richard Lawrence	S2c	15
Miller, Robert Thomas	TM2c/3c/S1c/2c	8-15
Mills, Lonnie Tolbert Jr.	F1c/2c	13-15
Mitchell, Harold Edward	S2c	15
Morris, Edward Lyle	FC2c/3c	10-15
Morris, Joseph Albert	S1c	9-11, 13-15
Nazay, George Gilbert	F1c/2c/3c	10-15
Naze, Donald Joseph	CTM/TM1c	11-15
Needham, George Melvin	EM1c/2c/3c	8-15
Nivison, Clinton LeRoy	EM1c/2c/3c	8-15
Page, Albert Francis	EM2c/3c	8-15
Page, Leonard Alton	MoMM2c	8-11, 13-15
Peterson, Elmer Norman	MoMM1c/2c/F1c/2c	8-11, 13-15
Politylo, Wasil (n)	EM3c/F1c	11-15
Rhoads, Guy Benjamin	MoMM1c/2c	9-15

Riggle, Mahlon Richard	TM2c/3c/S1c/2c	8-15
Rocaya, Saturnino (n)	OC2c/3c/StM1c	10-15
Rogers, Benjamin Franklin	CRT/1c/2c	8-15
Rosete, Tomas (n)	OCk1c	4-15
Sadler, John Colby	TM2c/3c/S1c/2c	8-15
Saint, James William	MoMM1c/2c	9-15
Steinbecker, Gerald Andrew	F1c/2c	13-15
Strausser, Clarence Elias	F2c	13-15
Underhill, William Hopkins	MoMM2c/F1c	12-15
Wall, Vernon Palmer	MoMM1c/2c	9-15
Warren, Thomas Wilson	TM2c/3c/S1c	8-15
Wiegenstein, Michael Paul	CMoMM/1c	10-15
Wyatt, David Bernard	S1c	15
Young, Robert Porterfield	EM3c	14-15
Zuel, Edward Andrew	EM2c/3c	11-15

U.S. ARMY PASSENGERS LOST:

Almero, Emiliano A.	Ts/Sgt.
Bueno, George B.	Sgt.
Cendonia, O. C.	T/5
Francisco, Alberto C.	S/Sgt.
Fria, A. B.	Cpl.
Hammill, Charles H.	S/Sgt.
Herbig, Robert P.	Sgt.
Ibea, Artemio I.	S/Sgt.
Kopp, Howell Stewart	Capt.
Miller, George F.	1st Lt.
Peralta, George E.	1st Lt.
Pugose, Emil L.	Sgt.
Ramos, O. B.	Pfc
Rimando, Juan F.	Pfc
Rodriguez, Ireneo R.	Sgt.
Ruiz, Ruperto R.	T/5
Wise, Braynard L.	CWO

1

"We're at War"

Manila Harbor *8 December 1941*

Bill Deragon was not expecting much to interrupt his evening duty aboard his submarine this night. Although war was imminent, two-thirds of the crew had been allowed to take liberty ashore on Sunday. Some of the men had come back aboard after the bars closed that night to sleep off the effects of the day's softball games. Others were still ashore in Manila, due to return back aboard by early Monday morning.

It was now the early morning hours of Monday, 8 December 1941, in the Philippine Islands. Lieutenant William Nolin Deragon had the duty watch since his skipper was asleep in his stateroom and the junior officers were off duty. As second in command of the United States submarine *Seawolf* (SS-197), Deragon would normally have his junior officers cover the watch schedules while the ship was at sea. Each officer stood four hours on, eight hours off, and another four hours on during each twenty-four hour period.

Seawolf was securely moored to the port side of her sister submarine *Searaven* (SS-196), which was in turned moored to the large submarine tender *Canopus* (AS-9) in Manila Harbor in the Philippines. Some 5,000 miles and an international dateline away, it was the morning of 7 December 1941 in Hawaii—where the Japanese Navy was just beginning to deliver the surprise attack that would end Lieutenant Deragon's quiet hours.

Of French Canadian ancestry, Deragon was fluent in French and had been found by the men he commanded to be calm in even the most dire of situations. One of his crewmen later described the Exec as "a tall, rangy man with a long face etched with two sharp lines from nose to mouth, and deep-set eyes."[1]

Deragon had joined *Seawolf* in late 1939 prior to her commissioning and had subsequently served as her first lieutenant and gunnery officer prior to advancing to executive officer, navigator, and first lieutenant. The first lieutenant was in charge of the maintenance and repair of the exterior of the ship, including the ship's boat, kingpost boom, winches and her mooring lines. This job would normally be passed on to a junior officer but *Seawolf* had to divide her duties among only four officers aside from the skipper.

Upon graduation from the Naval Academy in 1934, he had first served on the cruiser *Louisville* for a year and then the destroyer *Sands* for two years before reporting to the Submarine Base in New London, Connecticut, for training in submarines. Bill Deragon had then been assigned as engineering officer of the older *S-42* in the Panama Canal area from January 1938 through October 1939.

When war finally caught up to him, Bill Deragon was below in *Seawolf*'s wardroom when he got wind of something happening on the bridge around 0315 on 8 December. He moved quickly through the passageway past the chief petty officers' stateroom and the captain's stateroom before hopping through the oval bulkhead doorway into the control room. He grabbed the steel rungs of the perpendicular ladder and swiftly climbed it into the conning tower and then moved up another ladder onto the bridge of his submarine.

There he found the ship's leading signalman, Frank Franz Jr., acknowledging a message from the tender *Canopus*. Franz was well experienced. He had been aboard *Seawolf* since she was commissioned and had previously served on the submarine *Shark*. With a series of flashes from his signal lamp, Franz was through and he excitedly related to his Exec the urgent plain-language message addressed to all ships and stations he had just received: FROM COMMANDER ASIATIC FLEET . . . TO ASIATIC FLEET . . . URGENT . . . BREAK . . . JAPAN HAS COMMENCED HOSTILITIES . . . GOVERN YOURSELF ACCORDINGLY.[2]

The message had originated from four-star Admiral Thomas Charles Hart, a 1897 graduate of the U.S. Naval Academy who was commander of the U.S. Asiatic Fleet. Just after 0230 on Monday, 8 December, a radio operator at Hart's headquarters in the Marsman Building at Pier 7 in Manila received a transmission from Admiral

Husband E. Kimmel's headquarters in Honolulu. "Air raid Pearl Harbor," was the message. "This is no drill." Hart rushed to his operations room and soon the word flashed to his Asiatic Fleet.[3]

From Hart's headquarters, the word was quickly passed to Captain John Wilkes, Commander of Submarines, Asiatic Fleet, who was stationed aboard his flagship, the sub tender *Holland* (AS-3). Wilkes' chief of staff, Commander James Fife, next awakened Cdr. Stuart S. Murray, Commander of Submarine Division 21. "Get up and get the submarines started going on patrol," Fife ordered.[4]

Based out of the Philippines, the U.S. Asiatic Fleet's Submarine Force comprised four divisions of boats. There were 29 total submarines, including six older S-boats and 23 newer fleet boats. Monday, 8 December, found some on patrol, others still en route to their designated stations and another group in Manila Bay for reloading or repairs.[5]

Holland blinkered the first message of the Pearl Harbor attack on to the tender *Canopus* and followed with other messages as the news unfolded. *Canopus* had in turn then signaled the other nearby ships, including Frank Franz and Bill Deragon on *USS Seawolf*.

Seawolf was attached to the four-submarine Division 202 under Commander Willis Merritt Percifield. All four "Sea" boats—*Seawolf*, *Seadragon*, *Sealion* and *Searaven*—were of the same *Sargo*-class and the crews had become familiar with each other. The submariners of these boats, in fact, casually referred to each of their sisters boats by dropping the "Sea" prefix. The four were thus known as "the *Wolf*," "the *Dragon*," "the *Lion*," or "the *Raven*."

The *Wolf* had just returned from Shanghai, having escorted the first shipload of Marines out of China back to Manila. The first message from *Canopus* had clearly spelled out to those awake on the *Wolf* that the war was on. Deragon quickly called down to the control room, where electrician John Bilkey was on duty. "Bilkey, wake Captain Warder!"

"Aye, sir," replied Bilkey.

John Howard Bilkey was no newcomer to the Navy, either. "I had been on the old S-boats awhile," he later recalled. "I had even been aboard the *Squalus* before she was lost, just long enough to learn how to run the electrical controls." An open valve had caused *Squalus* (SS-192) to sink off Portsmouth, New Hampshire, in 1939 with a loss of 26 lives. Incredibly, 33 *Squalus* submariners were rescued with an experimental diving bell and their submarine was later raised off the ocean floor. On 13 September 1939, Bilkey and others who were busy fitting out *Seawolf* in the Navy Yard stopped to watch as the refloated *Squalus* was towed up the Piscataqua River.

With the knowledge of how easily a sub could be lost even in peacetime, John Bilkey knew that war would bring even greater risks. "I was on duty when I found out about Pearl Harbor," he later said. The surprise attack was certainly shocking, although America's entry into the war was not a foreign thought to Bilkey. "The war had already been going on for others for some time."

Seawolf's skipper, Lieutenant Commander Frederick Burdett Warder, was on the bridge with Bill Deragon and signalman Franz momentarily. He ordered the crew to be awakened and had the word spread throughout the ship. Chief of the boat Eddie Souza made the rounds through the ship, instilling urgency in his men. "There's a war on!" he prodded them.[6]

Shortly after receiving the first blinkered message, Frank Franz copied the signal from *Canopus* of another directive from Admiral Hart: SUBMARINES AND AIRCRAFT WILL WAGE UNRESTRICTED WARFARE.[7]

It was 0345 and throughout the ship, there was suddenly the lively sound of men hustling to prepare for war. Gunner's mate first class John Nelson Bennett was quick to action. He and his assistants broke out the machine guns and ammunition from the lockers below deck to man the topside bridge guns. Known to all as "Gunner," Bennett had served on the old *S-46* before putting *Seawolf* into commission in 1939. After two years of peacetime drilling, Gunner Bennett was ready for his weapons to see some real action.

Monday morning, 8 December 1941, found *Seawolf* anchored 600 yards from the Cavite Navy Yard in Manila Bay. The numerous submarines operating from this division were serviced by three submarine tenders stationed in Manila Bay: *Holland*, *Otus* and *Canopus*. These enormous ships carried spare torpedoes, provisions, stores, and countless spare parts to refit and reservice submarines. On the morning of 8 December, *Holland* was behind the breakwater in Manila Harbor, *Canopus* was anchored off the Cavite Naval Station, and *Otus* was at Mariveles, the small naval base on the Bataan peninsula. The aircraft carrier *Langley* and her escorts were also stationed in Manila Bay this morning.

The *Wolf* had been awaiting her turn to go into overhaul at the Cavite Navy Yard. Two of her sister subs, *Seadragon* and *Sealion*, were ahead of her going through complete overhauls. The process could take as long as eight weeks to remove the ship's engines, tear down her electrical systems, and then rebuild the ship. The *Dragon* was nearly complete, but the *Lion*'s engines still lay on the dock.

The *Wolf* was scheduled to go into the yard on Thursday. She had spent time in the roadstead of Tawi Tawi Island—the most southwestern island in the Sulu archipelago—just 40 miles from the tip of Borneo. "We had returned to Manila because of typhoon danger and an expected navy yard overhaul," recalled quartermaster Henry Thomson. Some of *Seawolf*'s torpedoes had been offloaded in anticipation of the overhaul, along with other gear.[8]

Word of the Japanese attack on Hawaii put an immediate end to the overhaul plan. Those sailors who had been ashore walked, ran, or caught rides back to the dock, to catch the launches out to where *Seawolf* was anchored. Machinist's mate Pete Lober was ashore for the weekend. "Word got out about Pearl Harbor, so we had to hurry back," said Lober. He made his way aboard a bus and returned to *Seawolf* to help expedite her getting under way. Another shipmate, machinist's mate Lucien Thomas Rajotte, was on the same bus. "I was on liberty at the time Pearl Harbor was attacked," said Rajotte. "I made it back to the ship about six in the morning and found out we were at war while I was on the bus."

Pete Lober always found downtown Manila a great place to go for a drink or a steak while on liberty. When *Seawolf* had first been transferred to the Asiatic fleet the previous year, he had found a little extra pocket money to spend in the Philippines. "We had a big acey-deucy tournament on the way from Honolulu to Manila, and I won it," he said.

Thoughts of any more leisure time ashore were quickly erased as Lober made his way back to the *Wolf*. In addition to being a member of the auxiliary gang, he held the position aboard ship commonly known as the "fuel king" or "oil king." He was responsible for keeping the ship's engineering officer updated daily on the fuel levels of the ship's tanks. "I had to go down into the hull every morning to check each tank by the tubes that went into them," he recalled. "You could read how many gallons remained in each tank and then make your reports."

In different areas of Manila, submariners got the word on Pearl Harbor in their own ways and hurriedly made their way back to their boats. *Seawolf*'s senior radioman, Joseph Melvin Eckberg, sat down for coffee at the Plaza Cafe in Manila, trying to shake off the effects of the past night's drinking. "The place seemed to be seething with excitement," Eckberg recalled. Something had obviously happened which he was unaware of. One of the Filipino workers finally told him, "You no hear Japs bomb Pearl Harbor?" Eckberg suddenly "felt the tension" and knew it was time to race for his ship.[9]

Fireman first class Paul Zimmerman, a native New Yorker who was a member of the *Wolf*'s auxiliary gang, was also on leave with a number of his buddies. "About 4 a.m., this air raid alarm went off and we all scuttled back to our ships in Cavite's naval yard." The city had quickly come to life. "The activity along the water was different," said Zimmerman. "Everyone was armed and boats were carrying everyone back to their ships out in Manila Bay."

The first day of war would be a busy one for *Seawolf*'s crew. On the bridge, signalman Franz stayed busy reading messages from other ships. Another of his messages from *Canopus* called for all submarine captains to come aboard the sub tender for a conference.[10]

Seawolf got underway at 0825 to move away from *Searaven* and her tender. She anchored at 0833 in Canacao Bay and Lieutenant Commander Warder turned his ship over to Lieutenant Deragon. Warder then took a launch over to *Holland*, the flagship of Captain John Wilkes, the commander of submarines at Manila. The impending war and then the Japanese attack brought about changes for Wilkes. He was due for rotation back to a new assignment and his planned replacement was Captain Walter Edward "Red" Doyle, who had no experience commanding submarines in Asiatic waters yet. Admiral Hart informed Wilkes that he would not be returning to the States just yet. Wilkes would instead remain as "special advisor" to Red Doyle, in effect retaining his billet as Commander, Submarines Asiatic Fleet.[11]

Lt. Cdr. Warder found the conference of skippers busy this morning. In addition to senior sub leaders Wilkes and Doyle, their chief of staff, Commander Jimmy Fife, and operations officer Stuart "Sunshine" Murray were present. Commander Murray served as commander of Submarine Division 21, which included *Salmon*, *Seal*, *Skipjack*, *Sargo*, *Saury*, and *Spearfish*. As the sub captains mustered in

Wilkes' quarters on *Holland*, the first air raid siren—thankfully a false alarm—sounded at 1000 over Manila.[12]

Murray handed out secret operations orders to each submarine skipper. Wilkes felt that the first patrols made by these American sub skippers would be "the most dangerous and would also be the most informative as to enemy methods on formations, convoying, [and] anti-submarine warfare." The first patrols would therefore be limited to three weeks in duration in order that the lessons learned by these sub crews could be quickly disseminated for the benefit of future patrols.[13]

Warder was ordered to be sparing with *Seawolf*'s torpedoes. John Wilkes had full faith in the secret magnetic exploder that was included in *Seawolf*'s warheads. For merchant ships, the skippers were urged to use only one torpedo if at all possible. Capital ships warranted as many torpedoes as the skipper saw necessary. The priority for attack was capital warships first, loaded transports second, followed by light forces, transports, and supply ships in ballast.

While at sea, Warder and his fellow skippers would receive daily communications from Captain Wilkes and Admiral Hart from Manila. At night, while the subs charged their storage batteries on the surface, high-frequency radio broadcasts would be transmitted from Cavite or Corregidor. *Seawolf*'s radio operators could also be reached up to about 1,000 miles away during the day while submerged at 50 feet or less with low-frequency radio.[14]

Captain Wilkes advised Warder and the other skippers to use caution on their first patrols. Sunshine Murray was blunt. "Listen, dammit," he said. "Don't try to go out there and win the Congressional Medal of Honor in one day."[15]

Seawolf and her sister submarines were part of a limited number of combat vessel afforded to the Asiatic Fleet of the Philippines. They were in the most advanced position to intercept Japanese ships and their survival was critical.

"Your crews are more valuable than anything else," Murray concluded. "Bring them back!"

Lieutenant, junior grade (jg) Richard Holden was the duty officer when Lt. Cdr. Warder's launch returned from the meeting on *Holland* after 1030 on 8 December. He watched as his skipper climbed back aboard carrying a sealed envelope that held his secret operations orders.

An energetic, black-haired 1937 Annapolis graduate, Dick Holden was well qualified to command men. This Rutland, Vermont, native's family was one of the oldest military families in America, with officers in constant service dating back to Justinian Holden, who was a captain in the Massachusetts militia when the Peacock War started in 1632. In the 300 years following this, the proud Holden family had honorably served in every notable conflict the nation had seen.[16]

As a midshipman, Dick Holden participated in track for four years at the Academy. His first assignment after graduation was that of a junior gunnery officer aboard the cruiser *Vincennes*. He had then served aboard the cruiser *Salt Lake City* until beginning submarine school in New London in June 1939. He completed submarine training in December and reported aboard *Seawolf* as her fifth officer just in time for her commissioning that month in 1939.

Two years later, Holden was a lieutenant, junior grade, and was the third senior officer aboard his submarine. Although war was believed to be imminent, it seemed the farthest thought when Holden and his fellow officers had enjoyed the traditional Army-Navy Day festivities at the Los Tamoros Country Club in Manila on 29 November 1941. The Army had competed against the Navy in a "polo" match. Instead of horses, the men used little Filipino bicycles and enjoyed liquor chuggers from little tables set up all around the playing field. By the time the event was completed, most were too drunk to care who had won and most of the borrowed bicycles were in shambles from head-on crashes.

During the afternoon, Brigadier General Jonathan Mayhew "Skinny" Wainwright had kidded Lt.(jg) Holden about his home state Vermont, a former free republic before joining the United States. War was on the horizon and Vermont boldly declared war on Japan long before the U.S. became involved. "I understand that Vermont is offering a $500 bounty to anybody who signs up," chided Wainwright.

"Sir, that's only for the enlisted men," Holden retorted.

Two weeks later, *Seawolf* was scrambling to get underway for war. Dick Holden was a large man, towering over his crew with his six-foot-four-inch height. He also had a deep bass voice which allowed him to bark authoritatively when necessary. Preparing his boat for war this day required no challenge, however. The men of the *Wolf* had been expecting war and the urgency of getting their boat underway for war patrol moved them.

Dick Holden was *Seawolf*'s engineering and electrical officer as well as her battle stations diving officer. Although every department aboard ship was stressed with preparing to ship out to war, he had a special dilemma to contend with. One of his engines was completely disassembled on the main deck in preparation for the *Wolf* going into overhaul in two days.

Fortunately, *Seawolf* had an able assortment of machinist's mates who immediately attacked the problem. Holden had four chief petty officers in his engineering department: David Butler, Carl Henry Enslin, William T. Metz, and Lloyd W. Sandridge. All four were plankowners, men who had put *Seawolf* into commission. At 32 years of age, Chief Enslin was an old-timer as compared to some of his teenage firemen and oilers. Born and raised around Lake Arial, Pennsylvania, Enslin was known to all aboard as "Swede." Enslin had joined the Navy in 1927 and completed submarine school at New London in 1929, before most of the *Wolf*'s crew were even in high school. He had since been assigned to the tender *Holland*, *S-29*, *S-36* on China Station, *R-11* at New London, and *S-40* before going to Portsmouth to join *Seawolf*.

Sandridge was the senior engineman while Dave Butler and Swede Enslin each headed up one of the two engine room gangs. *Seawolf* had two main engine rooms, one forward and one aft. In the forward room, Enslin and Sandridge could count on able plankowners like Albert Hershey, John Street, and Lucien Rajotte to get the job done this day. The second senior man in the forward room was machinist's mate first class Otis Charles Dishman. At 38, he was twice the age of some sailors aboard the *Wolf* and he looked the part of the old salt. Dishman's body and muscular arms sported various tattoo designs, from trains to flowers to lovely ladies.[17]

Chief Dave Butler from Brooklyn was in charge of the after engine room with Machinist's Mate 1/c Salvatore "Sandy" Randazzo— another plank owner—as his able assistant. Collectively, the firemen, oilers and enginemen who lived in the heart of *Seawolf*'s engineering plant called themselves the "black gang."

In addition to the machinist's mates who ran the ship's diesels, there was a small specialty group attached to Lt. Holden's engineering department known as the auxiliary gang. "There were only a few of us auxiliarymen," explained fireman Paul Zimmerman. "I stood duty in the control room, even at battle stations. We took care of the compressors, the air conditioning equipment, refrigeration equipment, and the hydraulic equipment." Loading the ship and getting the engines back together was quite a task, but the *Wolf*'s crew was all over it. "We were well trained and had all been together for more than two years prior to Pearl," Zimmerman related. "So, we had no problem getting her ready for war."

Chief Bill Metz was the auxiliary boss over Zimmerman, MM1c Pete Lober, and MM1c Maurice "Red" Jenkins. "We all worked together," stated Lober. "There were only four of us in the auxiliary gang. I also worked in the engine rooms, too. I had to take care of the injectors that made the water." As the "oil king," Lober was also responsible for monitoring the ship's fuel consumption with chief engineer Dick Holden.

The black gang and auxiliarymen worked tirelessly on 8 December to reassemble one of the *Wolf*'s partially disassembled engines for patrol. Another important task was bringing aboard live ordnance. The billet of gunnery and torpedo officer was that of *Seawolf*'s fourth officer, Ensign Douglas Neil Syverson from Michigan.

Fresh from the Naval Academy class of 1939, 25-year-old Syverson was the newest face in the wardroom, having arrived aboard just two weeks prior on 17 November. He had served as communications officer aboard the battleship *Pennsylvania* and as a gunnery officer on the destroyer *Kennison* before being sent to Mare Island, California, in 1941. He there served as communications officer of the cruiser *Augusta*, which was undergoing an extensive overhaul. Doug Syverson escaped this duty after four months by applying for submarine school in New London.

In his role as gunnery officer, he would take over Bill Deragon's old post on the torpedo data computer for attacks, although the Exec would continue to work with him for a while. Syverson was also *Seawolf*'s communications officer, in charge of her radio gang and decoding all messages.

Immediately after returning to *Seawolf*, Lt. Cdr. Warder spoke quickly with Lieutenant Deragon and announced that the *Wolf* was to make immediate plans to go to sea. Deragon called up Chief Souza and had him organize his work parties to begin loading stores, ammunition, and live torpedoes.

"We had to stop at Corregidor and load torpedoes because we didn't have any aboard," recalled Paul Zimmerman. "The first couple of days after we had come back from Shanghai, we had been busy unloading equipment. All of our engine parts were scheduled to go over to the Cavite Navy Yard for overhaul."

Zimmerman and his fellow crewmen were now charged with bringing everything back aboard that would be necessary for a war patrol, including the torpedoes. Ensign Syverson worked with Chief Souza's deck gangs and his torpedomen to unload their practice torpedoes and stow 24 live warheads aboard, each complete with the new magnetic exploder. The torpedo gang rigged their booms and swung the huge warheads over from a *Canopus* boat, easing them down through the hatches.

With his resounding voice, Chief Souza instructed the boom workers to swing over the war fish. *Seawolf* could carry 20 warheads below decks, split between her two torpedo rooms. An additional four torpedoes were stowed in the *Wolf*'s superstructure. By opening a section of the topside decking, the torpedo gang used the king post and boom to jockey torpedoes into storage tubes located below the main deck just forward of the conning tower. Using chain hoists they were gingerly placed in their special holding tubes. Now fully loaded with two dozen torpedoes, *Seawolf* would have to fire some of her war fish before these four spares could be transferred below while at sea. That would be another tricky operation in itself.

Even as the loading progressed during the day, the war moved closer to *Seawolf*. Radio news commentator Don Bell, broadcasting from KZRH in Manila, was a well-known voice to the men of the Asiatic Fleet. He was on the air early on 8 December announcing that the Japanese had attacked Pearl Harbor. His reports by mid-day were far more alarming. Japanese aircraft were overhead and were bombing the Army's Clark Field and nearby Iba Field. From atop one of Manila's tallest buildings, Bell would have a bird's-eye view of events over the coming days.[18]

Bell's broadcasts were picked up by *Seawolf*'s radiomen and played for all as they sweated to load their ship with supplies and strip it of unnecessaries. Forward of the radio shack and the control room, yeoman first class John Edward Sullivan sat in the tiny little ship's office, typing an updated sailing list of the crew of *Seawolf*. During war patrols, Sullivan would be charged with maintaining the ship's detailed war diary, a concise record of all attacks, course changes, and maneuvers during patrol. When in port, the yeoman served as the boat's chief record keeper. A big, blond-haired thirty-two-year-old Irishman from New York, "Sully" was also the paymaster in port. Sullivan's yeoman shack was right beside the officers' wardroom and contained his little desk, typewriter, and file cabinets for the ship's records. In his off time, Sullivan had a special knack for baking cakes in the galley.[19]

On 8 December, Sullivan received the papers for one new hand who reported aboard from *Canopus*, ship's cook second class Harold Lankford Wright of St. Louis. "Gus" Wright was a welcome addition to the *Wolf*'s little galley, where SC1c Auston Baker and his assistant SC2c Bill Mallory were charged with continually feeding the crew of 66 officers and enlisted men. "Our cooks made the most wonderful bread," said Paul Zimmerman. "The submarine force always had the best rations." Baker's cooks were assisted by two Filipino officers' stewards, OS3c Mariano Bugawisan and OS3c Brigido Tamayo. Charged with serving the submarine's officers and maintaining a little pantry just off the wardroom, both men worried about their families ashore in the Philippines as the radio brought news of more bombings.

In the crew's mess room, the cooks put on extra coffee to help with the hangovers of those who had enjoyed themselves ashore a little too much the previous Sunday evening. Senior cook Baker was a plank owner, and he had taken all of the crew's taunts and jests about his cooking in stride for two years.

Mallory, on the other hand, was still coping with the crew's occasional lack of concern for his feelings. Once, one of his shipmates returned back aboard in the early morning hours after a long night of drinking. James "Red" Snyder, a member of the after engine room, came down the hatch and made his way back toward his bunk. Mallory had just gone into the reefer storage below deck for provisions without installing the restraining straps across the passageway. Snyder came stumbling through, completely oblivious to the open plating in the floor. Although observers yelled, "Watch it!" Snyder stepped right on the cook's head at deck level as if it were something he did routinely and nonchalantly moved on. By the time Mallory emerged with a tirade of profanities, Snyder was gone toward the wash room and the cook was faced with only the roar of laughter from his galley. "The breakfast crew was in stitches of laughter," recalled crewman Hank Thomson.

Good food and good-natured chiding was all part of the daily life of a submarine cook. During the morning of 8 December, it was all business as launches from *Canopus* brought out fresh food and dry provisions for the *Wolf*'s stores. The provisioning of the ship for war patrol was a duty that fell to her most junior officer, Ensign James Mercer, a slim Reservist from White Plains, New York, with thick black eyebrows. Mercer had attended the University of Michigan and was the only non-Academy officer aboard *Seawolf*. He had enlisted in the Naval Reserve as a seaman, first class, in 1935 and had completed that enlistment in 1939. After his University of Michigan schooling, Mercer made a practice cruise on the battleship *New York* and then completed midshipman's school at Northwestern University in early 1941. He was then commissioned an ensign in the Reserve and ordered to submarine school at New London.

Twenty-five-year-old Jim Mercer was assigned to *Seawolf* on 24 August 1941. As "George," or the lowest ranking officer, he was assigned as *Seawolf*'s commissary and assistant engineering and electrical officer. He had been aboard ship for three months and shared the engineering officer's stateroom with Dick Holden. Located just beyond the officer's wardroom, the engineer's stateroom had two bunks that could be turned down for sleeping.

Ensign Mercer worked with Auston Baker's galley gang to store all the items coming down the hatches. Eager hands carried the goods below and stowed them. A hand brigade was formed to receive each box and pass it through the hatch below decks and down to its proper storage area. Canned meats, sacks of coffee, rice and beans, milk, and other goods were passed below toward the galley.

From below, items not needed for a war patrol were passed back over the side: cans of paint and various tools that would not be needed. "The *Wolf*'s spit-and-polish days were over," noted radioman Mel Eckberg.[20]

The *Wolf* had been waiting for war for two years.

The crew was well versed in submarine service. Most had been in the Navy for years. Many of the older men were married and had started families. Many had enlisted in the Navy at age seventeen or eighteen and had known no prior job. The 50 per cent additional pay for sub duty had attracted most. The quarters were tight and the work was hard, but the men developed a love for submarines that could not be taught.

8 December 1941 found *Seawolf* with 66 hands—61 enlisted and five officers. This included ten chief petty officers (CPOs), one for about every five enlisted men under their direct supervision. This was an unusually high allotment of CPOs. "Our boat only rated three to four CPOs," said crewman Bob Hanson. "When the war broke out, we had several who had just rated but had not yet been transferred."

The *USS Seawolf* had been launched on 15 August 1939, at Portsmouth Naval Shipyard in Kittery, Maine. Her keel had been laid down on 27 September 1938, by the Navy Yard. Following this, the newest United States submarine had come to life at Flatiron Pier on the Piscataqua River at Portsmouth, New Hampshire. Soon after her launching, work stopped at the Portsmouth Navy Yard on 13 September 1939, as the sunken *Squalus* was towed back into the yard and placed in a nearby drydock. Some of the earliest sailors to report to *Seawolf* were there as the bodies were removed from *Squalus*, a grim reminder of the dangers of their business.

Among the first of her crew to arrive at Portsmouth prior to her commissioning was Lieutenant Frederick Burdett Warder, destined to be her skipper. A small, trim man with blue eyes, Warder was born in Grafton, West Virginia, in 1904. He had been a boxer at the Naval Academy. Following graduation in June 1925, he spent two and a half years on the cruiser *Milwaukee*, but he was eager to have his own ship.

"I wanted an early command," he later stated. "There was no future that I saw in the surface Navy—not fast enough." He asked for and received permission to enter submarine school in New London in January 1927. His first submarine assignment was on *S-16* down in Panama. With no air conditioning, "I had no trouble keeping my weight down." Warder worked his way up to engineering officer and then Exec of *S-16*, being aboard her through July 1932.[21]

"Next, I went to postgraduate school for two years in marine engineering," he related. He attended Naval Postgraduate School at Annapolis and continued his marine engineering at the University of California at Berkeley, earning his Master of Science degree. "Then I went to China with the *S-40* for about a year as Exec and Navigator." Warder's next stint was as squadron division engineer for Submarine Division Ten. In 1936, he was given command of *S-36* and held that command through May 1937.

Freddie Warder was then assigned as shop superintendent at the Navy Yard, Portsmouth, New Hampshire, from July 1937 through October 1939. At that time, he was ordered to assist in the fitting out of the new submarine *Seawolf* at his shipyard and he was thrilled to find that he would be her prospective commanding officer. It meant moving from the old 200-foot S-boats with no air conditioning to being the prospective commanding officer (PCO) of a new *Sargo*-class boat, which were 310 feet long and had air conditioning. He would take command of the last so-called "Sea-boat." *Seadragon*, *Sealion*, *Searaven*, and *Seawolf* were all commissioned within five weeks of each other.

"I was the youngest of the submarine commanders" to get a Sea-boat, Warder recalled. Freddie Warder's stateroom had a little desk, depth gauge fitted into the wall, a gyrocompass repeater, and a night bell to call for his mess attendants. His stateroom was right beside the officers' wardroom and next to the CPO quarters. Warder was happy that the *Raven* and the *Wolf* had the new Winton engines installed. "The *Dragon* and the *Lion* had the horrible HOR engines." He felt the HORs—eventually cursed as "whores" by the enginemen who struggle to keep them repaired—were a poor engineering concept that "should never have been bought." The HORs "smoked like hell on the surface."[22]

The commissioning pennant was run up *Seawolf*'s mast on 1 December 1939, and the members of her commissioning crew officially became known as "plank owners," with each actually being given a small piece of her wooden decking. Lt. Freddie Warder published his orders as commanding officer and the first watch was posted, as SS-197 was officially installed as a unit of the United States Navy. The first *Seawolf* had been authorized in 1909, although her name was changed to USS *H-1* in 1911, prior to her being placed in commission. *H-1* operated with the Pacific Submarine Fleet from

USS SEAWOLF (SS-197)

Launching of the *USS Seawolf* on 15 August 1939 at Portsmouth Naval Shipyard in Kittery, Maine. *Official U.S. Navy photo.*

Port side view of *Seawolf* (SS-197) taken August 1940 shortly after her completion. The 197 numerals on her bow and conning tower were painted over before she entered the war in 1941. *Official U.S. Navy photo.*

1913 to 1917, before being based Stateside. She grounded off Santa Margarita Island on 12 March 1920, and sank twelve days later during salvage operations.[23]

Warder's new *Seawolf* was impressive. She weighed 1,480 tons and was 310 feet long. Her living spaces were air-conditioned and she could make 20 knots surface speed. She was of the *Sargo* class of submarines, capable of a test depth of 250 feet due to her 11/16-inch steel pressure hull. *Sargo* boats were considered to be improved S-class boats. They were slightly wider and longer than the previous *Salmon*-class submarines which had been approved in 1936. She had four forward torpedo tubes and four after torpedo tubes. She could carry 20 torpedoes internally plus another four externally in special tubes in her superstructure.

Seawolf and her sister *Sargo* boats had seven watertight compartments in addition to her conning tower. Her four Winton diesels could generate 5,000 horsepower on the surface. She also had two auxiliary generators and four electric motors which could generate 2,660 horsepower while submerged. She ran on two 126-cell batteries which were capable of driving her submerged for 48 hours at 2 knots. At full speed, her batteries could be depleted in just over an hour. *Seawolf*'s range was 10,000 miles on the surface at 10 knots with 109,025 gallons of diesel fuel. She could carry enough food and provisions for her crew of 66 to patrol for up to 75 days.

Seawolf was the last of the ten *Sargo* boats and Freddie Warder was the youngest skipper of one of the last four "Sea" boats. The Electric Boat company built *Sargo* (SS-188), *Saury* (SS-189), *Spearfish* (SS-190), *Seadragon* (SS-194), and *Sealion* (SS-195). Mare Island Navy Yard in California built *Swordfish* (SS-193). The other *Sargo*-class boats were built at Portsmouth: *Sculpin* (SS-191), *Sailfish* (SS-192), *Searaven* (SS-196), and *Seawolf* (SS-197).

Freddie Warder actually came close to losing command of *Seawolf* on the brink of war. While overseeing her construction at Portsmouth, he injured his knee in a fall from a ladder. He escaped talk of a transfer to *Searaven*, but by November 1941, he was suffering from a bad knee, a stomach ailment, an ear infection, and aching teeth. Contemporary Dick Voge was considered as a replacement to take command of the *Wolf* while his own boat *Sealion* was undergoing an overhaul. Warder wrote home to his wife Mary that he was "getting more shots than ever before" in his life. He managed to get his boat back out on maneuvers the next week and retain his command. He later commented that removal from *Seawolf* for his health reasons "would have broken my heart." In spite of the bad knee that would sometimes slow him in clearing the bridge, Warder stayed on as *Seawolf*'s skipper and division commander John Wilkes complimented him on a "well running boat," no small praise coming from this tough leader.[24]

Freddie Warder's original second-in-command was Lieutenant James Alvin "Caddy" Adkins. "He was very knowledgeable," said Warder. "The crew admired him and respected him and all he was trying to do was take the load off me." One crewman, Bob Hanson, later said of Adkins, "I thought he was great but tough. Half the crew hated his guts—the other half liked him."[25]

After two years as *Seawolf*'s Exec, Adkins was detached to his own command in 1941 before the outbreak of war. He took command of the old S-boat *S-21*. His boat broke down on patrol in the Atlantic, but Warder later helped Adkins secure his own command again. In command of *Cod*, Caddy Adkins was officially credited postwar with sinking four ships.

Opposite page: Officers and crew taken on deck of the USS *Seawolf* (SS-197) 5 August 1940 at Portsmouth, New Hampshire. *First row (l-r):* CTM William S. MacDowell, CMM Ray H. Wilcoxen, Lt. William T. Kinsella, Lt. James A. Adkins, Lt. Cdr. Freddie Warder, Lt. Bill Deragon, Lt. Dick Holden, EM1c Evan E. Watts and CEM Clinton Jobe.
Second row: MM1c David Butler, TM2c George S. Tremblay, TM1c Robert Langford, S1c Albert Delnigro, F1c William T. Metz, EM1c Edmund C. Capece, TM2c Joseph Perry Jr., MM1c Albert E. Hershey, GM2c John N. Bennett, and FC2c George D. Leffingwell.
Third row: MM1c Lloyd W. Sandridge, F1c Peter N. Lober, SC1c Auston Baker, SM2c James V. Carney, F2c Orval C. Cross, MM2c Asa O. Apperson, Y1c John E. Sullivan, TM1c Clifford E. Harris, S1c Arthur E. Lamberson, RM1c Joseph M. Eckberg, and SM2c Frank Franz.
Fourth row: TM3c Arthur T. Campbell, TM1c Paul Connolly, MM2c Maurice Jenkins, RM2c Robert L. Hutchison, S2c Joseph R. O. Gervais, F1c William R. Butler, MM2c Lucien T. Rajotte, EM3c Mason Poole, and SC3c Carleton R. Lipham.
Fifth row: MM1c Carl H. Enslin, MM2c John E. Street, TM3c Clarence V. Kibbons, EM1c Alexander P. Mocarsky, Matt2c Melton Evans, EM3c Lawrence W. Crane, F1c Roy W. Bateman, F2c Frank C. Dvorak, MM2c Salvatore Randazzo, F1c Glenn K. Hickman, MM2c Otis C. Dishman, MM1c Edward K. Roszel and EM3c Lee Bob Parden.
Courtesy of Pete Lober.

30   War of the *Wolf*

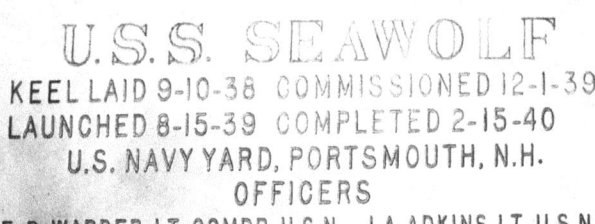

U.S.S. SEAWOLF
KEEL LAID 9-10-38   COMMISSIONED 12-1-39
LAUNCHED 8-15-39   COMPLETED 2-15-40
U.S. NAVY YARD, PORTSMOUTH, N.H.

OFFICERS
F. B. WARDER, LT. COMDR. U.S.N.   J. A. ADKINS, LT. U.S.N.
W. N. DERAGON, LT. JG. U.S.N.   W. T. KINSELLA, LT. JG. U.S.N.   R. HOLDEN, ENS. U.S.N.

CHIEF PETTY OFFICERS
McDOWELL, W. S., CTM.   WILCOXEN, R. H., CMM.   JOBE, C., CEM.

CREW

APPERSON, ASA O., MM2C
BAKER, AUSTON, SC1C
BATEMAN, ROY W., F1C
BENNETT, JOHN N., GM2C
BILKEY, JOHN H., EM2C
BRENGELMAN, HENRY B., EM3C
BUTLER, DAVID, MMIC
BUTLER, WILLIAM R., F1C
CAMPBELL, ARTHUR T., TM3C
CAPECE, EDMUND C., EM1C
CARNEY, JAMES V., SM2C

CASLER, JAMES B., QM1C
CRANE, LAWRENCE W., EM3C
CRAWFORD, CHESTER, SEA1C
CROSS, ORVAL C., F2C
DELNIGRO, ALBERT, SEA1C
DISHMAN, OTIS C., MM2C
DVORAK, FRANK C., F1C
ECKBERG, JOSEPH M., RM1C
ENSLIN, CARL H., MM1C
EVENS, MELTON, MATT2C
FRANZ, FRANK JR., SM2C
GERVAIS, JOSEPH R. O., SEA2C
HARRIS, JOHN S., TN3C
HARRIS, CHARLES E., MATT1C
HARRIS, CLIFFORD E., TM1C

HENSON, DUDLEY A., PHM1C
HERSHEY, ALBERT E., MM1C
HICKMAN, GLENN X., MM2C
JENKINS, MOURICE, MM2C
KIBBONS, CLARENCE V., TM2C
LAMBERSON, ARTHUR E., SEA1C
LANGFORD, ROBERT, TM1C
LEFFINGWELL, GEORGE D., FC2C
LIPHAM, CARLETON R., SC3C
LOBER, PETER N., F1C
METZ, WILLIAM T., F1C
O'HERN, JOHN P., RM2C
PARDEN, LEE B., EM3C
PERRY, JOSEPH JR., TM2C
POOLE, MASON, EM3C

RABYK, ANDREW, SEA1C
RAJOTTE, LUCIEN T., MM2C
RANDAZZO, SALVATORE, MM2C
REICHENBACH, HARRY A., EM1C
ROSZEL, EDWARD K., MM1C
SANDRIDGE, LLOYD W., MM1C
STREET, JOHN E., MM2C
SULLIVAN, JOHN E., Y1C
TREMBLAY, GEORGE S., TM2C
VAUGHAN, LEONARD B., RM1C
WATTS, EVAN E., EM1C

(Above) This *USS Seawolf* commissioning plaque from 1939 shows the names of her original crew members. *Courtesy of Carl H. Enslin Jr.*

(Left) One year later, this sketch helped the *Seawolf* crew celebrate the ship's first year in commission. *Kenneth Mallough collection, courtesy of Bobbie Mallough.*

With the departure of Adkins and Lt. Bill Kinsella prior to December 1941, the *Seawolf* wardroom shuffled. Dick Holden moved up to engineering and electrical officer. Lieutenant Bill Deragon, the original gunnery officer, moved up to be the new executive officer. Warder considered Deragon to be "a good man of French Canadian ancestry, born around Albany, New York." Another of the crew, Paul Zimmerman, said of Deragon: "He was a great guy, a bear of a man. He was the one who qualified us in submarines when we came aboard." Warder thought highly of Deragon—calling him a "tower of strength"—and would be pleased to be able to keep him as his Exec for the remainder of his time on *Seawolf*. One of his squadron commanders, John "Babe" Brown even told Warder that he thought Deragon was "the cream of the crop."[26]

During the early training days, Warder had encouraged his men to "learn the boat" and learn they had. In Building 150 of the submarine barracks, the crew had studied the *Wolf* for eight hours and more per day. Warder and his original officers—Adkins, Deragon, Kinsella, and Holden—assembled their crew into three watch sections and schooled them on every aspect of fighting their vessel to its fullest potential.

Seawolf made her first dives in February 1940 near the Isle of Shoals, where *Squalus* had sunk. In the weeks that followed, Freddie Warder drilled his crew on emergencies and situations of all sorts. Diving officer Bill Kinsella and his planesmen had tense moments when a practice of diving with hand-operated planes resulted in the bow planes jamming on hard dive. Chief Otis Dishman, the strongest man in the boat, could not budge the planes, even with the assistance of two other men. The fathometer showed *Seawolf* was only in 230 feet of water as she plunged toward the bottom out of control. She took a terrific down angle and went past 150 feet without leveling. Lt. Cdr. Warder ordered the safety tanks blown to increase buoyancy and shouted, "All back, emergency!" She passed 200 feet before her descent was checked. By blowing all of the water from her ballast tanks, *Seawolf* was now positively buoyant and she rose fast toward the surface, popping up like a cork out of a bottle. She was none the worse for wear, but it was a sobering experience for all hands.

After fitting out, *Seawolf* departed Portsmouth on 12 April 1940 for her shakedown cruise, which lasted until 21 June, taking her as far south as the Panama Canal. She stopped in New York, Tampa, Florida, Galveston, Texas, Corpus Christi, Coco Solo, Canal Zone, Cristobal, and Annapolis, Maryland, along the way before moving back up the East Coast again. Returning to Portsmouth, *Seawolf* had then gone through post-shakedown overhauling to correct errors and defects.

In Portsmouth, the workers corrected minor errors, fixed leaks, and tuned up what the *Wolf*'s machinists could not fix at sea. She left for Newport, Rhode Island, in September 1940, and there went through extensive torpedo training for six weeks. The torpedo crews fired practice torpedoes at dummy ships. At New London, the crew spent hours practicing in the escape tank with Momsen lungs—a special device designed to help them escape from a sunken submarine if necessary.

In October, *Seawolf* was assigned to the Pacific Fleet in San Diego. When she arrived from New London, her men were given five days leave. Three days later, the leave was cut short by a notice that *Seawolf* was being reassigned. The *Wolf* departed California on 16 November 1940 for Pearl Harbor.[27]

Seawolf reached Oahu five days later. While at Pearl Harbor, Chief Sullivan and Lieutenant Deragon acquired a washing machine for *Seawolf*. Sully had brought the idea to Lt. Cdr. Warder and he agreed that it could be purchased with the ship's slush fund. In the local Hawaiian department store, Sullivan selected a gleaming white enamel electric washer which was delivered by boat to the *Wolf*. Getting it below decks proved to be no simple matter. And then there was the chore of finding any free space to install it. Sullivan and Chief Metz's auxiliarymen went to work rebuilding the crew's washroom over the next three days to make space for the new washer. The crew nicknamed her "Baby" and many a pair of greasy dungarees would be cleaned by her.[28]

Warder approved of the full-size washing machine but found that the washbasins located in the washroom were in the way. The crew hoped to have them moved to the forward bulkhead to allow room. Warder supported the idea and so requested the wash basins be moved at *Seawolf*'s pending overhaul period. He found that "forward authorities disapproved request," but kept his Baby aboard.

The third day at Pearl Harbor, chief of the boat Eddie Souza supervised the loading of live torpedoes aboard the *Wolf*. The forward torpedo room was presided over by Chief Torpedoman Robert "Squeaky" Langford, a 35-year-old Iowan plankowner with a high-pitched voice. His forward gang—Arthur Lamberson, Keith Bjerk, Rudy Gervais, and Wilbur Chubbuck—carefully helped swing the booms with the torpedoes. The forward gang was tight, and most had nicknames, like "Pinky" Bjerk, a ruddy-complected sailor from Minnesota of Norwegian ancestry. In addition to "Pinky," those who could not pronounce his name were also prone to call him "Beejerk."

The torpedoes were moved down the skids into their tubes and racks in Langford's forward room. During battle stations, Langford lost the use of torpedoman Gervais, who was assigned to the conning tower. "He was so good at driving the boat that the captain assigned him as our permanent battle stations helmsman," said Bob Hanson.

The process was just as demanding to load the other live warheads into the after torpedo room. This gang was presided over by TM1c John Gibson, who had come aboard in the Philippines from one of the older P-boats. Gibson's after gang—Bill Reiland, Clarence Kibbons, and striker John Neil—was equally talented and equally tight.

Before departing Pearl, *Seawolf* received two new sailors on 30 November 1940: RM3c Paul Leroy Maley and F1c Edward Chapman. Chapman was a large man who was full of humor and energy. An Oklahoma native, he was part American Indian and quickly became known to his buddies as "Chief." Chapman would remain aboard *Seawolf* for all of her patrols. Maley, transferred from the submarine *Dolphin*, was well experienced on the sound gear. Maley bonded instantly with his senior radioman, Mel "Red" Eckberg, and took the bunk opposite from Eckberg's in Langford's forward torpedo room.

Eckberg had joined *Seawolf* during her construction in Portsmouth in August 1939, from previous service on *Plunger*. A husky man with piercing blue eyes and red hair, Eckberg already had 12 years of service in submarines, having joined the Navy at age 18 to follow in the footsteps of his older brother. During battle stations, Maley and Eckberg would track enemy ships together. For covering the full 24 hours of sound watches, RM2c Joe Ferguson from Houston, Texas, filled out the radio gang on the *Wolf*.

Yeoman Sullivan added an additional four men to his sailing list on 3 December just before *Seawolf* departed: MM1c Frederick Andrew Zirkel, F2c Paul Zimmerman, S2c Robert Wayne Lents, and QM2c Henry Hanford Thomson.

Zimmerman came over from the destroyer *Litchfield* and joined Chief Metz' auxiliary gang. Zirkel, a thin machinist whose trademarks were a corncob pipe and a pessimistic attitude, quickly became a favorite in Kelly's Pool Hall, *Seawolf*'s mess room. An old China Station sailor who asked to be called "Jew," Zirkel had plenty of sea stories from the Orient and even an illegal Chinese wife and a daughter.[29]

Eighteen-year-old seaman Bob Lents was received from the sub base at Pearl. He joined John Gibson's after torpedo room as a torpedoman striker. "We had air-conditioning in the crew's quarters aboard *Seawolf*," he recalled. "My bunk in the after torpedo room was right by the air-conditioning vent, so I'd have to have a big wool blanket on me."

Quartermaster Hank Thomson, like Paul Maley ahead of him, came from the older boat *Dolphin* (SS-169), replacing a sick signalman. He had previously served four years aboard the cruiser *Cincinnati* and found his only recourse to get off her was to enlist in the New London Submarine School. He found an open bunk below Big Swede Hanson and soon became close friends with this fun-loving man. Once aboard *Seawolf*, Thomson was motivated to earn his dolphins by becoming a fully qualified submariner.

> Qualification entailed drawing all the various systems and valves that made the submarine a ship. We also had to be able to make the torpedoes that were in the tubes ready for firing, be able to start and stop the diesel engines, put the generators on line to the maneuvering room, be able to dive the boat, and so on. Our drawings were checked by the executive officer and we were finally taken through the boat by the engineering officer to demonstrate that we could physically do all of the tasks. The procedure took a minimum of six months for enlisted men and one year for an officer.

Thomson became buddies with "Chief" Chapman, one of the new hands received at Pearl. "When he was mess-cooking, one had best not tarry or his plates would be snatched from him," recalled Thomson. "Chief was big enough not to dispute his ability to get the three messes served quickly."

Chapman and Thomson worked together on their qualification notebooks and went before Lieutenant Kinsella, the engineering officer, together for their physical interview.

Kinsella warned them sternly before starting, "One mistake or hesitation, and I am going back to the wardroom to write to my wife."

Chapman looked at Thomson and nodded confidently. The two aced their qualifications, earning their coveted dolphins.

Five days after pulling into Pearl Harbor, *Seawolf* departed Pearl on 3 December. In company with the submarine *Shark*, she had been ordered to join the submarine force in Manila under Admiral Hart. They were part of his Submarine Division 17. *Seawolf* proceeded to Manila Bay and operated from the Cavite Navy Yard.

Seawolf then spent the majority of 1941 conducting training cruises on gunnery and torpedo operations about the Philippines. "There were countless hours spent at sea with the submarine commander [John Wilkes] riding the tender and having our three divisions of submarines following astern on the surface by columns of ships by divisions engaging in surface tactics which were ordered by use of signal flags and had to be responded to be individual submarines," recalled officer Bill Kinsella. "In other words, it was the old 'squads-right and squads-left' type of drill."[30]

Lt. Cdr. Freddie Warder found it all thoroughly frustrating. Wilkes "had us running around like torpedo boats. It was also bad in avoiding collisions with your mates." Warder felt that most of the pre-war training days were wasted efforts for a submarine. He would much prefer submerging and working on attack procedures. "A shell through the superstructure would finish the submarine," he stated.[31]

Warder was thus not a proponent of gun actions with his little 3-inch deck gun. One of those who would later serve on the *Wolf*'s gun crew, Hank Thomson, recalled: "Warder never wanted to use the gun—except if a desperate circumstance occurred—and he didn't use it."

Lieutenant Kinsella felt that all the prewar training was "just one big bloody mess which never contributed a thing toward improving our readiness. All it did was frustrate us and waste a lot of fuel oil."[32]

Seaman Bob Lents, a trainee in the after torpedo room, stood his share of lookout watches on the periscope shears and helmsman watches in the conning tower during these fleet drills. Lents later recalled some of his experiences during *Seawolf*'s Asiatic Fleet training periods.

> Our cruises were all down in the southern part of the Philippines. We had torpedo practice and gunnery practice. I was on the deck gun during gunnery practice. We all had an "E" for efficiency on our sleeves and painted on the side of our ship for our gunnery efficiency. For torpedo practice, we usually fired dummy torpedoes to pass underneath the old sub tender *Canopus*. If we had a night practice, they had some kind of a light overhead on the ship to show us whether they passed underneath the ship or not. There was a boat off the tender that would retrieve the fish for us.
>
> Captain Warder was a good man and I liked him. I stood a lot of wheel watches while he was up there. Once, we had been in drydock down in Olongapo to have the bottom scraped and repainted. We were going back to Manila and I had the wheel watch. The skipper said, "I want to see how fast this thing will go."
>
> He called for emergency speed and we were going real fast. Then, this sailboat runs out right in front of us. He said, "Hard right rudder!" And I rolled a hard right rudder on the wheel. All of the hatches were open and a lot of stuff was laying on the deck and it was rolling everywhere. It was about a 40 degree turn on the roll and a lot of stuff fell over the side. We just about took water on.
>
> The skipper turned to me and said, "I didn't tell you to sink the damn thing!"
>
> I just said, "You called for hard right rudder. That's what I did."
>
> I met up with Warder after the war and he still remembered me. He remembered that hard turn and we laughed about it.

Lt. Cdr. Frederick Burdett Warder in a view taken in 1942 while on war patrol aboard his *Seawolf*. His heroics earned him the famous nickname "Fearless Freddie," a monicker which he did not allow the crew to use around him.
Photo courtesy of Hank Thomson.

The pre-war training helped mold the *Wolf*'s fire control party, the men who would be responsible for directing her torpedoes against enemy ships. The biggest problem was that they were never allowed to actually fire live torpedoes. In fact, not even the skipper had actually witnessed a live torpedo explosion. In short time, he would come to believe that the inability to fire actual live torpedoes would save many a Japanese ship during 1942.

Freddie Warder was a salty-tongued commander who was well respected by his men. He would become quite vocal on his opinion of ordnance performance in due time. He was equally vocal in defending his crew. In the days prior to the Pearl Harbor attack, he had stood up to Admiral Hart for one of his enlisted men and won.

The incident involved enginemen—tall, husky John Street and little Red Snyder—who had gotten into trouble following a baseball game and beer party ashore on 1 July 1941. "John Street got drunk and hit a shore patrolman, a court martial offense," recalled one of his shipmates. The MPs took both inebriated *Seawolf* sailors to the brig. Upon release, Lt. Cdr. Warder ordered Street back to the boat and reprimanded him in his own way. Snyder, on the other hand, had unloaded a torrent of profanities at the shore patrol and physically resisted his arrest. At captain's mast, Warder found Snyder guilty of the offense, confining him to the ship for ten days and docking his pay forty dollars.

With most skippers, these actions could have warranted a court martial or being thrown off the ship at the very least. Snyder was disciplined and Street, normally a good sailor, was merely threatened. When word of this soft discipline reached one of Admiral Hart's staff members, this officer objected strenuously.

"The whole event was blown out of context," recalled quartermaster Hank Thomson. Freddie Warder ended up going before Admiral Hart in his Manila office. The admiral was livid and threatened Warder with his command if he did not discipline his men more severely. Warder accepted the challenge and stood up to the admiral, saying that he had handled the situation in the manner he saw fit.[33]

The nickname "Fearless Freddie" that would later be hung on Warder certainly included his desire to take on Navy brass as well as enemy shipping. In the face of Warder's adamant stand, Hart finally backed down.

"Admiral Hart could not follow through on his threat because of the imminent war situation," said Thomson. "In the end, he told our skipper something like, 'You'd better sink a lot of ships!'"

"Big Swede" Hanson made it back aboard *Seawolf* on 8 December 1941 in time to go to war. Like Snyder and Street, seaman first class Henry Howard Hanson Jr. had already given Lt. Cdr. Warder a few premature gray hairs. Aboard ship, he was an able sailor who was striking to be a gunner's mate. Off ship, he was known to enjoy his liquor and had many escapades associated with his love for having fun. Sometimes he would return from leave in quite a stew and sometimes he would forget to return on time.

Big Swede Hanson would have his fair share of run-ins with Freddie Warder along the way. On one occasion, he even returned aboard ship with a monkey.

Seawolf had a pair of brothers aboard for her first nine patrols. Seen here on liberty are *(left)* Bob "Little Swede" Hanson and *(right)* his older brother Henry "Big Swede" Hanson. *Courtesy of Bob Hanson.*

"Swede brought aboard this monkey from Zamboanga, when we were operating down in the southern Philippines," recalled Paul Zimmerman. Henry Hanson was ashore with several buddies and his brother, Robert Norman Hanson, a fireman who was part of the ship's black gang. In the pre-war Navy and even the early days of World War II, it was not uncommon for brothers to be stationed aboard the same ship.

Henry Hanson, big and muscular, was known as "Big Swede," while Bob Hanson became known as "Little Swede." Bob had actually gotten on the ship first on 1 April 1941, received aboard from the tender *Canopus* in the Philippines. His older brother had joined the Navy in Minneapolis seven months ahead of him in 1939, but Bob was followed by Henry from *Canopus* to *Seawolf* a month later on 1 May 1941. With a Swede Enslin already aboard, they found they were not the only "Swedes" aboard the *Wolf*. "When you're in the service and your name is Hanson, Olson, or Peterson, they all call you 'Swede,'" Bob related.

The younger Hanson was there for the monkey episode. "My brother bought the monkey off one of the street vendors in Zamboanga." Once Henry smuggled his monkey aboard the *Wolf*, trouble started. "That thing got loose in the superstructure," recalled Zimmerman. "The captain found out and he told Swede, 'Either get ahold of that monkey and get rid of it, or you go! In fact, I think I'd rather keep the monkey!'"

Once the monkey was captured and returned to the street vendor, everyone had a good laugh over it. "Everybody was working on catching him," said Little Swede Hanson. "It took all day."

Freddie Warder would write to his wife during these prewar days that he had "good men in the ship but their activity on the beach has been a bit trying on me—and no damn good on my service reputation." Warder sounded off on such men like an angry father, but he wrote that most took his chastisements in stride. "You know my disposition is not of the sweetest and I think they like my sour moments very well indeed."[34]

Some of his crew were mere boys. Seaman 2c Charles Alfred Johnson reported aboard *Seawolf* only seven weeks before the war began. A seventeen-year-old who had joined the Navy from Kansas City, Johnson became known to his engine room buddies as "Johnny." Having lost his own dad at an early age, he found Freddie Warder to be a father figure to him. Johnny Johnson would later say that he would have gone to war in a rowboat if Captain Warder had asked him to.[35]

Thankfully, Warder may not have heard of another of Big Swede Hanson's memorable returns to the ship after partying ashore some time after the monkey episode. "He returned from leave one day and came across the planks to get aboard ship," said Paul Zimmerman. "He kept walking and walked right off the other side of the ship into the ocean!"

Henry Hanson made it across the deck and down below just fine on 8 December 1941. He had been transferred to the tender *Otus* a month before for treatment while the *Wolf* had gone to sea again. He was happy to return aboard this day to assist gunner John Bennett in preparing their guns for war.

Little Swede Hanson was equally busy with the black gang.

> The duty officer, Dick Holden told us, "Go to work. We're at war." We had one engine laying topside because we were going into the Cavite Navy Yard. We would be in there for months being overhauled. As soon as they said we were at war, we started putting the cylinder heads and pistons down below in the engine room. It was very rapid putting that engine back together again.

Bob Hanson was the forward engine room's oiler. Each engine room had an oiler and a throttleman constantly on duty. "The oiler took all the readings and checked all the equipment," said Hanson. "The throttleman stood between each of his two engines to control speed with the two throttles. He was also in charge of shutting the exhaust valves, air induction and securing the engines when we dived." After the ship submerged, the oilers went to the control room to finish out their shift by rotating in on the bow and stern planes.

Throughout the boat, the men had trained together for many months and knew their jobs well. Key chief petty officers managed the men within their respective divisions. The *Wolf*'s crew was ready. By late afternoon, the ship had been stripped off all restricted or confidential materials except for the publications that were essential to communications officer Doug Syverson.

There were no air raids that afternoon in Manila Harbor, yet Gunner Bennett's men stood watches continually on the bridge machine guns. The loading continued throughout the afternoon. The cooks kept hot coffee and sandwiches on supply through the day. The *Wolf*'s two Filipino messboys kept the stations restocked and ran food to the officers. After lunch, an oil lighter came alongside and refueled the *Wolf*. Auston Baker's supper menu for the *Wolf*'s first day at war was steak, french fries, asparagus, and ice cream.[36]

Following dinner, Warder called his men to quarters on deck. He looked over his line of men and then gave them the news. *Seawolf* would be leaving port this night, heading into the war zone immediately as an escort for a convoy.

"Needless to say, you all know we're not playing any more," he advised his men.[37]

The "History of the *USS Seawolf*" gives a notion of the seafaring traditions that the *Wolf*'s submariners were preparing to join.

> Coins of all denominations flashed over the side and into the water, in keeping with one of the many traditions which are so deeply rooted and have such a part in the submarine service. Their luck so assured, the men of the *Seawolf* took her down, and went hunting.[38]

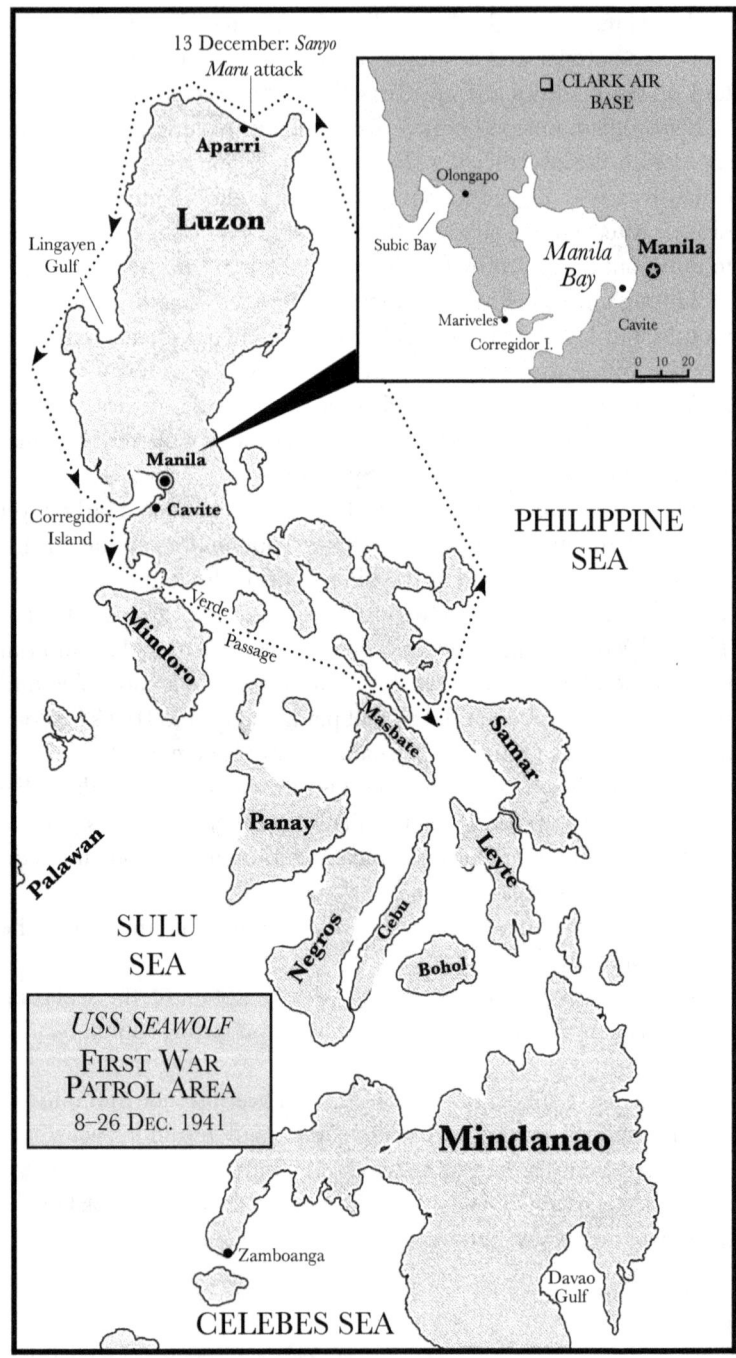

2

Faulty Fish

First War Patrol   *8–26 December 1941*

Freddie Warder——a man who routinely got by on two hours of sleep in peacetime——was eager as he stood on the bridge of his submarine. With lookouts stationed on the periscope shears above him, the *Wolf*'s diesel engines were rumbling and sputtering out blue exhaust smoke. It was still 7 December 1941 in Hawaii but it was 1700 on 8 December in Manila Harbor. Her deck crew was scrambling to stow the mooring lines as the ship began to get underway.

"All back one-third," called Lt.(jg) Dick Holden, the officer of the deck (OOD), from his perch on the bridge.

With *Seawolf*'s configuration, there were engine and helm controls built into the topside bridge superstructure. The duty helmsman repeated the order and the *Wolf* spun toward the harbor entrance. Once orders came to dive, the duty helmsman would secure the bridge wheel and move to the wheel in the conning tower.

Lt. Cdr. Warder still clutched the secret operations orders he had received from the Asiatic Fleet submarine commander, John Wilkes. "It was the first one I'd ever had," he later stated. "Just four or five lines. He said to go to Albany Gulf, which is east of Luzon, and resist Japanese landings. And there were amplifications that I had to go out there with a convoy at night." Warder could not share his submarine's actual destination with his crew until they were safely out to sea. The only other people in the know at departure time were his navigator, Bill Deragon, and his assistant navigator, Jim Casler.

The *Wolf* departed Manila Bay in company with sister submarine *Sculpin* and they rendezvoused off Sangley Point with the aircraft tender *Langley* (AV-3), tankers *Pecos* (AO-6) and *Trinity* (AO-13) and

Two views of *Seawolf* taken prewar. Above, she is seen from another ship as she submerges. Below, the *Wolf* runs on the surface with her prewar 197 numerals still painted on the conning tower. *Courtesy of Hank Thomson.*

the destroyer tender *Black Hawk* (AD-9)—all headed south toward Borneo. By 11 December, 22 of the Asiatic Fleet's 28 submarines would be out to sea, including *Seawolf*. The subs were deployed to protect against rumors of a possible approach of Japanese forces. Palau, lying to the eastward of Luzon in the Philippines, was believed to be a heavily fortified base, so five submarines were sent to be stationed along the eastward approaches to Luzon. These boats were *Seawolf, Sculpin, Skipjack, Tarpon,* and *S-39*.[1]

As the ships headed out from Luzon, the men continued to take in the news that came over the radio from Manila news commentator Don Bell. The Japanese had bombed Davao on the island of Mindanao. They had bombed Zamboanga on the southern tip of the Philippines and they had also bombed Luzon, Hong Kong, and Singapore. More news continued to come in about the surprise attack on Pearl Harbor.

War had suddenly hit home with the men of the *Seawolf* and they were in the thick of it. Scuttlebutt was rampant. The most common gathering place for off-duty men to speculate on the upcoming patrol was the mess hall, which had become known as "Kelly's Pool Hall" to her men. Three tables were set for meals, each top surface containing a rim to keep dishes from sliding off as the boat rolled. There was also a bulletin board for the Station, Lookout, and Watch bills, as well as any other relevant notices that may need to be posted.

Just aft of the mess hall was the galley compartment, where Auston Baker, Gus Wright, and Bill Mallory cranked out the meals. There were two ovens, a four-coil electric stove, a huge coffee urn, a sink, and a Mixmaster for pastries, pots, and pans, which were packed into shelves. The cooks and the two Filipino officers' stewards, Bugawisan and Tamayo, would serve three meals a day from this small cramped space, plus sandwiches, snacks, and an endless supply of hot coffee for watch standers. In the scullery adjacent to the galley, those assigned as mess cooks pealed potatoes and scrubbed dirty dishes.[2]

"Junior members of the crew did mess cooking until they later went on to make higher rates," said Paul Zimmerman. "There was usually somebody from the torpedo gang, someone from the black gang, and so on. If you were junior, you always had a turn at mess cooking." There were games of pinochle, checkers, acey-deucy or blackjack in Kelly's Pool Hall. It was the place to discuss the latest rumors about the ship's heading. The scuttlebutt now dealt with the Pearl Harbor attack and the Japanese moving into the Philippines.

The ship was all business, and yet there were the little personal touches throughout. For some it was the pictures of their wife or child pasted to their locker. In the mess hall, books by Jack London, Zane Grey, and numerous others had filled the double bookshelf. Engineman Otis Dishman had brought aboard a portable phonograph which he kept on a workbench in the engine room. "Dishman loved country and western music," said Bob Hanson. "He had quite a collection of records that he brought aboard before the war." Once *Seawolf* submerged and shifted to battery power, he would start up his record player. "We had one guy who would sneak into the radio room and switch Dishman's record phonograph into the ship's speaker system," related Hanson. "It usually didn't take long before the officers put an end to this."

Torpedoman Squeaky Langford brought aboard a miniature Chinese carved teakwood chest he had acquired in Tsingtao.

Machinist's mate Al Hershey, the boat's top softball player, had a bagful of bats, balls, and gloves which was well used during R&R.[3]

The crew's softball games now seemed like only a distant memory. Reality lay in the war at hand. *Seawolf*'s mission for her first war patrol included seeing the *Langley* task force safely through the narrow Verde Island passage south of Corregidor. The surprise attack on Pearl Harbor was just hours into the history books as the *Wolf* entered the minefield outside of Manila's harbor.

On the bridge, Lt. Cdr. Warder, Lt.(jg) Holden, and their bridge lookouts found that the Army's searchlights were annoying and dangerous. Sweeping searchlights alternately lit navigational buoys, the sky and vessels. "This procedure blinded bridge personnel and could have revealed [the] presence of darkened ships" to enemy aircraft or ships that might be present, Warder noted in his patrol report.

Within the hour, *Langley*'s convoy had cleared the minefield as the submarines followed the carrier tender through the mines. She headed toward Verde Island Pass, with *Sculpin* and *Seawolf* continuing to trail astern of their convoy. Although a half moon kept the sea well illuminated, the convoy kept all of its navigational lights burning brightly, a move Warder found to be "entirely unnecessary."

Seawolf and *Sculpin* submerged at dawn on 9 December with Verde Island 2 miles south of them and left the surface ships on their own. The *Wolf* was at war now and could not afford to be caught on the surface by Japanese aviators. The *Wolf* cruised submerged at 120 feet during the daylight hours on account of rough sea conditions and remained there except for occasional trips to 55 feet for radio sweeps. She surfaced that night at 1808—a half hour after sunset—in complete darkness in a choppy sea.

Warder proceeded immediately on three main engines at 80 per cent power, making 16 knots. His men commenced a battery charge and air charge. The cracking of the conning tower hatch brought a rush of fresh air which blew loose papers and trash toward the engine rooms as the hungry diesels began to suck oxygen.

While submerged, the *Wolf* was driven by powerful battery-powered generators. Her crew included nine electricians mates plus several strikers who were responsible for this equipment. The electrical gang was presided over by three chief petty officers. Clinton Jobe, a six-year Navy veteran from Kansas City, and Edmund Capece, a seven-year vet from Brooklyn, were both *Seawolf* plankowners. Alexander "Pop" Mocarsky, the third chief electrician's mate, only had five years in the Navy but had more years in general life than his companions.

At age 43, his gray hair helped the East Hartford, Connecticut, native earn his nickname. "There were many people in the boats and surface craft whose formal education was limited," said shipmate Hank Thomson. "Mocarsky couldn't pass the written exam for chief before the war but he had managed to repair our gyrocompass just by feeling when the experts from the tender *Sperry* couldn't fix it. We called him 'Pop' because he was older than Captain Warder."

Three of *Seawolf*'s electricians had seen prior experience on older submarines before putting SS-197 into commission: EM3c Henry Brengelman aboard *S-45*, EM1c Larry Crane aboard *R-11*, and EM2c Mason Poole aboard *S-42*. Crane, a New Yorker, was rarely seen without his signature tobacco pipe in his mouth. Hank Brengelman, who became an integral part of the ship's fire control party, was described by one shipmate as "a roly-poly German with pale blue eyes and a love for books."[4]

EM1c John Bilkey had joined the Navy in 1934 and soon chose his specialty. "I went through electrical school in San Diego and then they sent me down to Panama for four years before I came back to the States to help put *Seawolf* into commission," he related. Like the Hanson brothers, EM2c Lee Bob Parden from Nebraska had served on the same ship with his older brother George on the four-stack destroyer *Pillsbury* (DD-227) in the Asiatic Fleet in the 1930s. Later assigned to a sea-going tug, Parden found the easiest escape from that unpleasant assignment was to volunteer for submarine school and the extra pay involved.

The electrical gang was responsible for boiling the iron and other impurities out of the water that they used to service their 252 battery cells. The evening recharging of the sub's batteries was essential in order to keep her properly powered for submerged running during the daytime. John Bilkey later related the length of charging time.

> The battery charge depended on how hard we had pushed them. If you could get enough charge to 'em, you could do it in about five or six hours. If we happened to be in a position where we had to dive, we just cut the charge off in the middle and went ahead and used the batteries anyway. You never took the batteries bottom to top any time. If you took 'em bottom to top, it would take as much sometimes as

Ensign James Burr "Jughead" Casler (left) with Lt. Cdr. Warder in 1942. Casler served as plotting officer for torpedo attacks and was promoted from chief quartermaster into the wardroom. *Courtesy of Hank Thomson.*

twelve hours to recharge them because you had to dribble into them a little bit instead of a full charge.

While the electricians took care of the important task of charging batteries during the night, the ship's navigators were also busy. Lt. Bill Deragon and his senior quartermaster, CQM Jim Casler, took their navigational sights at night using the sextant to take careful altitude readings of the most prominent star formations. When daylight readings were permitted, the navigation team took evenly spaced altitude readings of the sun to help plot the ship's position accurately. The afternoon sun lines, when used in conjunction with the nightly star sightings, could more accurately help the Exec and his assistant with their important task.

Casler was more than just the average assistant navigator. He handled more than his fare share of the duties. "Casler had an eighth grade education and trouble at home as a kid," recalled quartermaster Hank Thomson. "He lied about his age and joined the Navy. He was my boss when I transferred to the *Seawolf* in 1940. He took an interest in me and improved my performance."

Born in Fresno, California, James Burr Casler was a descendant of Anthony Wayne and Aaron Burr. He had joined the Navy in September 1927, so had already been in the service 14 years when the war broke out. Casler had served in the quartermaster department aboard the tanker *Neches* from 1927 through 1938. Aboard *Neches*, he earned the nickname "Jughead," which was often shortened to simply "Jug" by his buddies. Casler switched to submarines, spending more than 18 months aboard *Shark* before joining *Seawolf* shortly before her commissioning in 1939. The men found that he was an expert navigator but a poor cribbage player.[5]

Jug Casler's lack of formal education did not inhibit his ingenuity. In fact, he was credited by the Navy with inventing a special sextant telescope for night vision that was used on all U.S. submarines during World War II. "Jug broke a 5x35 binocular in half," recalled Hank Thomson. "He had gotten up to the *Canopus* optical shop to get it aligned in the brackets that were for the star finder scope, which was a three-power." Casler's crude invention served its purpose. "One had to turn it upside down to bring the star down to the horizon and then make the measurement," said Thomson. "It did bring the horizon a little closer on poor visibility nights."

As *Seawolf*'s leading quartermaster, Casler was also over the other three men in his department. He had a first class signalman, Frank Franz, whom he trusted from their previous service together on the submarine *Shark*. During the night, Frank Franz stood the 0000 to 0400 watch. Casler found Thomson—who held the 0800 to 1200 watch—to be very efficient and now worked on training his striker, Norm Kisver. Casler put Kisver on the 0400 to 0800 watch. When diving officer Dick Holden made his trim dive before dawn and eventually took the boat down for the day, this allowed Chief Casler to be on hand to supervise Kisver closing the hatch properly.

Kisver, born in Brooklyn and raised in upstate New York, had joined the Navy from the Civilian Conservation Corps. "I thought I'd like to see the world, you know," he said. "So, I joined the Navy since conditions weren't too good then." Prewar, he had served aboard both the *Searaven* and the old *S-14*. Reporting aboard *Seawolf* on 26 May 1941 from the cruiser *Indianapolis*, Kisver quickly made friends with a fellow Brooklynite, electrician striker Robert Koehler— —with whom he would make many liberties ashore.

Throughout the day, Kisver, Franz and Thomson covered the quartermaster watch during the 24 hours. Each man stood four hours on and eight hours off. They kept the quartermaster notebook, routinely logging in any course change or speed change. The quartermasters frequented the bridge during the night to record navigational data for their log books.

Also allowed topside at night were the mess cooks, who came with weighted garbage bags to discard the previous day's trash. With the hatch open and fresh air flowing through the boat, the cooks did their hot cooking of breads, meats and baked goods. The cooking smells had the opportunity to escape through the vacuum created by the hatch and the diesels and moved along by the blowers.

During *Seawolf*'s second night at sea, Warder communicated his position, course and speed to the nearby *Sculpin* and then increased his speed during the night hours in order to pass *S-39*'s assigned area, in and near San Bernardino, safely. *Seawolf* raced through San Bernardino Straits during the early hours of 10 December. The *Wolf* dived at 0454 and proceeded towards her operating area.

During the morning hours that day, sound operator Paul Maley twice reported hearing sonar soundings being taken at irregular intervals by two ships. Radioman Maley reported that he believed the two ships were talking by sound. Nothing was in sight and no propellers were audible. If sound conditions were right, the fathometer's pings could be heard more than 30 miles away.

Seawolf's sonar could not pinpoint these ships but it soon became apparent that these were two Japanese submarines moving through San Bernardino Strait. Freddie Warder later recalled his first brush with enemy submarines.

> They were both heading north and so was I. I didn't know what to do with them. We didn't have any sonar that has since evolved and you couldn't see them. They were just two submarines coming into the Philippine waters. They were pinging on the bottom because it was goddamn shallow and I was pinging too, except that I was keeping my pinging down to a minimum, in order to get the hell out to the east, where I was ordered. These Japs were being a little more liberal with their pinging but I could not fix them or frame them for torpedo attack.[6]

Warder did not attempt to further develop this "scanty contact" because of flood currents moving into the straits and "because I judged my mission to call for early arrival in patrol area." That night, *Seawolf* received a dispatch directing her to shift to Babuyan Channel. As *Seawolf* ran toward her patrol area, the radio brought disturbing news from Manila. The Japanese bombed the area as *Seawolf* was running through San Bernardino Straits toward Luzon. Her destination was Aparri, a town some 250 miles from Manila on the northeast tip of Luzon. Don Bell was picked up on the *Wolf*'s radio, announcing the destruction of Cavite by Japanese bombers.

Seawolf had narrowly missed being caught. The bombers pounded the naval installations and the Navy Yard, destroying *Sealion*. *Seawolf* had missed her own destruction by 48 hours!

En route to Luzon, *Seawolf* rolled considerably in the choppy seas. On war patrol, everything was conserved, including the water. Men soon began growing beards and would keep them throughout the patrol. "There was no water for shaving or for showers," recalled Hank Thomson, "just enough to wet our toothbrushes."

The men off duty found entertainment in listening to the San Francisco short-wave station KGEI or to the propaganda of Tokyo Rose as she came in over the radio antenna. She made threats and taunts about the string of Allied defeats, describing how the U.S. fleet was at the bottom of Pearl Harbor. Mel Eckberg, among those in Kelly's Pool Hall, later wrote that the card players and mess cooks countered back to Tokyo Rose's threats with retorts that "were unprintable."[7]

Tokyo Rose also gave out specific names of United States ships and their tonnage, listing the dates and places where they had been lost. For the untrained ear, it might sound as though the U.S. Navy was practically down to its last ship. Machinist's mate John Street––the tall, slow-spoken Coloradian who was in charge of the *Wolf*'s No. 2 engine and the source of the run-in with Admiral Hart's staff months before—was always prepared. Street loved numbers and trivia and had brought a number of books aboard ship, including the *World Almanac*, a *Universal History*, and *Jane's Fighting Ships*. When Tokyo Rose's boasts went too far, Street could quickly prove that some of the "lost" U.S. ships she had named never existed at all.[8]

During the early morning hours of 12 December, *Seawolf* encountered seas that grew progressively violent. Warder took her down for the day at 0457, where he stayed except for hourly observations at 63 feet. "By the time we got to the east side of Luzon, the seas were twelve to fifteen feet high and there was a storm blowing," Warder later stated. "Those transports could not have landed troops and we couldn't defend." Even submerged, the stormy ocean rocked the *Wolf*. "Ship not controllable at 55 feet, vertical antenna depth," logged Warder. "Force of sea appreciable at 120 feet."[9]

The morning periscope watch sighted the coast of Luzon at 0903. *Seawolf* proceeded northward along the eastern coast of Luzon toward the port of Aparri, where she had been assigned to patrol. The Japanese had landed a small invasion force at Aparri on

December 10. The Japanese need Luzon for an important forward base for their aircraft.

Sound operator Joe Ferguson detected the pinging of a Japanese ship off Luzon that afternoon. Lt. Cdr. Warder soon found an *Amagiri*-class destroyer near Cape Engano but was unable to get close enough to this warship to attack. He surfaced that evening and proceeded toward Aparri.

The *Wolf* submerged at 0447 on 13 December. Lt. Dick Holden had the duty watch in the conning tower, making careful periscope sweeps off the coast of Luzon. With five officers, two of whom did not stand deck watches, Deragon broke the boat up into three watch sections covered by the three junior officers. Holden covered the 0000 to 0400, followed by Ens. Jim Mercer with the 0400 to 0800 watch and Ens. Doug Syverson with the 0800 to 1200 watch. The next 12 hours repeated with Holden covering 1200 to 1600, Mercer covering 1600 to 2000, and Syverson manning the 2000 to 2400 shift. Each watch team worked four hours on and eight hours off.

Holden, making his periscope sweeps, soon spotted a Japanese destroyer guarding the mouth of the harbor of Aparri.

"Call the captain," said Holden.

The duty messenger in the control room ran to alert Lt. Cdr. Warder, who appeared in moments dressed in shorts and sandals. Warder called for, "Up periscope!" and rode the handles up from the well. After a long sweep of the target ship, he called, "Down scope. Sound battle stations."[10]

The raucous blasts of the battle stations alarm rang throughout *Seawolf.* Men tumbled out of their bunks or raced from Kelly's Pool Hall to their assigned station. The war was less than one week old and the *Wolf* was ready to make her first attack.

In the tiny conning tower, the plotting party gathered to work out the course and speed of any potential target ship. Grouped at or near the plotting table were Ensign Jim Mercer, CQM Jug Casler, and Hank Thomson. Torpedoman Rudy Gervais, a brown-eyed young man of French descent, took over as the battle stations helmsman. Warder would make the periscope observations while Bill Deragon would serve as his assistant approach officer. "We had been practicing firing torpedoes for over a year and Warder had never missed," recalled Thomson.

In the control room, chief of the boat Eddie Souza stood sentinel at the colorful "Christmas Tree," the panel of red and green hull safety indication lights. An all green board while *Seawolf* was submerged indicated that all hatches were securely dogged and safe. A red light upon submerging could spell disaster. Near the depth gauge, two planesman sat before the two giant chrome wheels which operated the bow and stern planes that balanced the *Wolf* underwater. Diving officer Dick Holden stood watch over the planesmen. Gunner Bennett was on the stern planes and signalman Frank Franz was on the bow planes during battle stations.

Nearby were the high-pressure manifolds, whose high pressure air could blow thousands of gallons of water out of the ballast tanks to bring *Seawolf* to the surface on command. Auxiliaryman Paul Zimmerman had the watch over the air manifold during torpedo attacks. In one corner of the control room was one of the four heads aboard ship. In the opposite corner, in its own little compartment, was the radio shack, a six by eight station with just enough room for three people. Here, Mel Eckberg and Paul Maley manned their sound gear to track the attack. As senior radioman, Eckberg's job was to monitor the target ship during torpedo attacks while Maley kept track of escort vessels or any other ships on his sound gear.[12]

As *Seawolf*'s progress continued, Warder kept up a running commentary of what he saw through the periscope. Unnecessary—but appreciated by all—was the fact that he allowed his voice to be broadcast over the 1MC for all hands to hear. He described the destroyer he was watching above. Warder contemplated attacking the tin can but decided that the man-of-war was likely protecting something of value deeper in the harbor.

Warder stayed low and allowed the destroyer to pass on by, her pinging closely monitored by his sonarmen. Warder "endeavored to gain attack position but was never able to get range under 12,000 yards." The target destroyer was headed for the western entrance to Babuyan Channel. Warder secured his crew from battle stations at 0749 and resumed his course for Aparri.

During the day, *Seawolf*'s periscope watch spotted two aircraft as she reconnoitered Aparri and moved within the 30-fathom curve. Warder found only one small single-stack freighter in the anchorage, but she was 4,200 yards inside the 30-fathom curve. He chose to pass this small game by due to the risks of bringing his submarine into dangerously shallow water during the daylight to make an attack. That evening, a ComSubs dispatch alerted *Seawolf* that a large expeditionary force was in the South China Sea.

The seas were heavy again this night and Warder drew away from the bay to allow his men to rest. Squeaky Langford has his torpedo crews inspect their torpedoes this night, as action was expected the next morning. The heavy 20-foot warheads were slid on tiny rollers from the loading rack by strong arms. *Seawolf* moved from Aparri toward the Luzon's Port San Vicente during the early morning hours of 14 December.

As dawn approached, Warder had his OOD, Ensign Jim Mercer, practice a crash dive. At 0458, Mercer cupped his hands and yelled, "Clear the bridge!" He pulled the diving klaxon as the lookouts piled down the ladder into the conning tower. Chief Jug Casler stood ready to assist in the conning tower as quartermaster striker Norm Kisver dogged the hatch behind the lookouts and Mercer. Kisver had the hatch securely dogged before the engineers below had secured their diesels. The diesels pulled heavily on the available oxygen in the boat, thereby creating a powerful vacuum with the hatch closed. "This makes for a quieter dive and I believe it to be a safer indication of hull tightness," noted Warder. Mercer's section effectively cleared the bridge and dived the boat in 45 seconds.

Fifteen minutes after submerging, *Seawolf*'s sound operators picked up enemy pinging as the ship moved into Port San Vicente. Freddie Warder was more than a little surprised. "We were told by our intelligence that the Japanese destroyers had no echo-ranging sonar," he later stated. "But, boy, there were pings coming off that destroyer! He didn't pick me up, this son of a bitch, probably because he was too close to the ground."[12]

In the shallow waters, Warder ordered silent running, shifting steering machinery to hand. All blowers, fans, air conditioning, and refrigerator machinery was shut down and the temperature within the boat began to rise. "When we were at silent running, everything was shut down," recalled Little Swede Hanson in the forward engine room. "It got up to 115 degrees or more in the engine rooms." *Seawolf* was plagued by the destroyer's pinging for the next four hours until she effectively dodged the searching surface vessels.

Having shaken the searchers, the *Wolf* proceeded slowly into Port San Vicente, drawing within the 100-fathom curve. The port's waters became shallower as the *Wolf* proceeded in. Regular periscope observations were made and Warder retired to his cabin for some of the inbound run. Lieutenant (jg) Dick Holden had the watch at 1334 when he spotted the mast of a ship at anchor ahead.

Holden called for the skipper as *Seawolf* moved in closer. Holden also believed that he had seen a submarine to the westward of the ship when first spotted. He and Warder quickly decided that the vessel was a large ship, distance about 5,500 yards.

"Battle stations!" went the call over the intercom, followed by the three blasts of the alarm sounding through the boat. The red emergency lights came on as men raced to their assigned stations. In the conning tower, Rudy Gervais took his post at the helm. Quartermaster Hank Thomson stood behind the skipper wearing phones and a chest telephone. He was Warder's battle stations talker and would relay orders to other compartments.

Warder called out details of the vessel as she came closer into view. Casler, working the plot at his little table in the conning tower with Ensign Mercer, warned that the water was too shallow for the *Wolf* to completely reach this distant ship safely. Casler did find that the water would be deep enough to at least reach firing distance.[13]

As the ship came closer into view, Warder called out that he had a Japanese seaplane tender at anchor, a large one at that. The ship had guns fore and aft with two mast cranes visible. Using the recognition books, the approach team identified the ship as *Kamikawa Maru*, although Warder later felt that the vessel may have been the fleet supply ship *Mamiya Maru*. She was anchored in 7 to 10 fathoms of water. Postwar research would show that this target was the seaplane tender (AV) *Sanyo Maru* of 8,360 tons, a prime target.

"Open outer doors," Thomson called to Squeaky Langford's forward torpedo room. Once the forward room acknowledged the order and performed the task of opening their torpedo tube doors, Thomson repeated back their, "Forward tubes ready, Captain."

At 1420, Warder's boat was well within the 30-fathom curve and he coached Rudy Gervais on the helm into his final approach course with, "Come to oh-eight-one."

Gervais made a slight twist on his wheel as his skipper took his last periscope observation. The final data was called down to Doug Syverson on the torpedo data computer (TDC) in the control room. Exec Bill Deragon, Captain Warder's assistant approach officer, dropped down to the conning tower to assist Syverson with this first attack. Deragon was the *Wolf*'s original torpedo officer before being promoted to second-in-command.

While these two officers were the resident experts on the TDC, *Seawolf* had a fire controlman to keep it in order, 26-year-old George

Darrell Leffingwell. Born and raised in Mason City, Iowa, he had traveled after high school as a magazine salesman before moving to San Diego. One of six kids, Leffingwell and two brothers ended up joining the military—one in the Air Force and two in the Navy.

"As fire controlman, my main duty involved working with the torpedo data computer and the deck guns," related Leffingwell. "It was not so much fire control as it was making sure the weapons performed as designed." George had been with *Seawolf* since her commissioning and had since acquired the nickname "Lefty" from his buddies. During torpedo attacks, he would be on hand to assist the TDC officer. "The target bearings were supplied down to us," recalled Leffingwell. "I was there to assist Lieutenant Syverson in any way he needed. It was usually offering my opinion on what should be done in a certain situation."

Also in the crowded control room during attacks was the ship's paperwork king, yeoman John Sullivan, who kept *Seawolf*'s war diary. This log included all attack and course change data and was the basis for the skipper's detailed patrol report which he typed prior to the ship returning to port from any patrol.

The generated range was 3,800 yards. Chief Langford's torpedomen were ordered to set the depths on their warfish at 40 feet for the first two tubes and 30 feet for the last two, so the torpedoes would pass under her keel and actuate the magnetic exploder.

"Fire one!"

One week after Pearl Harbor's day of infamy, *Seawolf* fired her first torpedoes in anger. The *whoosh!* of rushing air and shudder announced to all that a war fish had been fired. The air pressure within the ship jumped up with each firing and the pressure was felt in the men's eardrums. Langford's torpedomen quickly flooded the tube after firing to compensate for the thousands of lost pounds.

Seawolf hissed and jarred three more times as all four bow tubes were fired at the seaplane tender. The four torpedoes were fired to strike *Sanyo Maru*'s stack, forward kingpost, mainmast, and finally at her stack again.

"Full right rudder," called Warder. "Take her to 90 feet."

Warder did not want to risk his periscope being spotted to watch these torpedoes. Eckberg and Maley on sound tracked the torpedoes in toward *Sanyo Maru*. Three minutes after firing, sound reported two muffled explosions, which were likely the sounds of warheads exploding against the beach. *Seawolf*'s first shots had failed.

"Jesus, they missed the target!" Warder shouted. "Oh, hell! Make ready the after tubes."[14]

Squeaky Langford's torpedomen set to reloading their four tubes and completed this task in about seven minutes. In the after room, Chief John Gibson's men opened their outer doors and stood ready to fire. A fat tender still sat an anchor, ripe for the plucking. At the moment, there was no time to decide what had gone wrong.

Seawolf eased forward at 1/3 speed as Warder made final observations. His first torpedoes had been fired at high speed, so Warder had Gibson's crew set the stern torpedoes on low speed with a 15-foot depth setting. Warder noted that the Japanese tender was busy now. "I could see people on the deck, walking up and down the gangway," he later said. Men raced about on her deck as she apparently was making preparations to get underway. From a range of 4,500 yards, Warder fired all four of his stern tubes as he exited with his periscope just breaking the water from his running depth of 63 feet. As he withdrew, he could only see the top of *Sanyo Maru*'s stack.[15]

Leading soundman Mel Eckberg called out, "All torpedoes running hot, straight and normal, skipper."

Watching through the periscope, Warder saw no flame, no smoke, nor no explosion. His engine room gang, however, reported hearing four muffled explosions as *Seawolf* withdrew. "We had never heard a torpedo go off in our practice prior to Pearl Harbor," admitted Paul Zimmerman. "We had heard a lot of explosions, but as far as hearing an explosion after we fired a torpedo—we couldn't be sure from these explosions if we had hit her or were being depth charged."

If any of the torpedoes hit, their exploders must have failed. Postwar research would show that *Sanyo Maru* was indeed struck by a dud torpedo which caused only slight damage. "My sound reported an explosion, but apparently the only good I did was to get one dud into that ship forward of the gangway on the starboard side," Warder later recalled.[16]

Seawolf's skipper felt that better training in the months prior to war would have paid great dividends. "I'd never listened to a live torpedo explosion in my life and I didn't know what a live one sounded like," admitted Warder. "This is one of the things that John Wilkes did wrong in the warming up to war. Never were we allowed to fire a live warhead because it cost too goddamn much money."[17]

Warder felt that the pre-war submarine training should have included at least some live torpedo exercises against rock cliffs or der-

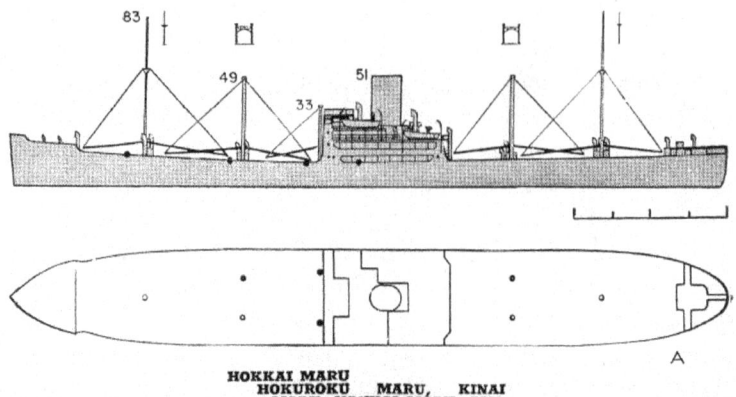

HOKKAI MARU
HOKUROKU MARU, KINAI
MARU, NANKAI MARU, SAN-
YO MARU, TOKAI MARU

Seawolf's first wartime torpedo target was the 446-foot, 8,360-ton seaplane tender *Sanyo Maru*. *ONI-208J.*

elict ships and that the submarines should have gone through depth charge indoctrinations to help all hands adjust to the explosions. Had live torpedo firings been practiced in 1941, Captain Wilkes' skippers would have quickly found just how poor their torpedoes were.

U.S. torpedoes had been manufactured at a single site at Newport, Rhode Island since 1908. *Seawolf* was carrying the newer Mark 14 torpedoes, which Newport had started sending to the fleet in 1938 to replace the aging World War I-era Mark 10 torpedoes. The newer Mark 14s were more than four feet longer at 20 feet, 6 inches and 1,000 pounds heavier at 3,209 pounds each. Each Mark 14 warhead also weighed 643 pounds versus the previous 500 pounds of the old Mark 10s.[18]

The problem with the "improved" Mark 14s was that submarine doctrine at the start of World War II called for the skippers to fire their torpedoes with magnetic exploders versus exploding on actual contact. A large metallic object such as a ship creates enough magnetic disturbance that a compasslike detector installed in a torpedo could be calibrated to the existing magnetic field. Torpedoes equipped with magnetic exploder held promise that they could be detonated beneath a ship's unprotected keel at the point of greatest magnetic influence, breaking the ship's keel.[19]

A large influence for the protocol to use a magnetic exploder was the need to conserve war fish. Each Mark 14 produced at Newport cost the Navy $10,000, or $142,857 in 2006 dollars, adjusted for inflation. Newport torpedo engineer Ralph Waldo Christie's requests for live firing tests were refused by the Chief of Naval Operations, Admiral William H. Standley, until Newport could guarantee that the test target would not be destroyed. This ridiculous request, and the extreme cost of torpedoes, meant that the U.S. Navy did no live firing tests of its important weapons in the years leading up to World War II. Boats such as *Seawolf*, which left port on the day of Pearl Harbor's attack, were going into combat with a weapon whose effectiveness was completely unknown.[20]

Only time, complaints and the compilation of patrol reports which documented U.S. torpedo failures would begin to force the issue to naval brass. Another of the boats deployed from Manila, *Swordfish*, had two torpedoes explode prematurely during her early attacks in December 1941. Another Manila boat, Lieutenant Commander Tyrrell Dwight Jacobs' *Sargo*, had her first Mark 14 torpedo—fired on 14 December—premature 18 seconds after firing. Jacobs and his leading officers conferred and decided to break standing orders by deactivating the magnetic influence features on *Sargo*'s tin fish. In the ensuing attacks *Sargo* made on her maiden patrol, Lt. Cdr. Jacobs was frustrated to find that his submarine's Mark 14s were also running deeper than their settings. Jacobs broke radio silence to inform John Wilkes that in six attacks with thirteen torpedoes fired, none of his war fish had exploded. *Seawolf*, *Sargo*, *Swordfish* and many of their sister subs were thus conducting the Navy's first live firing tests of the Mark 14 torpedoes, and the results were less than satisfying.[21]

The German Navy was far ahead of the U.S. Navy in figuring out their own faulty fish. Kommodore Karl Dönitz, commander of the German U-boat forces, quickly found from his submarine commanders that the magnetically-detonated torpedoes had problems and routinely ran 6.5 feet deeper than their settings. As early as January 1940, Dönitz was already aware that "at least 25 percent of all shots fired have been torpedo failures."[22]

In an attack very similar to Freddie Warder's Aparri attack, famed U-boat skipper Günther Prien had taken *U-47* into Bygden Fjord off Norway in April 1940. He found fat transports and cruisers lying at anchor and fired all bow tubes and all stern tubes from close range. All ran too deep and missed, however. Such continued failures drove

Dönitz to making improvements in the magnetic and contact pistols of German torpedoes. By late May 1942, U-boats began using the improved torpedoes and results improved. It would be early 1942, however, before the Germans solved their other big torpedo problem, the depth-control issue.[23]

The U.S. Navy was not aware that they had a very similar and very serious torpedo problem. Freddie Warder and his *Seawolf* men were among the first to make this discovery on 14 December 1941.

Freddie Warder was fit to be tied. He had bravely conned his boat into an enemy harbor past a pinging Japanese destroyer only to have eight torpedoes fail him against a defenseless enemy ship.

Warder withdrew from the harbor now that the Japanese were thoroughly aware of his position. He called for Ensign Syverson to discuss the torpedo failures. Syverson explained that all of the fish had been properly prepared and that he had personally inspected them prior to the attack.[24]

Warder decided to run out of Port San Vicente to the northward as sound reported pinging in the distance. At 1453, with the enemy destroyer pinging on the *Wolf*, the first depth charge exploded on the starboard quarter, well distant. Warder called for silent running and depth charge stations as he continued to clear the area submerged. He had Dick Holden take her to 200 feet and make full speed ahead. Electrician John Bilkey later stated:

> At silent running, you shut down everything except the screws. You run as silent as you possibly can and run as slow as you possibly can. The normal procedure when your battery runs low is to slow down. The batteries threw out a lot of heat and it made it pretty hot. We had the air conditioning shut down.

Seawolf's hull was test rated for 250 feet, but Warder rarely took her below 200. In theory, her hull would hold even up to 350 feet, but test dives in the past proved that other equipment and pipes were prone to leak from the intense pressure below 200 feet.

For depth charge attacks, Warder would order the personnel from that more exposed compartment and seal it off. The conning tower

Faulty Fish

In the tight living quarters, men slept among the torpedoes while on patrol. TM1c Art Lamberson is seated and MoMM1c Bud MCoy is in his bunk, right foreground. *U.S. Navy photo, courtesy of Mike McCoy.*

included port holes to the sea where the duty quartermasters could see the condition of the sea. "When we were submerged, you could see fish swimming when the porthole covers were open," said Hank Thomson. For depth charge attacks, these were covered over, just as they were at night to prevent light from shining from them.

Although this depth charge attack was not considered severe by later standards, it was the first for her crew. They had not even been allowed to receive a practice depth charging during peacetime just for indoctrination. The *Wolf*'s crew was indoctrinated under fire.

"I was scared shitless," recalled Paul Zimmerman, who was on duty in the control room. Earlier, he had been unsure of the muffled explosions he had heard in the distance, whether they were torpedo or depth charge explosions. This time, he had no doubt what was exploding above him.

Over the next ten minutes, five or six more ashcans were dropped at irregular intervals, all distant. The Japanese destroyer never came close enough for the sound of its screws to be detected.

By 1515, *Seawolf* had secured from depth charge attack quarters and eased back to 120 feet. She surfaced three hours later in perfect blackness and spent the night maneuvering in the Babuyan group while waiting for the moon to rise.

Seawolf commenced a west-east patrol of the northern approach route from Takao to Aparri on 15 December, but found no enemy targets. Visibility was poor the next day and rough seas caused the skipper to run at 120 feet with only hourly periscope observations. Warder scoured Aparri after dawn the next morning, and submerged at 0507 on 17 December at a point 13 miles from Aparri Harbor. The seas were heavy again this day, making depth control difficult at the 63-foot periscope observation depth. Bill Deragon and Jug Casler were unable to fix the *Wolf*'s position by landmarks due to the prevailing 1,000-yard visibility so they finally had to employ the ship's fathometer to establish her position that night.

Lt. Cdr. Warder began writing a letter to his wife Mary that night. Due to censorship, he could not give specific details of his Aparri torpedo attack but he did offer: "I can say that we had a splendid baptism and have had success at the northern latitudes."[25]

Throughout 18 December, numerous sound contacts and even propellers were heard without an enemy ship ever being sighted. The *Wolf* closed to within 7 miles of Aparri the following day, noting an increase in Japanese flying boats from the airfield but no shipping targets to pursue. *Seawolf* reconnoitered Port San Vicente again on 20 December but this harbor was now void of shipping.

The war was not quite two weeks old and the *Wolf*'s crew was already frustrated with faulty fish and a lack of Japanese targets.

Once the moon had disappeared that night, *Seawolf* performed the dangerous procedure in enemy waters of striking two torpedoes below from her deck tubes.

This procedure was tricky enough in daylight within the safety of a harbor. At night, within enemy waters, it was doubly dangerous for the ship and the men working topside. The tubes within her superstructure forward of the conning tower held four spare torpedoes. For this task, torpedo officer Red Syverson and his senior torpedomen—Eddie Souza, Squeaky Langford, and John Gibson—had their ablest men ready to move fast.

With a king post and boom rigged and the angled torpedo hatch open, the men pulled two of the spare torpedoes from the extra tubes under the deck gratings and gently snaked them along the main deck. From there, each 3,200-pound torpedo was skidded down in the forward torpedo room with more sweating, cursing, and careful maneuvering of chain hoists over the course of three and a half hours. A crash dive at this particular moment would have been fatal for a number of men. Should a Japanese patrol plane or destroyer find the *Wolf* during this tricky process, the results would not be pleasant.

Warder ordered all stop and the ship lay to on her battery power only southwest of Babuyan Island. The ship's little motor launch, located below near the spare torpedo tubes, added to the difficulty of transferring live weapons at sea. "If boat had not been aboard we could have had them all below during this time," Warder wrote.

By 0330 on 21 December, Warder had had enough for one night and *Seawolf* was back underway en route to examine Port San Pio Quinto—having only moved two of the four spare fish.

While the torpedomen moved two warfish below, radioman Eckberg also went topside to fix a radio insulator which had been smashed. He could see "the thin outline of a small island less than half a mile away" as he climbed up on the bridge to repair the insulator which held his radio antenna. "This was the first time since Manila that I had been topside, under the sky," Eckberg wrote. When he finished the repair, he found half a dozen other sailors crowded around the base of the ladder in the control room just enjoying the fresh air that was coming down the hatch.[26]

Seawolf found no shipping at Port San Pio Quinto during the day on 21 December and returned to the area off Babuyan Island that night where she had worked the previous night. Once again, *Seawolf*'s engines were stopped to secure more torpedoes below decks. This night, the torpedomen were able to move the other torpedoes in just over an hour and get their ship back underway.

After another fruitless day of patrolling, a radio dispatch at about 2200 on 22 December reported that a Japanese convoy of transports was moving into Lingayen Gulf under escort of aircraft and destroyers. Warder patrolled off Cape Bojeador the following day but found only pinging and propeller noises heard by his sound operators. Orders came from ComSubAF at 2014 on 23 December for *Seawolf* to return to Cavite.

Warder increased speed to 17 knots and set course for Manila. The Japanese had many ships about the area and returning to Manila would be no small task. Warder had to run the blockade of enemy vessels to return to the bombed port. The off-duty men shot the breeze in Kelly's Pool Hall or played cribbage. Others lay in their bunks reading magazines. The washing machine "Baby" purchased

Christmas 1941 aboard *Seawolf* during her first war patrol. Chief electrician John Bilkey *(left)* and chief yeoman John Sullivan pose with Sullivan's hand-made Christmas tree. *Courtesy of Hank Thomson.*

in late 1941 and installed aboard ship was put to good use keeping clothing clean. "The machine is the standard home size and takes up more room in the washroom than is desirable but this is more than overcome by the use made of the machine after a week or so at sea," wrote Lt. Cdr. Warder in his patrol report.

As *Seawolf* approached Cavite on 24 December, chief yeoman John Sullivan offered his own little Christmas celebration to the crew. En route from Aparri, he had started working on a homemade miniature Christmas tree. On a green monk's cloth in the yeoman's shack, he had built the tree from odds and ends. The trunk was a broom handle with holes Sully had drilled in it. With medical applicator sticks from "Doc" Loaiza, he had inserted branches into the holes. From signalman Frank Franz, Sully had added red and blue flag bunting and some of his own green and pink file paper. He made tinsel by gluing tinfoil from cigarette wrappers to paper strips and decorated the branches with that. His two foot Christmas tree also included a dozen flashlight batteries that Sully had painted red, green and silver and strung about on a dry-battery circuit.[27]

Hank Thomson recalled that Sullivan was always an inspiration to the crew.

> Before the war, he had been in the lighter than air dirigibles. When not busy on the *Seawolf*, he would assist the

enginemen, torpedomen, and anyone else just for fun. He would go into the galley and bake a cake. When I hurt my arm before the war, Sully would send short messages for me on the signal light. He couldn't receive light, but could acknowledge at my word that I got the message.

John Sullivan's little tree became the curiosity of the boat on Christmas Eve 1941. Freddie Warder stopped in to admire it and approved the use of the camera to record the moment. Sullivan spread cotton batting about the base of the tree for snow and constructed a cardboard fireplace behind the tree.

In addition to the tree, the after engine room displayed two immense stockings made of bunting, one bright red and the other sewn in white. Each stocking was filled with goodies ranging from a can of oil to a Stilsen wrench to a pair of pink silk panties. Machinists Red Snyder and Walter McCoy had inscribed each with "loving" messages to each other. Like Snyder, McCoy had been in the Navy nearly three years. Tall and thin, McCoy was 27 years old. He went by "Bud," but shipmates also called him "Bones" for being so thin. Bud McCoy was the oldest of five brothers, all of whom joined the Navy. He served with brother Ralph in 1939 aboard the battleship *Oklahoma* before Bud entered the sub service and joined *S-41*.[28]

Seawolf made her approach to the Philippines on Christmas Day 1941. Auston Baker and his cooks—Gus Wright and Bill Mallory—celebrated the holiday by baking twenty hams all on the previous night. Wright, a thin, buck-toothed man of about 28 years, also announced to the crew in the mess hall that their hams would be served with mince pies.[29]

Seawolf approached Subic Bay around 2300 on 25 December. Sound picked up the pinging of the patrol vessel as she approached. With a bright moon, Warder dove his boat as a precaution until the moon had set. Several miles off Corregidor, she picked up a small white light which pointed toward her and blinked a message. Frank Franz replied with his blinker gun. The ship blinkered back that a pilot was being brought over to escort *Seawolf* into Manila.[30]

Shortly before midnight, a PT boat delivered a young pilot to the submarine. The pilot joined Warder on the bridge and guided the sub through Manila's mine field at 0430 on 26 December before departing again. Just before dawn, *Seawolf* pushed on in and submerged at 0512 off Mariveles Harbor in Manila Bay as directed.

Seawolf's first war patrol was officially concluded. It had been 18 days in duration, short by design from John Wilkes' orders. By being at sea, his submarines had managed to harass Japanese shipping and be further spared from the devastation that had hit Cavite. They had returned in three weeks to bring intelligence of the enemy's movements.

Freddie Warder's *Wolf* had already done her best to get in some revenge for Pearl Harbor. Unfortunately for the "Silent Service," the dud torpedoes which had allowed *Sanyo Maru* to escape death on 14 December were far from the last faulty warheads that U.S. submariners would be forced to contend with during the early months of World War II.

Shortly after his arrival at Mariveles, Lt. Cdr. Warder wrote to his wife Mary of *Seawolf*'s first war patrol. "I'm very proud of my crew," he wrote. "They are going to beat the hell out of these gents in time. In the meantime, they are going to be very annoyed."[31]

USS *Seawolf* First War Patrol Summary

Departure From:	Manila
Patrol Area:	Aparri, northeast Luzon
Time Period:	8–26 December 1941
Number of Men Aboard:	66: 61 enlisted and 5 officers
Total Days on Patrol:	18
Number of Torpedoes Fired:	8
Ships Credited as Sunk:	0
JANAC Postwar Credit:	0
Shipping Damage Claimed:	one dud hit on *Sanyo Maru*
Return To:	Manila

3

Special Missions

Second and Third Patrols   *31 December 1941–7 February 1942*

War was evident about Manila Bay when *Seawolf* returned from her first war patrol. She could not tie up at the docks during daylight, so Lt. Cdr. Warder kept his boat submerged throughout 26 December off Mariveles. During the day, the faint pounding of bomb explosions could be heard as the Japanese continued to come over Manila and bomb airfields and military installations.

Warder surfaced at dusk and ran with his decks awash. At 1900 on 26 December, orders came for *Seawolf* to move into Corregidor. She eased up alongside the dock, where a few men were allowed to go ashore—warned about the chances of expected air raids. Trucks brought supplies to the dock and working parties were formed to stow goods below decks. That night, the radiomen heard Tokyo Rose calling for the surrender of the men on Bataan, warning them that they were encircled and should "give up now without dishonor."[1]

Sister submarine *Swordfish* came in alongside the dock that night also to replenish supplies. Before dawn, both boats eased back out into Manila Harbor to submerge before the Japanese bombers made their daily visits. Once again, the dull explosions of bombs from Japanese planes could be heard from underwater as Manila Harbor's shipping was dive-bombed. *Seawolf* and *Swordfish* resurfaced at dusk and returned to the docks they had occupied the previous night.

Knowing that the sun would rise around 0500, the men worked quickly in order to get their ship loaded. Warder would prefer to get out past the mine field on the surface while it was still dark. By midnight, oil lines were hooked to *Seawolf* as oil king Pete Lober and engineer Dick Holden oversaw the transfer of thousands of gallons of fuel oil. There was no R&R period ashore for the *Wolf*'s crew following her brief first patrol. Ashore, war was evident everywhere.

Things had deteriorated so rapidly about Manila that Admiral Hart and Captain Wilkes decided it was time to abandon Manila as an advanced submarine operating area. While taking on provisions at Corregidor on 28 December, *Seawolf* received three new men from the destroyed *Sealion* who were happy to come aboard and leave their hellish world ashore behind. They were GM3c James Edward Larson, S2c Vernie Marion Fuller, and Ensign William Alexander Whitman. Leading gunner's mate John Bennett was thrilled to add another experienced gunner to his team. Larson became known to all as Swede and he joined the Swedish ranks of Swede Enslin, Big Swede Hanson, and Little Swede Hanson aboard the *Wolf*.

For 22-year-old Reservist Bill Whitman, reporting aboard *Seawolf* was reunion time. He was friends with Jim Mercer and had attended submarine school with Red Syverson. Raised in Seattle, he had graduated from the University of Washington with a degree in mechanical engineering. "Whit" was commissioned an ensign on 1 July 1941 and then went immediately into submarine school at New London. He reported aboard *Sealion* at Cavite in November 1941 and became her assistant first lieutenant.

Like Larson and Fuller, Ensign Whitman was damned happy to come aboard *Seawolf* on 28 December 1942. Her wardroom was short one officer and he was short one submarine to serve on. Whitman had been aboard *Sealion* when the Japanese bombers struck Cavite Navy Yard on 10 December. Stranded in the yard, the *Lion* was struck by two bombs which smashed her pressure hull and killed four sailors in her engine room. Fragments of one of these bombs ripped through the nearby *Seadragon*, killing one of her junior officers. The damage was severe and *Sealion* settled to the bottom as Whitman and the other survivors scrambled out of the hatches that were still above water. With Cavite's Navy Yard destroyed, ComSubsAsiatic John Wilkes ordered *Sealion* destroyed on Christmas Day to prevent her from falling into Japanese hands. Three depth charges were rigged in her compartments and *USS Sealion* became the first United States submarine casualty of World War II.[2]

On Christmas Eve, General Douglas MacArthur had ordered all troops around Manila to retreat to Bataan Pensinsula, a 50-mile peninsula on the western side of Manila Bay. MacArthur moved his headquarters to heavily fortified Corregidor Island, known as "the Rock." Manila was declared an open city on 26 December and the city would be completely taken over by the Japanese in another week. In addition to MacArthur's staff, Corregidor was soon crowded with planeless aviators and subless submariners. Ensign Whitman joined *Seawolf* as Doug Syverson's assistant communications officer.

When *Seawolf* returned to the Corregidor dock on the night of 30 January, John Wilkes had decided to abandon the Philippines as an operating base for his Asiatic submarines. The sub tenders *Holland* and *Otus* had already been sent south to Darwin, Australia, to set up a new operating base under Captain Red Doyle. There were now ten submarines left in Manila Bay plus the tender *Canopus*, which had been damaged by air attack. Captain Wilkes decided to withdraw his submarines to Surabaya and Darwin, leaving the old and slow *Canopus* behind. All of his available submarines would be used to carry men from the Philippines. Wilkes then divided his own staff between two submarines, *Seawolf* and *Swordfish*.

Admiral Thomas Hart and two of his staff officers had left "the Rock" on 26 December aboard *Shark*, bound for the Dutch naval base at Surabaya, Java. Captain Wilkes decided that each of his ten submarines in Manila Harbor could evacuate 25 men. Wilkes, division commander Willis Percifield, Sunshine Murray, and others boarded *Swordfish* to follow Hart to Surabaya.[3]

As the ComSubsAsiatic staff prepared to come aboard from Corregidor, *Seawolf*'s crew was stowing aboard boxes of highly confidential papers. The most precious of these documents were stowed in locked positions throughout the boat. Gunner John Bennett came aboard with four yellow rectangular cans of dynamite explosives. In the event that the Japanese captured the *Wolf*, Bennett was responsible for destroying all gear, including the torpedo data computer and all radio and sound gear. Each dynamite box, stowed near classified equipment, had a separate five-foot fuse Bennett would attach only in the case of actual need.[4]

At midnight, the 1MC speaker system called for men to go topside in preparation for *Seawolf*'s departure. Shortly thereafter, visitors began climbing down the hatch into the *Wolf*. One of them was Captain Jimmy Fife, who would serve as Chief of Staff, Submarines, Asiatic Fleet. He wore a tan field jacket, khaki trousers, and brown

Army regulation shoes, attire that at once announced that he was not of the crew.

Scuttlebutt ran rampant throughout the *Wolf*. Among the passengers were Don Irish and Charles "Duke" Woodard, two radiomen who had fought to keep their radio going on Corregidor during the past month of almost constant bombings. Both men were thin from lack of good provisions and ample water supply.

Woodard had joined the Navy in 1934 at Bremerton, Washington. "I wasn't a hero for volunteering," he recalled. "I was just tickled to death to get in and have a job." Prior to the war, he served as a radioman aboard the submarine *Shark*. When the war broke out, "Woody" Woodard was aboard the tender *Canopus* at Manila. "They came aboard *Canopus* looking for radiomen to go ashore onto Corregidor," said Woodard. "They took me, and I joined the flag there."

The only medical help aboard ship was Chief Pharmacist's Mate Frank Loaiza of Puerto Rican ancestry. Full of nervous energy, Loaiza was an able medic who took the crew's taunts of "Doc" and "the Quack" in stride. From his locker cabinet in the after battery compartment, he passed out vitamins to those who had been without regular meals for many days. "Doc" Loaiza took in radioman Woodard to look over an ulcerated leg wound that he sported.[5]

Seawolf's deck log shows that her ComSubsAsiastic passengers came aboard ship at 0100 on 31 December 1941. They were: Cdr. James Fife Jr., Cdr. Eliot Hinman Bryant, Cdr. Joseph Anthony Connolly, Lt. Cdr. Ralston B. Vanzant, Lt. Cdr. Morton Claire Mumma Jr., and Lt.(jg) H. V. Combs Jr. Four enlisted men also came aboard as passengers: RM1c Donald W. Irish, Y1c A. A. Borowski, RM2c Charles C. Woodard, and OS2c A. Llanes. Cdr. Jimmy Fife became the senior officer on board *Seawolf* for this patrol.

Warder was pleased to reunite with one of his old Academy classmates, Mort Mumma, who had been captain of the Academy rifle team. He had taken command of the old *Squalus*, which had been raised from the bottom of the ocean and renamed *Sailfish*. Mumma had taken *Sailfish* out of Manila on 9 December, one day behind *Seawolf*. On 13 December, Mumma attacked a Japanese ship and his boat was pounced on by the escorts. In the ensuing depth charge attack, Commander Mumma lost his composure and asked his Exec to take command. "Mumma washed out," Warder recalled of his classmate. Mumma decided that he was not fit for submarine com-

Cdr. James Fife Jr., chief of staff to submarine squadron commander John Wilkes, was among the ten staff members evacuated from Corregidor by *Seawolf* on her special mission. *U.S. Navy photo.*

mand and returned *Sailfish* to Corregidor by 16 December, where Lt. Cdr. Dick Voge took command of his ship.[6]

Seawolf was underway in the early hours of 31 December 1941. She cleared the minefield outside of Manila at 0244 and commenced a zig-zag pattern once in the open waters. She ran at periscope patrol through the daylight hours and resurfaced at 1840 to proceed. During the night hours, *Seawolf* commenced an investigation of Palauan Bay at full speed on four engines. No shipping was found so *Seawolf* eased back to 13 knots to proceed on her route.

Shortly after departure, Warder was directed to attack enemy shipping in Lingayen Gulf, where the Japanese had started landing in late December. Communications officer Doug Syverson received the coded message in the night and woke the skipper once he had decoded it. "I got this general directive from John Wilkes, 'penetrate Lingayen Gulf and shoot everything you can hit,'" said Warder.[7]

Jimmy Fife was in Warder's upper bunk in his stateroom. He shook him awake and shared the report with him. Based on *Seawolf*'s southerly course toward Australia, this would have meant heading back up north of Corregidor. Not only did this countermand the current objective, but Warder found the patrol orders vague.

"I will not go into Lingayen Gulf on these orders unless you countermand them, because you are senior submariner present," Warder stated. "I'm happy to go into Lingayen Gulf and try if you give me area assignments, where I'm sure I'm not shooting at my compatriots and not interfering with any of their endeavors."

Without hesitation, Fife replied, "Freddie, I agree with you. This doesn't say. Let's keep on to Darwin unless we get some area assignments."

The rush of U.S. submarines into Lingayen on Luzon's northwestern coast concerned Fife and Warder, although Warder silently felt that Fife should have set up a patrol area for *Seawolf*. Enemy shipping was known to be plentiful in this area as Japanese naval forces supported their landing ships in Lingayen Gulf. On 22 December, *S-38* torpedoed and sank a large troop transport in the gulf. This was the only Japanese landing ship sunk in Lingayen during December. The submarines of the Asiatic Fleet, in fact, had very poor luck in the first month of war against the plentiful Japanese targets. In 45 separate attacks, they had fired 96 torpedoes, which sank only three enemy ships that could be confirmed in postwar records.[8]

Dick Holden conducted his trim dive at 0532 on 1 January 1942 at a point 24 miles from Ambulong Island. *Seawolf* continued her uneventful patrol, passing through Basilan Strait the next night without any further contacts. Cdr. Fife pored over the *Wolf*'s charts during the voyage. He and his staff took up residence in the wardroom. Fife himself rarely roamed the boat. The *Wolf*'s two Filipino stewards, Bugawisan and Tamayo, had to set up three eating shifts to handle everyone: three breakfasts, three lunches, and three dinners. Baker, Wright, and Mallory now had more than 80 men to feed. Their days and nights were steady between cooking, cleaning, washing, and cooking again, and filling the ten-gallon coffee urn. The crew routinely went through 30 gallons of the coffee a day.[9]

Beyond the control room, the after battery compartment was where most of the crew slept. Others would hot bunk in the forward torpedo room, which was also air-conditioned. Six bunks, three on each side, were suspended on heavy chains from the ceiling above the torpedo racks. In the center of the room were four jump bunks, which could be dismantled and stowed away at a moment's notice.

Scuttlebutt got around that the *Wolf* was taking the Corregidor men to Australia. This was exciting news in that the crew might finally be able to cable their families back home to let them know that they were safe. In the radio shack, Paul Maley and Mel Eckberg helped some of the crew prepare their cables, which could be sent once they reached Australia.

Seawolf's crew would rather be sinking ships, but they realized the importance of taking on a special mission to evacuate these key people. A submarine was best suited for running the Japanese blockades. Freddie Warder found the peninsula of Zamboanga—located on the western side of Mindanao Island— blacked out as his submarine moved through Basilan Strait during the early morning hours of 3 January. Rough waters compelled him to run at 110 feet this day, but the weather improved throughout the day. *Seawolf* changed course that night in order to examine the southwesterly exists from Davao Gulf. As her voyage toward Australia continued, Warder investigated Balud Island of the Sarangani group, Talaud Island, and Beo Bay of Karakelong during the next two days.

No enemy targets presented themselves and *Seawolf* was able to average 220 nautical miles per day en route to Australia. By 6 January, the *Wolf* was approaching the equator. She submerged 43 miles west of Kasiruta Island at 0443 and Lt. Cdr. Warder elected to let his men have a traditional Navy equator-crossing ceremony while submerged this day. Among those initiated included some of the *Wolf*'s special passengers. "The skipper and our executive officer had been initiated before," related Chief Lucien Rajotte. "We also had aboard our division commander, who had not been initiated, so we really gave him the works."

Along with Rajotte, the "Royal Court" included primarily chief petty officers and older Navy veterans such as Pop Mocarsky. As the most senior shellback present, Lloyd Sandridge was Neptunus Rex, complete with a hand-fashioned trident. Among his Royal Court was the next most senior line-crosser, John Sullivan as Davy Jones. Sully was later in charge of typing up official shellback cards for each polliwog that survived his initiation. As one of the largest men aboard, machinist Otis Dishman wore nothing but a giant diaper in his role as the Royal Baby. As the Royal Prosecutor, chief of the boat Eddie Souza had the list of all those in need of initiation. *Seawolf*'s initiates included the youngest seamen to some of the oldest veterans. Swede Enslin, with more than 14 years in the Navy, was still a polliwog.

Each of the Royal Court members was dressed in the most outlandish pirate garb that they could find. As *Seawolf* reached the equator, King Neptune Sandridge told Lt. Cdr. Warder that he would have to take control of his submarine in order to cleanse it of unworthy polliwogs. Warder graciously turned his boat over to the commander of the seas and Dick Holden took over the diving duties in the control room once the boat was leveled off underseas. Sandridge's Royal Court then proceeded to initiate those who had never crossed the equator.

The three most junior officers—Ensigns Jim Mercer, Doug Syverson and Bill Whitman—were subjected to the punishment as well. Deragon and Holden had been with the *Wolf* since her commissioning, so they were part of the shellback group. Warder noted that "48 polliwogs were initiated into the mysteries of the deep when the ship crossed the line submerged."

Each man was brought before Dishman, the Royal Baby. "He had a big stomach with grease all over it," said Bob Hanson. "You had to kiss it." Each victim was blindfolded and made to believe he was kissing the baby's royal bottom. "They also cut our hair and fed us rotten stuff that made me gag." Quartermaster striker Norm Kisver—who enjoyed baking as one of his hobbies—did not find the cooking to his taste this day. "They gave us some rough treatment. Our baker, who was actually named Baker, mixed up some stuff in the galley," said Kisver. "I don't know what the hell it was, but it tasted awful."

Everyone had grown rough-looking beards in the past month, so these became a target for the Royal Barber. "A lot of the passengers were shaved at the same time," said Paul Zimmerman, "because King Neptune didn't approve of their looks." Hank Thomson added, "All polliwogs were given one-half of a haircut: half their beards, half their mustaches, and half their hair."

Each polliwog was read a formal sentence of his crimes and then assigned tasks to repent for his sorry ways. Each was subjected to mind games of various sorts. "They made some of us think we were about to be thrown overboard into the sea, but we were submerged 60 feet deep at the time," said Zimmerman. "We were loaded up with survivors from Corregidor, officers and men. So we initiated everyone. Our passengers were a little more scared than anybody."

Some men protested who had actually crossed the equator previously. Lt. Cdr. Warder, however, was adamant that each man who had no proof of such crossing would have to be initiated. Motormac Bud McCoy had crossed aboard *USS Chaumont* on 3 July 1940, and even had a card to prove it—but not with him at the time. One of the passengers, radioman Charles Woodard, had joined the Navy in 1934 and had also crossed. "I crossed the equator on a surface ship a long time before the war," Woodard said. "I didn't have any proof of it, though, so they initiated me again. That was no fun! The skipper made it a point to put it in my record this time."

Sullivan and Sandridge signed "Ancient Order of the Deep" initiation cards for each new shellback. This proved that the cardholder had "been duly initiated into the mysteries of the deep in keeping with the tradition of my realm" at Latitude 00° 00° Longitude 126°28° on 6 January 1942.

Some of *Seawolf*'s new shellbacks could hardly wait for the next time their skipper would allow them to initiate some unworthy bunch of polliwogs.

THE CREW: Bearded *Seawolf* officers and crew in early 1942 photos taken while on patrol. *Photos courtesy of Hank Thomson.*

Lt. Bill Deragon was Exec, navigator and assistant approach officer for the first eight war patrols.

Lt. Dick Holden, the *Wolf*'s engineering and diving officer during her first seven patrols.

Lt. Doug Syverson, torpedo and gunnery officer and later Exec of the *Wolf*.

CMM Otis Dishman, senior engineman of the forward room.

CPhM Frank "Doc" Loaizo, the ship's pharmacist's mate.

CEM Henry "Hank" Brengelman, part of the fire control party.

CMM Alexander "Pop" Mocarsky made seven runs on *Seawolf*.

Fire controlman George Leffingwell, gun crew member, future officer.

RM1c Paul Leroy Maley made nine runs on the *Wolf*.

CGM John "Gunner" Bennett was *Seawolf*'s chief of the boat during her tenth run.

CMoMM Albert "Al" Hershey made the first eight war patrols.

CMoMM Lucien T. Rajotte made the first ten patrols.

Motor machinist Walter "Bud" McCoy *(left)* was on *Seawolf* from the start of the war until her loss, eventually becoming the senior enlisted man aboard.

TM1/c Arthur Earl Lamberson *(right)* was a senior torpedoman in the forward room.

Top three photos courtesy of Hank Thomson; lower five photos courtesy of Harold and Jeff McCoy.

Special Missions

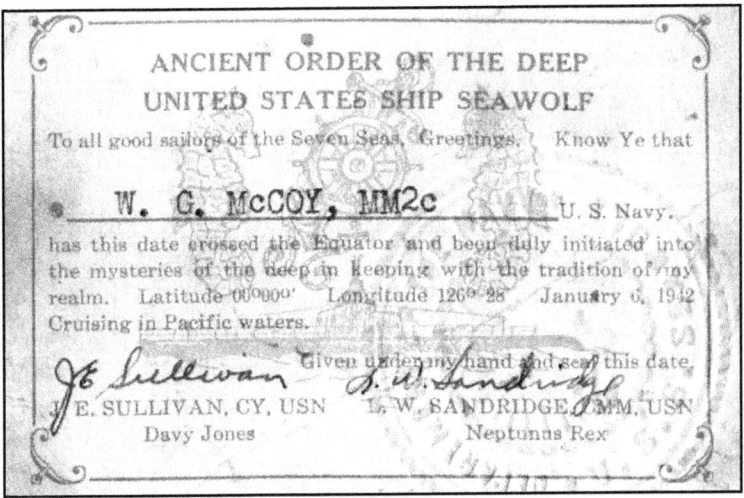

This card was awarded to *Seawolf* polliwogs such as Bud McCoy, who were initiated into the "Ancient Order of the Deep" upon crossing the equator on 6 January 1942. *Courtesy of Harold and Jeff McCoy.*

After running through the night, *Seawolf* patrolled submerged again on 7 January. She followed the same routine the next day, surfacing only shortly before sunset. In their first month of World War II, the *Wolf*'s personnel had lived in virtual darkness. "Officers and crew are somewhat in need of rest, sunlight, and exercise after 31 days of submerged operations (all surface running was done at night except for about 11 hours)," wrote Warder.

As *Seawolf* approached Australia, she stopped her zig-zag pattern and ran surfaced through the night. At 0047 on 9 January, Dick Holden called for battle stations when the port lookout announced an unknown ship on a parallel course and same speed 3,000 yards abeam. Duty quartermaster Frank Franz attempted to challenge this ship. Upon receiving no reply, Holden called for a dive at 0049. Sound could pick up no propellers. After staying down until 0110, *Seawolf* resurfaced and proceeded at 18 knots. "What had been thought to be a ship was a small dark cloud, now lifted from the water and still on about the same true bearing," noted Warder.

With the available moonlight, navigator Bill Deragon was able to definitely establish landfall on Cape Fourcroy at 0248. *Seawolf* met a British patrol vessel at 0620. This boat's maximum speed was 9 knots and Warder found that the British boat broke down twice. He finally offered the pilot to come on board *Seawolf.* He obliged and "left his boat to shift for itself. We offered to tow him but he declined."

With the pilot aboard, *Seawolf* moved through an 18 mile searched channel into Port Darwin, passing the entrance buoys at 0906 on 9 January. The *Wolf* moored alongside the familiar tender *Holland* at 0940 and her uneventful second war patrol ended. Tanless, bearded sailors who had seen no daylight since departing Manila Bay on 8 December came topside to bask in the warm sun. Later that day, the submarine *Tarpon* limped into the bay and tied up. She had weathered a typhoon, and had taken a large slug of water down the hatch while charging batteries. The water short-circuited her generators, and she had been forced to ride out the typhoon on the surface.

Seawolf's men were able to finally send out their Western Union cables to their families. Their loved ones could now know that they were safe, but still would have no idea where their sons or husbands were in the world. Red Eckberg found Darwin to be a "ghost town. When we got there, most of the civilians had fled to the interior." Many homes and stores were boarded up as if a hurricane was imminent. The heat was a blistering 110 degrees in the afternoon.[10]

Freddie Warder allowed his crew to go ashore and relax, keeping only a skeleton crew aboard to overhaul engines and carry out maintenance. The others polished their shoes, put on their whites, and took a liberty boat about a mile from *Holland* to town. They found that the center of town was a mile's walk from the dock over a dusty, red-clay road. They found milkshakes and ice-cream sodas instead of beer, and only one tough, middle-aged waitress to admire.[11]

By the afternoon, many of the men had had enough and opted to return to their boat. As the tender approached, Bill Deragon could be seen smoking a cigarette on the conning tower. "What are you doing back here?" the Exec called down to his crew.

"Lieutenant," called back Paul Maley. "There's nothing in port. No beer, no girls, no nothing!"

Deragon said he would see what he could do and soon had a dozen cases of beer from a transport. Gunner Bennett dashed to the gun locker to retrieve bats, balls, and gloves. The crew went ashore with the beer and found an empty lot for a baseball game. The tournament pitted the "Winton Wizards" against the "Deck Apes," the *Wolf*'s engineers versus the deck crew. Beer was consumed between each inning until the score of the game really did not matter.[12]

Special Missions 79

Seawolf remained in Darwin for one week preparing for her next mission. The men passed their free time wandering about the half-deserted town, looking at the uniforms of soldiers from other Allied countries who were also searching for entertainment. Yeoman John Sullivan had some personnel changes on 15 January prior to setting out on the third patrol. S2c Vernie Fuller was transferred to *Holland* immediately before *Seawolf*'s departure from Port Darwin. In return, Sully received MM2c Kenneth Cronk from the tender *Otus* for the engine rooms and S1c Francis James Cashero from *Holland*, both on 15 January. The son of a Kansas coal miner, Jim Cashero had been born in Colton, Oklahoma, a little town near Henrietta that no longer exists. "I was a senior in high school when I joined the Navy at age eighteen," he recalled. He enlisted in Kansas City, planning to finish high school later. At age 20, Cashero had completed sub school and was in Manila Harbor the day the war started.

Days later, he was under attack. "I was aboard the *Holland* when they came over and bombed the Philippines," he said. "Then they moved our ships down to Darwin, Australia, and I ended up being transferred aboard the *Seawolf*." Cashero joined Chief John Gibson's forward torpedo room as a torpedoman striker, and found a bunk slung above the torpedoes that would be his new home. It was tight quarters with a full load. "We had eight torpedoes in there, plus the ones in the tubes," he said.

On 15 January, Lt. Cdr. Warder received his next patrol orders from ComSubsAsiatic, Captain Wilkes. These were delivered to him by ComSubDiv 203, Cdr. Eliot Bryant, who had just made the trip aboard *Seawolf*. Wilkes' orders were for *Seawolf* "to proceed via Manila Straits, Sibutu Pass, and Apo East Passage using best sustained speed to Manila carrying .50 caliber anti-aircraft ammunition for the relief of Corregidor." Bryant also ordered that the *Wolf* would offload all but eight torpedoes and other ammunition that would enable her to carry up to 40 tons of .50-caliber. The final weight would be decided by just how much weight *Seawolf* could haul while still being able to safely dive with her safety tank full. Warder's boat was further detailed to bring back a full load of torpedoes and "such personnel as had been designated by the Commander in Chief, U.S. Asiatic Fleet to the Commandant Sixteenth Naval District."

The afternoon, chief of the boat Eddie Souza made the rounds and told his men to be ready for hard labor that night. The *Wolf* was making ready to head out on another special mission. Two dim lights were rigged up on the conning tower that evening. Around 2000, a string of motor launches came out of the darkness. They were fully loaded with ammunition which was to be brought aboard the *Wolf.*

The *Seawolf* sailors set to work loading the .50-caliber shells aboard, some cursing all the while that their boat was becoming nothing more than a transport vessel. Word eventually made it around the ship that they were loading anti-aircraft and machine-gun ammunition to take to Corregidor, known to the crew as "the Rock." *Seawolf* would run it through the Japanese blockade that would not allow a freighter to deliver such valuable cargo.

"We packed ammunition until it almost oozed out," recalled Mel Eckberg. "We thought the cases would never stop coming down." Shells were stacked in both torpedo rooms and throughout the ship in every conceivable nook and cranny. In the crew's berthing quarters, cases were stacked seven feet high. Men had to crawl over ammunition to go to sleep. The only eight torpedoes left aboard *Seawolf* were those in her tubes.[13]

"We had ammunition stored in the torpedo room. It was in the battery compartments and it was on the decks," recalled Bob Hanson. "We were walking all over it."

The surplus ammunition loaded aboard *Seawolf* at Darwin only amounted to a drop in the bucket: 675 boxes of .50-caliber ammunition and 72 3-inch anti-aircraft shells. "The way they were expending ammo up there," Warder recalled, "it amounted to about one day's supply."[14]

While the loading was going on, Warder noted Fred Doyle, division commander aboard *Holland.*

> I was alongside the tender *Holland* in Darwin for four days after I brought Jimmy Fife and his division commanders and staffies down there. I would see Doyle walking on the foscle of the *Holland* and he would see me doing the same thing on the *Seawolf.* We were overhauling two engines and loading all this goddamn ammunition. We had to get the hell out of there to Corregidor. He waved to me and we were in talking distance. He'd ask me how things were going, but he could see my crew working like hell and he didn't want to bother me really.[15]

Special Missions

By the morning of 16 January, the exhausted *Seawolf* crew had loaded 178,875 rounds of .50-caliber anti-personnel tracer shells and 72 rounds of 3" .50-caliber anti-aircraft shells for the relief of Corregidor. The total weight was 72,585 pounds, or 36 and one quarter tons of additional ballast for diving officer Dick Holden to contend with. Freddie Warder, ever eager to get into the business of sinking ships, read his secret patrol orders with Exec Bill Deragon. In Admiral Wilkes' orders it did not say that *Seawolf* could not attack enemy shipping, so Warder decided to have this clarified. He went to Commander Bryant aboard *Otus* for clarification.

"Commander, my orders say nothing about attacking ships of the enemy. Am I to seek attack?" Warder queried.

Cdr. Connolly piped in with, "Certainly if you encounter a nice target such as a heavy transport or a carrier, you should knock him off. After all, that is up to the skipper."

"No, this mission is not entirely up to Freddie," advised Bryant. He contended that if Lt. Cdr. Warder was coming to his superiors for advice that they should try to give it to him.

Cdr. Vanzant spoke up at this point. "As I see it, your mission is to get to Corregidor with this ammunition as fast as you can. Nothing more is expected of you."

With staff opinions flying about, Eliot Bryant finally gave Warder his verdict. *Seawolf* was to "do nothing which might jeopardize the success" of her special mission or unduly delay it. However, if Warder were presented while en route with a "firing set-up which looked like a sure thing," then he should take the shot.

Warder departed *Otus* with his superiors' advice and made final preparations for getting his boat underway. Calling Bill Deragon to his stateroom, he relayed the essential parts of his fresh conversation. They had been given just enough rope to hang an important Japanese ship, should they encounter one. Just to cover his bases, Warder had Sully Sullivan type up the clarifications to his printed orders that had transpired before the Division 203 staff.

At 1249, *Seawolf* got underway from *Otus* and proceeded out the searched channel of Darwin for the 1,600-mile voyage to Corregidor. Warder had orders to conduct sound exercises with the Australian anti-submarine vessel *Warrego* at 1330. By 1355, *Warrego* was not in sight and the *Wolf* finally received a radio dispatch telling her that the training exercise had been cancelled and that she was to clear the swept channel at dusk.

Navigators Deragon and Jughead Casler cleared the swept channel that night "by guess and deck reckoning" in Warder's words. The *Wolf* ran at 17 knots through the night as she charged her batteries. The additional tons of ammunition aboard proved to be troublesome for diving officer Dick Holden and his control room watch detail. As *Seawolf* ran submerged on 17 January, the afternoon watch lost their depth control three times momentarily as they encountered pockets of water with lower density. From her cruising depth of 120 feet, the *Wolf* dropped ten feet before they caught her each time via the use of the bow and stern planes, increased speed, and pumps.

At 1415, *Seawolf* dropped again to 140 feet as she hit another low density water pocket. This time, she apparently scraped bottom as the QC-JK sonar head stopped functioning and the pit log began behaving erratically. Warder ordered his boat up to 63-foot periscope depth and ran shallower until he could safely surface after sunset.

Hopes of the officers altering the watch schedule were put to an end by a mishap involving Ens. Bill Whitman. Junior officers Holden, Syverson, and Mercer covered all watches in a 24-hour period with four hours on and eight off. Whitman had to go through a thorough qualifying process before Lt. Cdr. Warder would put him on as an OOD. During a timing drill to clear the bridge, Whitman tripped over the hatch and took a nasty fall, fracturing his ankle in the process. Doc Loaiza taped and splinted the badly swollen foot, but Whitman would not be able to scramble up and down hatches for weeks. Bill Deragon assigned him to more duty time in the wardroom decoding all the incoming messages from the radio gang.

During the week's run to Corregidor, the crew passed their off-duty time reading or playing blackjack, poker, and hearts in the messhall. The radiomen piped in two commercial radio stations —San Francisco radio or Tokyo Rose—each day for the listeners' entertainment. Radioman Eckberg found that each man took the new war in his own stride. Al Hershey, a Wisconsin farmboy and former Navy wrestling champ, listened optimistically to Frisco radio's claims of how the U.S. Navy would soon take care of the Japanese Navy. Fred Zirkel argued back that the United States was in for a longer fight than they bargained for. "We'll be fighting these Japs a hell of a long time from now, and when it's over we'll know we've been in one hell of a fight," Zirkel proclaimed.[16]

"Call the skipper!" exclaimed Red Syverson as he steadied the scope on bearing 350. "I've got a Jap destroyer at 14,000 yards."

Warder quickly arrived and took a look. He concurred with his OOD and sounded battle stations at 1802 on 21 January. Although his patrol orders called for the speedy delivery of his anti-aircraft ammunition, Warder and Deragon had decided that enemy warships most certainly constituted a green light to attack.

Seawolf was near Kema Island, where she picked up a portion of the Japanese Western Invasion Force bound for Balikpapan: a light cruiser, six destroyers, and seven Japanese transports. Balikpapan was a port on the east coast of Borneo off Makassar Strait. The sea was choppy with occasional rain squalls, ideal weather for a submarine periscope to slice around in. Within minutes, Warder was making his setup on what appeared to be a light cruiser of the convoy's port screen. By 1815, the range had decreased to 2,900 yards.

Warder came left with full rudder and full speed in order to make a bow attack on this warship from about 1,000 yards with a 90° ship track. He then slowed to one-third speed and at 1820, as he was preparing to attack, he made another observation. He found that the target "had zigged away during my maneuver and had slowed down" from its original 14 knots. Warder positively identified this ship now as an *Asashio*-class destroyer but found himself too far out of position to attack.

Seawolf tailed four destroyers, hoping to develop targets astern of the convoy. She moved in toward another mast sighted in the approaching dusk and rain. Five minutes later, a three-stack Japanese cruiser came barreling down on *Seawolf*'s position from 12,000 yards away. Warder ordered helmsman Rudy Gervais to come hard left on a collision course with the cruiser and called for full speed.

He ducked under for a few minutes to avoid feathering his scope. Once again, when *Seawolf* poked her scope up to see how well she had closed the cruiser, she now found a freighter-transport ship at about 7,000 yards. The cruiser had approached to about 10,000 yards. Warder changed course to give himself a shot at the light cruiser before dark.

The cruiser changed course moments later and began opening up the range. In the distance, another escort flashed his searchlights. Warder counted a total of seven freighter-transports bearing about northwest. "Are these 'empties' bound for Palau or are they taking station with cruiser and destroyers?" Warder mused in his log. "Rain squalls and darkness adding to confusion."

After dark, searchlights continued to flash about among the convoy as *Seawolf* hung on to her contact. Warder gave up the idea of continuing to pursue when a loud explosion erupted at 1918. Although this depth charge may have been a "practice drop," it got Warder's attention.

"Rig for depth charge! Take her to 150 feet."

Seawolf stayed at 150 momentarily, but Red Eckberg and Paul Maley could not make out any approaching propellers. Warder ordered his boat back up to 60 feet and made a quick sweep. No destroyers were visible, so he ordered *Seawolf* to the surface. The *Wolf* pursued the convoy at 15 knots while charging her batteries. Some men were relieved not to be making an attack. Stacked high with extra explosives, the prospect of being jarred and shaken by enemy depth charges did not seem inviting. Warder went to the radio shack and had Eckberg send off an urgent dispatch to submarine command, revealing the enemy task force that had been spotted.[17]

Warder hoped to intercept the Japanese cruiser during the night, but heavy rains persisted and made the intercept all but impossible. "Rain is going with us at about our speed," the skipper noted. Having received no word back from ComSubAF regarding modification of *Seawolf*'s delivery mission, Warder reluctantly broke off the chase at 2200 and reversed course.

Seawolf ran at 17 knots on the surface toward Bangka Passage. At 0005 on 22 January, a flash of light was spotted about 1 point on the starboard bow. Jim Mercer, the OOD, slowed to 10 knots to investigate and called the skipper. The sound shaft was also lowered to listen for enemy ships. Warder ordered battle stations. The rain was still coming down.

Lt. Cdr. Warder was looking out the forward port of the pilot house, endeavoring to find the light which had been reported when Ensign Mercer reported, 'Lights all round us!'

Warder jumped to the pelorus platform and was able to make out flashes of a light abeam to port, well above the water. By the time he made out how far away it was and which way the ship was going, the enemy ship had extinguished his light. "I could have fired a stern tube but felt sure it would be wasting a torpedo," he logged. The lookout and Mercer reported that there had been two ships signaling each other, one forward of *Seawolf*'s beam and one on her beam. Warder believed that this had been a Japanese destroyer and another ship in column at high speed, running to catch up to the convoy.

Seawolf, unable to attack the rapid ships in the darkness and rain, ran on the surface until dawn. She continued a periscope patrol throughout 22 January as the seas grew stronger. At 1550, Lt. Dick Holden sighted what appeared to be a large tripod mast. Eddie Souza called over the intercom for the skipper. Warder arrived in the conning tower and took a look. He agreed that there was some object on the horizon, but could not tell if it was a ship. *Seawolf* went to battle stations as she closed on the potential target at 6 knots.

The target became larger in the periscope. Warder still could not make her out but decided "there must be something 'fishy' about the target." The tracking party charted out the target for about half an hour as Warder cautiously approached. He finally muttered, "Wait a minute!" He then called, "Secure battle stations."[18]

"Dick, come over here a minute and take a look at this ship you sighted," he said.

Holden took the periscope and found that his mast was that of a sunken schooner whose deck was under water. The upper one-third of the mast was broken and hanging down. A small boom was sticking up aft. Two sea gulls were sitting on projections from the hull and they gave good imitations of tops of belled-out Japanese smokestacks. "Don Quixote and the windmills had nothing on us," Warder wrote in his patrol report.

The approach on the sunken schooner was a source of comedy for the crew for days. Men joked whether they should have fired torpedoes at the seagulls or shot them with their 3-inch deck gun.

Seawolf left the Sulu Sea and entered the South China Sea to head for Corregidor. She passed through Sibutu Passage on 24 January.

Her port screw became fouled with something in the water during the early morning hours the next day. The forward torpedo room reported that the ship had hit something. Lt.(jg) Syverson stopped and backed the ship and then worked her up to 18 knots and managed to free whatever had fouled the port screw.

Seawolf passed close enough to Panay on 25 January to make out a grass fire ashore which was first thought to be signalling a ship. That night, a ship was spotted and battle stations were called, but it proved to be nothing more than a sailing schooner running without lights.

Seawolf continued to make good time and Warder radioed command that *Seawolf* would arrive one day earlier than planned. He offered to use that day hunting unless otherwise directed. Warder received no response so he spent 27 January investigating bays along Luzon's coast. The periscope watch noted considerable activity on Corregidor during the early morning hours, including rapid successions of red rockets being fired around the bay. From Mariveles Bay, Warder noted the flash and noise of gunfire.

Seawolf continued to poke through Subic Bay and then Port Silanguin Bay and Nazaso Bay without sighting any enemy shipping. At 0400 on 28 January, *Seawolf* met a motor torpedo boat which led her in through the minefields to Corregidor. Once safely through the fields into the harbor, the *Wolf* submerged 1 mile east of the entrance for South Dock. *Seawolf* settled lightly on the harbor bottom in 135 feet to wait for darkness.

She surfaced at 1936 and moored to the starboard side of Corregidor's South Dock. As soon as the mooring lines were secure, the *Wolf*'s crew began unloading the 36 tons of ammunition with the aid of Army personnel. Forty minutes into the unloading, *Seawolf* received warning that an air raid was likely between 2100 and 2400. Warder got his boat underway and stood out about a mile from the dock, laying with his deck awash.

Shortly after midnight on 29 January, *Seawolf* received an all clear and eased back to the dock to recommence unloading. The army gratefully received the .50-caliber ammo but was disappointed with the 3-inch anti-aircraft shells. There were plenty on Corregidor and experience had proved that one in three was a dud.[19]

Electrician Bob Parden, sweating through the night to unload ammo, was surprised to receive a drink from one of the soldiers he was helping. The soldier handed him a drink in a shaving lotion bottle, which Parden gladly accepted. In return, Parden and some of his mates had purchased several cases of booze in Australia, which they handed out to the Corregidor defenders.[20]

In the early hours of 29 January, the ammunition offloading was completed. Some of the crew had a chance to go ashore then and take in the latest information. Red Eckberg, standing on the dock, could hear "the distant thunder of the Jap guns on Bataan, twenty miles away." Soldiers working to haul off the freshly delivered ammunition complained of the lack of good food and having to eat mainly rice. Typhoid was spreading and the smell of death hung in the air from bodies littering the ground in the vicinity of Corregidor. Eckberg, sporting a beard and wearing sandals, shorts, and shirt, walked up a dusty road to the communications center and got his first glimpse of the island fortress, 600 feet high, that split the Bay of Manila. He recalled that it "was then known as the biggest and most impregnable fort in the world."[21]

By 0543 on 29 January, the *Wolf* stood out from the dock. Doug Syverson's torpedo gang had managed to take aboard eight replacement torpedoes. The war fish were taken aboard by means of an ingenious skid devised by the torpedo officer of *Canopus*, Louis Darby "Sandy" McGregor.

Seawolf lay in Manila Bay at 128 feet all day until threats of bombers were gone with darkness. She moored to Corregidor's South Dock at 2017 and continued loading replacement torpedoes during the next hours, along with all the spare parts *Seawolf* could hold. Scarcely had the unloading completed than *Seawolf* became part of another mission in support of the Philippines. This time, Com16 had been ordered by Admiral Hart to have *Seawolf* take passengers to Surabaya before starting her next war patrol. At 0200 on 30 January, a military truck pulled up to the south dock and men spilled out. "Out of it came more than a score of the most disreputable human beings in Army uniform I had seen," recalled Mel Eckberg.[22]

Seawolf took aboard 25 passengers, designated by name, half by Admiral Hart in Surabaya and half by MacArthur. There were a British intelligence officer, 12 army pilots, six navy pilots, five navy enlisted men, and one navy yeoman. Bob Hanson recalled that there was no shortage of other passengers—including several divers from a submarine rescue vehicle—willing to try to sneak aboard to escape "the Rock."

There was Major Reginald Vance, air officer to General MacArthur's staff on Corregidor. There was also a British Secret Intelligence Service agent, Major Gerald Wilkinson. He had been in Manila before the war in charge of a Pacific-wide British trading company. Wilkinson had warned one of MacArthur's senior staff members in late November 1941 that war between Japan and America was imminent, per his intelligence from Singapore.[23]

By 28 December, Maj. Vance had only 18 serviceable aircraft on Corregidor and these continued to be lost. In the early morning hours of 11 January, Vance tried to fly out Maj. Wilkinson in the only North American A-27. Wilkinson was bound for Mindanao and then to Singapore in an effort to get help for the Philippines. After taking off, Vance's A-27 lost oil pressure and he was forced to make an emergency landing on Corregidor in the moonlight. He misjudged the runway in the dark and smashed the A-27. Fortunately, he and secret agent Wilkinson escaped with only minor injuries.[24]

The Army Air Corps pilots were a mixed bag. Their squadrons in the Philippines had been scattered during the chaos in December once the Japanese invaded. Captain Glenwood Stephenson was commander of the 16th Pursuit Squadron, which flew P-40s. Lt. William E. Eubank was acting commander of the 91st Fighter Squadron, one of two elements of P-39 squadrons that boarded.

First Lt. Jim McAfee was in charge of a small group from the planeless 27th Bomb Group. He had been in charge of 90 men and a few P-40s and P-35s on Nielson Field, located about 5 miles north of Nichols Field. McAfee inherited this job on 14 December when headquarters abandoned this exposed field for the safety of Fort McKinley. "I guess they figure that nobody but a low-ranking 1st Lieutenant would be dumb enough to stay on that field," McAfee wrote in his diary. "It's going to be bombed to hell."[25]

Lt. John Carpenter was a B-17 pilot of the 93rd Bombardment Squadron who was attacked by Japanese fighters on 8 December as he was coming in for a landing. He dodged enemy aircraft for hours and finally brought his bomber back to Clark Field, dodging bomb craters in the runway upon landing. The number of planes operable at Clark trickled over the ensuing weeks. Carpenter lost his B-17 when it was hit by a crashing P-40.[26]

Lt. Harl Pease, another B-17 pilot, had been involved in one mission from Clark Field on 10 December which proved to be the first

United States Air Force bombing attack of World War II. With four other pilots, Pease rained his load of 100-pound demolition bombs on Japanese transports unloading off Luzon.[27]

By Christmas Eve, the acting squadron commanders had received orders to move their men via boat or plane to Corregidor for evacuation. Lieutenant McAfee found everyone to be "running for their life," taking all the provisions they could carry before the Japanese took them. Captain Stephenson's men reached Corregidor carrying food, field kitchens, clothing, and camp equipment.[28]

The six Navy pilots and their enlisted men had also seen their share of action from Manila. Lt. Harmon Utter and Lt.(jg) Thomas Pollock were scheduled to fly Admiral Hart from Manila to Java, but before they could do so Japanese planes strafed and destroyed their PBY. Both men would receive commendations for their brave flying prior to the fall of the Philippines in early 1942.

In addition to the 25 passengers, *Seawolf* had crammed spare parts into every available inch aboard ship. Space was even tighter than it had been with the ammunition. Chief Souza worked out a hot-bunking arrangement with the crew and aviators. *Seawolf* got underway at 0318. After waiting for the moon to set, she cleared Manila Harbor's minefields at 0601 and set course for Surabaya, Java.

Pilot Jim McAfee wrote in diary for 30 January of *Seawolf*, "She's quite big but I bet we're cramped." He and his fellow aviators fell right in with the submariners. "Played poker till 1 with some Navy guys and I cleaned them. It's a good thing—they can't spend their dough and I'm broke."[29]

After his morning watch ended at 0800, Red Eckberg went forward to his bunk in the forward torpedo room. He noticed two of the aviators: "They were just youngsters, and they looked worn out. They were discarding some of their personal gear—gas masks, knives, and belts." He found that the fliers appreciated the good food and the endless supply of hot coffee aboard. Eckberg noted the scanty personal effects of one dirty, scarred flier and offered to share his cleaning-gear locker with him.[30]

"We were hot-bunking it during this time," said torpedoman Jim Cashero of the forward room. Engineman Bob Hanson found that the aviators were in poor shape. "When they came aboard, all they could bring with them was what they had on their backs. So, we shared all of our cigarettes and everything else with them. They

Military Personnel Evacuated from Corregidor
by USS *Seawolf* on 31 January 1942

British Army	Stafford, 2nd Lt. Robert F.
Wilkinson, Maj. Gerald	**U.S. Navy VP-101**
U.S. Army	Utter, Lt. Harmon T.
Vance, Maj. Reginald F. C.	Entire, Lt.(jg) R. K.
Eubank, Capt. William E. Jr.	Pew, Lt.(jg) L. A.
Shedd, Capt. Morris H.	Pollock, Lt.(jg) Thomas F.
Marrocco, Capt. William A. (MC)	Swenson, Lt.(jg) H. R.
Stephenson, Capt. Glenwood G.	Williamson, Ens. L. H.
Bender, 1st Lt. Frank Peter	Benedict, ACMM C. P.
Carpenter, 1st Lt. John W.	Kelley, ACMM D. M.
Croxton, 1st Lt. Warner W.	Payne, ACMM Earl D.
Hoevet, 1st Lt. Dean C.	Clark, AMM1c J. W.
McAfee, 1st Lt. James B.	Cook, AMM1c R. F.
Pease, 1st Lt. Harl Jr.	Seward, Y1c N. E.

were happy to be leaving Corregidor." Older brother Henry "Big Swede" Hanson did not smoke, but Little Swede shared plenty of his cigarettes. "When we had the aviators aboard, they would get four hours on the rack," he said. "Then, someone would wake them up and they'd go find another empty rack to crawl into."

The *Seawolf* crew soon came to appreciate what their passengers had been through. Some had been shot down more than once. Many had made their way through jungles and swamps before reaching the safety of the American lines. Baker, Wright, and Mallory filled them with steak, ham, lamb, pork chops, pies, and pastries. Fortunately, *Seawolf* had been able to add officer's cook first class Tomas Rosete, a Filipino received from the submarine detachment at Corregidor. With stewards Bugawisan and Tamayo, Rosete would help take care of the additional passengers. Doc Loaiza offered the Corregidor evacuees vitamin pills each morning to help restore their health.[31]

Flyboy Jim McAfee particularly enjoyed the good food.

> I went first to the galley and never will I forget the taste of a jar of sandwich spread I found on the table. I'm to sleep in one of the crew's bunks while he works. They have plenty of food. Coffee, too!
>
> The boys say this food is a little better than usual; they got it off a bombed ship in Darwin. Fine—I volunteered for a watch on deck when we surface on nights. It's fun and I

think I would like sub life. The captain and his crew are sure fine men. I've struck up a good frienship with the sec in command—Mr. Deragon. He's Annapolis '39. We know people. He says he was surprised I like subs. I'd like six months duty aboard but no more.[32]

"The crew adopted the men," recalled Eckberg. "We showed them how we made our leisure time go by in Kelly's Pool Room—cribbage, hearts, dice, acey-deucy. We let them wash their clothes in Baby, and they got a kick out of watching us in the evening, sitting on our bunks and darning socks or sewing buttons."

The crew had fun with the flyboys, too, as Eckberg related.

> Now and then, in the course of a routine test, we'd let a little air hiss out of a valve, and some wisecracker would whisper, loud enough for our passengers to hear, "Jeez, the Skipper's taking her down to 1,000 feet." The faces of the aviators would be a sight. The thought of going down 1,000 feet under the water didn't make them any too happy.[33]

Anxious to get back to war, Lt. Cdr. Warder ran his boat at 18 knots as much as possible en route. *Seawolf* ran submerged during the daylight hours and made speed during the night. She sailed south through the Sulu and Celebes seas, Molucca Passage, and thence to the Dutch port of Surabaya. As the *Wolf* moved closer to Java, the water temperature became warmer and the boat became warmer. The aviators were amazed that their new home could still feel so comfortable with its air-conditioning system.

During the evening hours, the crew listened to the radio to catch up on news of a naval battle playing out in Makassar Straits. "Hope we get there in time to help. We have plenty of torpedoes and nothing to use them on," Army Lieutenant McAfee wrote in diary on 2 February. He found Freddie Warder to be "a fine guy. He is teaching me how to run a sub. I read part of the time we were under water. They have a fine library. Most of the crew have beards as I do."[34]

Seawolf approached the straits on 3 February. When the pilots were not helping to stand watches, they passed time below playing cards. "We're getting a lot of bridge in," wrote McAfee. "Colonel Vance and I skinned Steve [Stephenson] and Major Wilkinson today."[35]

This squadron photo of the 27th Bomb Group shows three pilots rescued from Corregidor by *Seawolf*: 1st Lt. Jim McAfee *(first man on the left, second row)*, Capt. Glenwood Stephenson *(second man, second row)* and 1st Lt. Frank Peter Bender *(fifth on second row)*. Photo courtesy of G. Wayne Dow.

The *Wolf* stayed submerged during the day as she made her way through Makassar Straits and approached Surabaya. "We stay down about 60 feet all day and surface at night," McAfee wrote on 4 February. "The air gets rather thin just before we surface due to all the passengers. We sweat easily then and feel like gasping for breath. Actually the oxygen content is plenty high and we are nowhere near to suffocating." Auston Baker and the cooks even treated their aviator passengers to a turkey dinner en route to Surabaya.[36]

Seawolf arrived off Surabaya late on 6 February and made her way toward her rendezvous point during the early morning hours of 7 February. Although relieved to be off "the Rock," Jim McAfee had a new appreciation for Navy life on a wartime submarine.

> Everyone (all of us landlubbers) will be glad to get off and get a good bath. Too, we can't work all the valves and pumps on the head (toilet) very well. If you don't work it just right and twist valves and pump, why you're liable to have the whole mess blow right back out of the place. That would be rather untidy. Anyway, it has been something that money wouldn't buy, this trip has. I will hate to leave some of these fine fellows. They do a grand job and have my admiration—they aren't scared of the devil. Anyhow, land ho![37]

No enemy attacks had been made and the patrol was not deemed successful for combat insignia. It was just a mission that needed doing. It was not until February 1946 that Warder received a Navy Letter of Commendation for his third patrol. He was cited for having "penetrated extremely hazardous waters and delivered much needed supplies and ammunition to embattled forces on Corregidor."

4

"Into a Hornet's Nest"

Refit and Fourth Patrol   7–25 February 1942

The *Wolf* rendezvoused with her patrol boat escort at 0707 on 7 February and followed it from Madura Straits into the channel and through the two mine fields off Surabaya. Freddie Warder kept his decks awash during the morning as he awaited an all-clear to offload his passengers. At 1248, the radio announced an air raid warning. At 1310, two planes were spotted circling over Surabaya. Warder dived and ran a periscope patrol. When a Dutch ship was spotted entering the harbor that afternoon, Warder exercised Doug Syverson's TDC party on the departing ship with a simulated "attack."

Seawolf moored port side to Holland Pier at Surabaya at 2005. Although Surabaya was subject to air attacks, the 25 passengers who left the ship at 2030 were elated to be anywhere other than Corregidor. As they departed, the fliers shook hands all around. The happiest sight for the *Seawolf* crew was a Navy paymaster who came aboard and gave the *Wolf* crew a week's pay in Dutch guilders. Ship's cook Gus Wright returned to his mess hall with a 5-gallon can of fresh milk, the first seen aboard *Seawolf* in two months.[1]

The *Seawolf* sailors were allowed to go into town at Surabaya, which was quite a spectacle. At this time, Surabaya was the third largest naval base in the East, behind Singapore and Hong Kong. Oil tankers filed into this base routinely to fuel from a half-mile long jetty which extended along the harbor's entrance. Army and Navy men rode jammed trolleys through the streets of town, mingling with natives, people of multiple ethnicities, and British and Australian soldiers. In town, the crew found fresh fruit in the marketplaces and

MoMM1c Kenneth G. "Casey" Mallough had survived *Sealion*'s destruction in December 1941 before joining *Seawolf* and making six patrols aboard her. *Courtesy of Bobbie Mallough.*

plenty of bars to visit. The only downfall to Surabaya was that there was no mail waiting for the crew and no way for the men to communicate back home.

Yeoman Sullivan received aboard three new hands on 9 February, TM1c Claiborne Weade from *S-36*, and F1c Henry Kraght and MM1c Kenneth "Casey" Mallough from *Permit*. Twenty-five-year-old Mallough, like Swede Larson and Bill Whitman, had escaped *Sealion*'s destruction. "Casey had been on the *Sealion* at Cavite Navy Yard when she was being overhauled. *Sealion* was bombed and destroyed, so we picked up Casey," recalled fellow engineman Paul Zimmerman. "Later in the war, he served on *Sealion II*." Born Kenneth Geiser Mallough, he had picked up the nickname Casey as a schoolkid in North Dakota and it stuck with him. When *Sealion* was bombed at Cavite, Casey had just changed shifts and headed ashore to the civilian housing. He lost some of his close buddies when the explosion wrecked the *Lion*'s engineering spaces.

During the next weeks, Mallough made do with all he had left—a motorcycle he had acquired in the Philippines and some borrowed Army clothing—since all of his gear had been destroyed on *Sealion*. Casey used his motorcycle as a messenger runner ashore until he finally was able to flee Corregidor on *Permit*. Designed for 65, she had staff officers, crew, codebreakers, and PT boat crewmen stuffed aboard her when she departed "the Rock" on 17 March with 111 people on board. Kraght and Mallough were no doubt glad to be off of her when *Permit* reached Australia on 7 April.[2]

On the *Wolf*'s second day at Surabaya, the U.S. Submarine Commander, Captain John Wilkes, informed Lt. Cdr. Warder that his *Seawolf* crew could make use of a rest camp at Malang run by the local Dutch naval garrison. A tiny town high in the mountains about 50 miles south of Surabaya, Malang would be used by the crew in two shifts. Half would stay aboard to effect repairs while half went to the rest camp. After three days, the two sections would swap.[3]

On the morning of the third day in Malang, half the *Seawolf* crew headed to the nearby railway station. All dressed in their whites, they soon had to scatter as an air-raid siren pierced the morning air. Lt. Holden called for the men to evacuate the train station. The crew scrambled over a 20-foot embankment and plunged into a rice field, six inches deep in water. There they remained until the flight of 27 Japanese bombers passed overhead in V formation to drop their bombs near the waterfront.

"We enjoyed ourselves at Malang," recalled Mel Eckberg. "Jap bombing planes came over Malang daily, but they didn't bother with the little resort." Paul Zimmerman heard from the first group to Malang that "they didn't have any toilet paper up there. You did your job over a ditch with running water." When Zimmerman's second group went up, "we brought a couple of rolls of toilet paper."

The men were housed in barracks at Malang's camp. There, they could enjoy horseback riding along the little river in camp. Out riding along the river one day, a group of *Seawolf* crew came upon some local women who were washing their clothes and hanging them out to dry. At that moment, the horses took off and Chief John Sullivan was thrown from his right into the river. He was unhurt but received quite a ribbing from his buddies.

Back aboard *Seawolf*, the other half of the crew found the daily Japanese bombing raids to be troublesome. Air-raid sirens wailed at any given time. Long before the morning bombers would appear, the *Wolf* would nose out into the deep waters of the bay and submerge, either lying to or cruising about until dusk brought safety again.

Some of the more enterprising *Seawolf* crewmen learned to take advantage of the routine air raids. Paul Zimmerman relates how they managed to "acquire" some extra goods.

The Dutch Navy Yard personnel at Surabaya lived in shelters up above the yard. Some of our crew learned that when the air raid alarms went off, the Dutch would run for these shelters. While our boat went out and submerged in the harbor, we would leave one of our crew there to pick up provisions. Some of the cans they collected were printed with words they couldn't read. Most of the cans turned out to be full of stew. They also brought back bales of tea that they thought at first was coffee. Of course, the guys made out like mad when they thought they were taking coffee.

When the alarm ended, our ship came back and picked up the crew that had stayed ashore. Here they were, loaded with a couple of jeeps of stew and cans of things we had no idea of what was inside. We had spare torpedo tubes in the upper hull, but we didn't have torpedoes in these skids at this time. So we used the torpedo tubes to store all the liberated Dutch goods.

The second group to head to the Malang rest camp also dealt with the air raids en route. "On the way up there, the train stopped," recalled Bob Hanson. "Everyone jumped off and went into the bush. You could pretty much set your clock by when the bombers were going to go over, generally in the morning and midday." At Malang, Hanson found the rest camp "was run by the Dutch military, complete with beer halls, softball, volleyball" and other entertainment.

At long last, Lt. Cdr. Warder received new orders during the second week of February. Following his conference with Captain Wilkes, he came back aboard and *Seawolf* began at once preparing his boat for war patrol again. Her days of hauling brass, ammo, aviators, and papers appeared to be over at long last.

Seawolf's diesels rumbled to life in the afternoon of 15 February in Surabaya's harbor. The "deck apes" took in the lines as Lt. Cdr. Warder called for, "All back one-third" at 1450. *Seawolf* trembled as she backed into the harbor from Surabaya Roadstead and then twisted around toward the north. She moved out past the protective minefields with a Dutch pilot aboard to help with the conn. He stood on the bridge with Freddie Warder, guiding the submarine through the darkness out to sea beyond the wharves and jetties of West Gate.

At 1632, a small motor launch pulled up alongside *Seawolf* and she halted long enough for the pilot to climb over. With a wave, he was gone and the *Wolf* was headed out with a patrol vessel.[4]

Surabaya had been a good R&R for the crew, but it was not a safe area. Soon thereafter, the conditions in the area deteriorated. Four days after *Seawolf*'s departure, Dutch Admiral Karel Doorman led American, Dutch, British, and Australian ships in a battle against Japanese landing forces at Bali. The destroyer *Stewart* (DD-224) was damaged and returned to Surabaya for repairs in drydock on 22 February. There, *Stewart* was further damaged by bombs and would eventually fall into Japanese hands when Surabaya was captured during the first week of March.

Seawolf had thus far led a charmed life, slipping away from Corregidor and Surabaya before both fell into Japanese hands. Warder was eager for a fight, having spent three patrols either on limited time or special duty. His patrol orders called for him to cover the Java Sea and Lombok Strait area to help repel enemy invasion forces. In their quest to move closer to Australia, the Japanese had bypassed the Celebes Sea area. Their focus was on taking such key islands as Bali, Lombok, Flores, and Timor, which would serve as stepping stones to taking Surabaya on Java. *Seawolf*'s mission was now to impede the Japanese progress toward Australia with her full load of Mark 14 torpedoes.

Once past the mine fields of Surabaya, the patrol vessel signaled *Seawolf* to proceed independently at 1752. Warder ran his boat at 17 knots until clear of all danger areas and then dived to periscope depth. Bill Deragon and Jug Casler shot the stars during the night and found that the prevailing heavy head sea called for a slight course change. Running through the Java Sea on 16 February, sound reported a propeller count of 100 rpm around noon. The *Wolf*'s sound head was lowered but the noise faded out before a location fix was possible. Heavy rains had reduced visibility to about 1,000 yards, so Warder was not able to get a sighting on this vessel.

Seawolf surfaced at 1909 and started her battery charge. At 2134, Mel Eckberg picked up a coded dispatch directed to Lt. Cdr. Warder. His orders called for *Seawolf* to head east at once for Lombok Strait and intercept a Japanese invasion force that was preparing to land troops on the southeast coast of Bali Island. Warder ordered course 155° and made 17 knots for Lombok. Word spread that *Seawolf* was

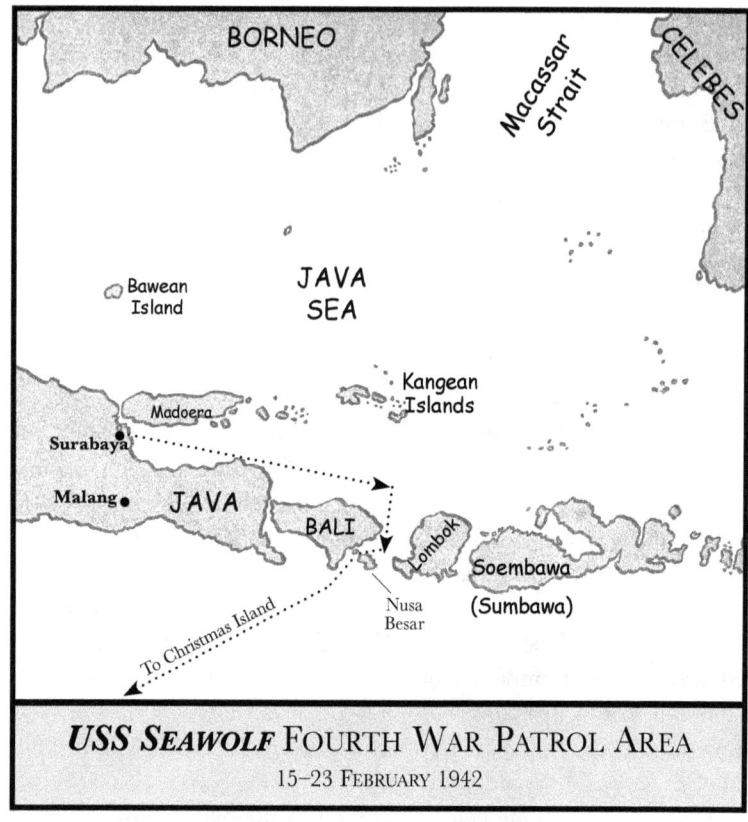

USS SEAWOLF FOURTH WAR PATROL AREA
15–23 FEBRUARY 1942

heading for Lombok Straits, a narrow passage between Lombok and Bali islands, 120 miles southeast of Surabaya.[5]

Seawolf cleared the 100-fathom curve during the early morning hours of 17 February and headed for the center of Lombok Strait. She submerged at dawn and had Lombok Island in sight through the periscope at 0705. Lombok Island had a 7,000-foot volcanic peak which was visible for many miles at sea. Vicious currents made headway a struggle at best as *Seawolf* fought a strong set to the southward. Bill Deragon and his assistant navigator fought the underwater currents which kept submerged progress minimal.

To keep his lookouts on full alert, Warder upped the ante. "Captain Warder had a motivational idea for the lookout who sighted an enemy ship that we successfully attacked," recalled quartermaster Hank Thomson. "This person would get a bottle of scotch when we got in from patrol."

In Lombok Straits, *Seawolf* had Bali Island to the west and Lombok Island to the east. War news told that the Japanese were focusing on an assault on Bali to take that airfield. A British sub was posted at the northern entrance to the straits. An old S-boat was assigned the central area, and *Seawolf* was given the southern entrance. *Seawolf* encountered heavy rain and a 20-knot wind on 18 February in Lombok Strait. Sound reported propellers at 0752 but poor visibility once again prevented the *Wolf* from sighting this vessel. Dick Holden sighted fires burning in the vicinity of Benoa at 2115 and Warder changed course to investigate. "Decided fires due to local destruction of oil and gasoline stowage in expectation of Jap landing," he logged. "Reversed course in order to regain patrol station and avoid revealing vessel in firelight."

The radio gang picked up an urgent dispatch for *Seawolf* during the early morning hours of 19 February directing the ship to close in on Badung Strait, located between Bali and the small island of Nusa Besar in Lombok Strait. Warder was to intercept a Japanese armada spotted racing toward this strait. *Seawolf* halted her battery charge and raced on the surface northward for the straits at 18 knots. She did not slow until Nusa Besar came into sight at 0124.

The contact report was for an important Japanese convoy that had sortied from Makassar on 17 February for Bali, Java. It included the troop transports *Sasago Maru* and *Sagami Maru* with part of the Imperial Japanese Army's 48th Infantry Division aboard. Close cover for the transports was provided by Destroyer Division 8—the destroyers *Asashio*, *Oshio*, *Arashio*, and *Michishio*. Providing more distant cover was the cruiser *Nagara* and her Destroyer Division 21 screen of *Hatsushimo*, *Nenohi*, and *Wakaba*. Command of the invasion force fell on Rear Admiral Kubo Kyuji, former skipper of the cruiser *Suzuya*, who flew his flag aboard *Nagara*. As the invasion force crossed the Flores Sea, Mitsubishi A6M "Zeke" fighters based at Makassar provided air cover. At 0200 on 19 February, the invasion force began disembarking on Bali with little to no resistance.[6]

As *Seawolf* approached Nusa Besar during the morning of 19 February, a flash of light could be seen on the horizon and she changed course to close on it. Warder flooded down his ship at 0206 to reduce her visibility to enemy ships while continuing to head for the flashing light. En route, sound reported high speed screws moving aft rapidly. By 0218, sound had two sets of screws. In the radio shack, Red Eckberg "heard pinging all around. We'd gotten into a hornet's nest." The swishing sounds of propellers filled his earphones as enemy ships approached. Paul Maley entered the subdued light of the radio shack, lit a cigarette, and waited impatiently.[7]

The pinging was first detected off *Seawolf*'s starboard bow and then another pinger became steady off the port side. The pinging came closer, so Warder made a sharp course change at 10 knots "to leave them a wake to ping on." By 0240, the pinger on the port beam was sighted as a destroyer approaching fast. Due to the poor visibility, the DD was at an estimated 1,000 yards when finally spotted.

"Left full rudder! Clear the bridge!" roared Doug Syverson as he pulled the diving alarm. The klaxon blared as Syverson and his lookouts scrambled down the ladder and secured the conning tower hatch. The *Wolf* dived and altered her heading. The destroyer had located *Seawolf* so Warder commenced evasion tactics. Eckberg could also hear the screws of two other ships. Warder elected to remain at periscope depth as he was uncertain of the depth.

By 0255 the pingers had apparently lost the *Wolf*. Warder eased down to 90 feet and made his way toward the likely landing place. He ordered his sound shaft and pit log brought in on account of the uncertainty of currents. The crew was secured from battle stations at 0400 as the periscope watch tried to keep the *Wolf* off the beach as she approached.

At dawn, *Seawolf* eased up to periscope depth. The first observation after daybreak showed no more enemy ships in sight. Warder conferred with Jim Mercer, poring over the charts. He figured that the enemy destroyers had spread out to run an entrance patrol, unaware that the *Wolf* was already inside the straits with them. *Seawolf* moved ahead at one-third submerged speed, conserving her battery as she searched for the enemy warships near Nusa Besar.[8]

Warder sounded battle stations again at 0531 as his boat closed in on the landing zone on Bali. He found a heavy rain killed his visibility, however, and he was forced to circle "to await improvement of visibility." At 0603 a light was sighted and the *Wolf* headed toward it. A minute later, an *Asashio*-class destroyer was sighted and Warder realized that he had successfully penetrated the landing zone.

Navigation had been difficult to pinpoint during the morning, so Warder worked with Jim Mercer at the plotting table. The water was shallow and treacherous with the currents. In the radio shack,

Eckberg recalled: "Over my phones I heard the roar of shallow water eddying and swirling around the high coral shoals. The *Wolf* was weaving her way with infinite care through a subterranean maze of jagged, razor-sharp reefs, any one of which could rip her hull from stem to stern."[9]

At 0622, with rain increasing in intensity and nothing in sight, Warder elected to continue circling the landing area until conditions improved. In the after battery compartment, Gus Wright had donned the battle phones to relay orders from the conning tower. In the wardroom, Bugawisan and Tamayo had the battle phones to do the same. At 0630, the *Wolf* lurched as her keel hit the beach at 63 feet. Lights flickered and men were thrown from their feet as a grating noise enveloped their ship. "We're aground!" someone shouted.

Freddie Warder had found himself trapped in a shallow coral cul-de-sac which had caused him to hit bottom. "All back emergency," he ordered. *Seawolf* shuddered and the bottom grated noisily against her forward hull as she backed free from the shallows. The *Wolf* reversed her course and eased carefully back the way she had come. Warder continued backing at full speed and then 2/3 speed until he felt sure he was clear of the beach.

Electrician John Bilkey later remarked, "Captain Warder used to complain about shallow waters. He would say, 'Submarines are made to go down, not hit bottom with!'"

Warder stopped the *Wolf* and had Dick Holden keep her at 55 feet. Shifting to low-powered periscope, he swept the seas above. Still the heavy rains obscured any view. Warder called for left full rudder and ahead standard. Steadying on course 090° he then ordered all ahead 2/3 speed. Several minutes passed quietly. *Seawolf* had successfully freed herself from the beach and was moving on.

Wham! At 0642, she suddenly lurched again as the boat slammed bottom at periscope depth for a second time. "Hit bottom violently, starboard side, entire length of ship," wrote Warder.

"All stop!" he called. A quick sweep of the scope showed nothing of alarm, so Warder ordered, "Surface!" Lieutenant Holden blew the safety tank and brought *Seawolf* up. Warder backed emergency quickly and then killed all headway. In the radio shack, Red Eckberg was fearful: "I went ice cold. For the first time in my life I think I knew absolute, craven fear. Here it was broad daylight, and Captain Warder was bringing us up in the middle of a Jap task force."[10]

As the ship broke the surface, the skipper and his lookouts scrambled up through the conning tower. "I went on bridge and found myself fenced in with coral patches and discolored water," wrote Warder. *Seawolf* had surfaced into a rain storm, which served to mask her surfacing so close to the enemy ships. Warder considered this "an act of God." The rain was so intense that his view was limited to 500 to 1,000 yards. After a quick sweep of the coral surrounding him, Warder chose his exit path. "Picked myself a hole and went ahead, starboard ahead standard, left full," he wrote. "Then port ahead standard, right full."

Seawolf weaved her way out of the dangerously shallow shoals and put two engines on the line to clear the area. By 0654, she had cleared the coral but a destroyer was still in sight about 6 miles away. "Don't think he can see me [on] account low visibility and land background," noted Warder. As the destroyer faded into the distance, Warder put his other two engines on battery charging.

Twenty minutes later, a four-engine plane was spotted flying about 2,000 feet above the waves. The lookout reported the aircraft broad on the port quarter, 5,000 yards away. Lieutenant Holden cleared the bridge and took *Seawolf* down to 120 feet. Fortunately, the pilot had not spotted the submarine and no bombs followed.

Seawolf commenced a periscope patrol and the weather began to clear over the next hour as the morning sun burned off the clouds. Mel Eckberg, listening on sound for the enemy ships, reported that the number two listening head was dead. The projector on the end of the second sound shaft had snapped off when the boat ran aground. *Seawolf* was down to only one sound device to use to find the enemy, look out for ships, and track the sound of torpedoes.[11]

The clearing weather allowed Warder's navigational team to identify landmarks. Deragon, Casler, and Mercer pored over their dead-reckoning and plot information and decided now that *Seawolf* had struck bottom off Serangan Island. Casler's job as chief quartermaster was that of plotting officer, working closely with Mercer to maintain an exact plot of the enemy forces.

At 0855, smoke was sighted on the horizon and *Seawolf* headed for it. Within ten minutes, the smoke was found to be coming from a destroyer at 14,400 yards. Warder sounded battle stations, but the destroyer circled to starboard and withdrew from sight to the north during the next quarter hour. Warder pursued and had the tin can

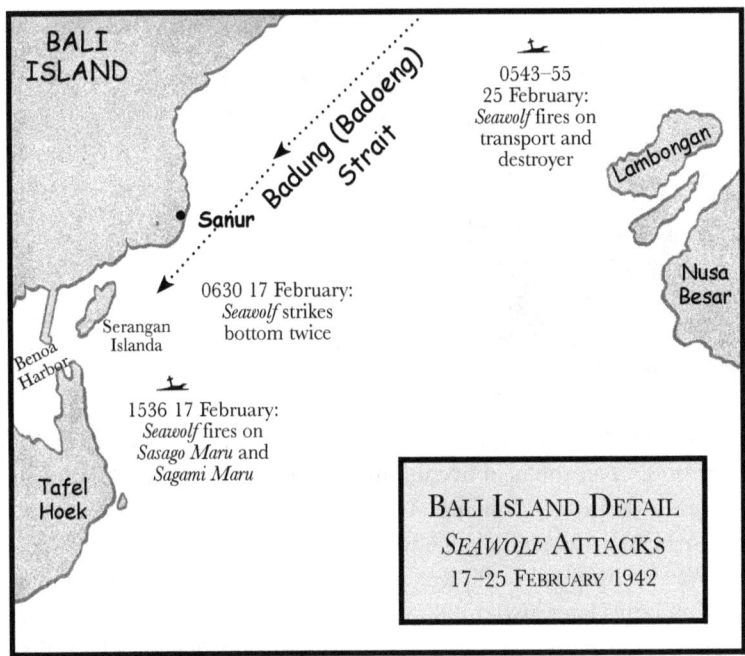

BALI ISLAND DETAIL
SEAWOLF ATTACKS
17–25 FEBRUARY 1942

in sight again by 0946, but could not overtake her. Another destroyer was sighted a short time later, also heading on a northerly course.

"Realized that we were not getting in fast enough so increased to 4.5 knots except for periscope observations at about 15-minute intervals," Warder recorded at 1151. He found that his boat was bucking a current this morning, whereas the *Wolf* had been riding with the currents the previous day.

Warder was moving in close to the beach at Bali again from a deeper water angle. By 1256, Red Syverson had sighted masts in Sanur Road. *Seawolf* moved toward the masts and soon sighted an *Asashio*-class destroyer lying to at about 16,000 yards. Battle stations were sounded again at 1350 for the third time of the day. The target ships became more clearly defined as *Seawolf* approached. She had found two destroyers and two troop transports. Using *Booth's Merchant Ships* recognition book, Warder and his conning tower team identified the transports as being similar to the freighter-transport ships *Akagi Maru* and *Kinai Maru*. These transports were actually the *Sasago Maru* and the 7,000-ton *Sagami Maru*, which had sortied with Rear Admiral Kubo Kyuji's force on 17 February for Bali.

Chief Radioman Joseph Melvin "Red" Eckberg was the key source for a 1945 book about *Seawolf*'s first seven patrols. *Author's collection.*

One of the destroyers commenced pinging at 1417 off *Seawolf*'s port bow. Over the next five minutes, Warder spotted a total of four different destroyers "milling about at low speed on various courses between APs and strait." The freighters were still 8,000 yards distant so Warder dropped to 90 feet and increased his speed to 4.5 knots. Five minutes later, two distant depth charges were heard by sound.

Making a periscope observation at 1442, Warder found another destroyer at the mere range of 1,800 yards, "lying to with anti-aircraft machine gun batteries manned, and all his people looking up, or to shoreward. He looked too good so I turned left at full speed to give him straight bow attack."

"Open doors to tubes 1 and 3," he ordered.

Seawolf slowed to 1/3 speed and steadied on the course of the target warship. On his next observation, Warder found that the destroyer "had kicked ahead, range now 2800, angle on bow starboard and large." Undeterred, he turned his attention back toward the troop transports. Sound reported another couple distant depth charges as the *Wolf* closed in.

Seawolf move into shallow water as she approached the two transports *Sasago Maru* and *Sagami Maru*. As the distance closed to 3,400 yards, Warder could see that the northernmost freighter had her port gangway down. He ordered his own sound head and pit log rigged in due to the shallow waters and increased speed to 6 knots. At 1511, five depth charges rumbled a little closer in the distance. A minute later, Warder noted "five more depth charges, still closer." The Japanese were obviously aware there was a fox in their henhouse, but had not made solid contact with the *Wolf*.

Warder began his final firing observations at 1536 from a range of 2,300 yards. The northern freighter was observed with her "anchor chains straight up and down; 1 DD crossing between me and them at dead slow speed." *Seawolf*'s No. 1 and 2 torpedo tube doors had been open for 45 minutes since the approach on the destroyer. Warder now ordered tubes 3 and 4 opened and prepped for firing. He decided to fire tubes 1 and 3 at the first freighter and tubes 2 and 4 at the second.

At 1536.50, Warder commenced firing. Warder passed course changes to Rudy Gervais, the battle stations helmsman.

Seawolf fired all four forward tubes. Eckberg on sound reported the fish to be running, "hot, straight and normal." On the scope, Warder observed the smoke from his first two torpedoes pass to the left, or ahead, of the southernmost freighter. He could not see the wakes due to the water being "much riled up"with considerable floating debris. Sound reported all four fish running normally.

"Oh, hell!" snapped Warder, his face glued to the periscope. "Here comes a destroyer turning left and heading toward us. Down periscope. Right full rudder. Rig for depth charge attack!"[12]

The destroyer sailors had spotted the wakes or the smoke from the Mark 14 torpedoes and turned their ship toward the source. Gervais swung *Seawolf* hard to the right as she headed for 90 feet at standard speed. Warder did not have time to observe the last two torpedoes. His main concern was just how much water he had under his keel. "Beach looked very close," he noted.

Throughout the boat, every bit of machinery was turned off for silent running, including the air conditioning. The cooks secured pots and pans that could potentially crash to the floor. The heat increased throughout the boat as men waited silently for the first enemy depth charges. *Sasago Maru* and *Sagami Maru*'s destroyer screen—*Asashio, Oshio, Arashio* and *Michishio*—moved in to attack.

Paul Maley and Red Eckberg listened to the *swish, swish, swish* of the destroyer's screws as the first one moved in closer. The range narrowed from 3,000 to 1,000 yards, and then under 500 yards. After reaching 90 feet at 1543, Warder called for his speed to be slowed to 1/3. One minute later, the enemy's ashcans began to explode.

Click—*wham!* Click—*wham!* Click—*wham!*

Two strings of three charges each were dropped just forward of *Seawolf*'s starboard beam and fairly close. "If I had made full speed from the firing point they would have had us," admitted Warder. "The conning tower was doing a dance by the time I got out of it."

Seawolf's first close depth charges were alarming. In the radio shack, Mel Eckberg was thrown from his seat.

> It was the loudest sound I had yet heard. It was as solid as a blow on the skull. It was like a thunderclap between my ears. I found myself on the floor, my stool upturned. Maley was on the floor beside me, scrambling to his feet. We were in a snowstorm—paint chippings and cork from the bulkheads filled the air.[13]

Photographs and books tumbled to the floor. Light bulbs shattered in their sockets. The hull buckled in from the enormous underwater pressure and then flexed back out again. Sailors dripping with sweat picked themselves up and silently dusted off the cork and paint flecks as the enemy destroyer wheeled around for another run.

Click—*wham*!

This one was more violent than the first, exploding just off the *Wolf*'s beam. The ship lurched heavily as the DD passed along the *Wolf*'s port side. "The charges were dropped so close that counting was impossible," wrote Warder. His estimate was that over the next 18 minutes of depth charging, *Seawolf* was jolted by about 43 ash cans. "They really rocked the hell out of us!" Warder later stated.[14]

"Lieutenant Syverson, assisted by me, was trying to plot our dead reckoning position," quartermaster Hank Thomson later wrote. "The pencils drew long lines with the explosions as they pushed the hull in. And then the hull would come out, depressurizing the air and causing a fog. We laughed, not realizing our predicament."

During depth charge attacks, *Seawolf*'s conning tower was abandoned and sealed to prevent leaks. Freddie Warder chose to sit on a stool in the control room to wait out the enemy's counterattack. He listened closely to the relay coming from his two radiomen and then called out evasive orders when he felt that the destroyer was preparing to drop more ashcans.

Warder dropped the *Wolf* down to 110 feet as soon as the DR showed that he could safely do so with some bottom to spare. Diving officer Dick Holden found that with the accumulation of bilge water and the encounter of a fresh water stream at this point that the *Wolf* sank slowly to 170 feet. Warder felt this "was probably a good thing as [the charges] gave the impression of going off over us."

Motormac Charles "Johnny" Johnson, making his fourth patrol on *Seawolf*, described the tension of such a severe counterattack: "Depth charges damaged machinery, destroyed light fixtures and various gauges, and forced prolonged operation at silent running." In the confined area of the *Wolf*'s engine room, Johnson found the depth charge attack drove "temperatures and humidity to dangerously high levels, as all air conditioning and ventilation was secured."[15]

In his patrol report, Warder describes his ship's pounding.

> My face was stung with flying cork and paint chips, my ears were deafened, and my eyes were kept busy watching the gauges and manifolds dance. On one blast the overhead appeared to come in six inches. This ship is strong. At no time was depth control lost in the wrong direction. Bow planes went out once and main motor relays blew and had to be reset. Conning tower door operating mechanism backed off enough to allow door to show open. Battery ventilation exhaust sprang a leak. No serious damage.

None of the 43 explosions on 19 February came close enough to finish off the *Wolf*, but she took her share of close charges. By 1643, Warder's periscope sweep found three destroyers, followed by one transport and a fourth destroyer. They were heading toward the north entrance of Lombok Strait at about 10 knots. One transport was missing. The remaining transport, standing out with the DDs, "appears to be down by stern and listing to starboard," he thought.

The destroyer just forward of the transport was seen to be firing in the direction of *Seawolf*'s periscope. Lt. Cdr. Warder dejectedly noted no possibility of further attack due to "excessive range, large target angle, and low battery." He took *Seawolf* out of the strait to the southward to rest his men.

Warder released his crew from battle stations and ordered Dick Holden to have the mess cooks run coffee throughout the ship to all hands. Enginemen Fred Zirkel, Otis Dishman, and Red Snyder made the rounds, inspecting carefully for damage to their precious diesels. The men crawling about reported that a few air and water lines had sprung small leaks, but no major damage was found.[16]

Seawolf stayed down until after dark to avoid patrolling Japanese planes. Officer of the deck Dick Holden surfaced at 1943 to charge the batteries and send off a contact report to submarine command.

Despite Warder's bravery, no Japanese ships were sunk at Bali and the Japanese took the island and its airfield with ease. North of Bali, the British submarine *HMS Truant*, under Lieutenant Commander Hugh A. V. Haggard, also attacked the Bali invasion force on 19 February. *Truant* fired on the flagship cruiser *Nagara*, but missed with all six torpedoes and received a working over by her destroyers.[17]

Chief Metz's auxiliary gang toiled all night repairing the leaks. *Seawolf* recharged her batteries and heading back for Lombok Straits. As she neared the southern entrance of Badung Strait after 2300, a searchlight swept the area and stopped momentarily on *Seawolf*, fully illuminating her. Officer of the deck Jim Mercer called for battle stations and prepared to dive, but the light soon resumed its sweeping.

Seawolf proceed around Nusa Besar Island on 20 February, due to the absence of Jim Dempsey's *S-37* in this patrol area. She fought the heavy currents throughout the morning hours once again. "Currents are swift, erratic," noted Warder. "Tide rips and whirl pools make this strait a very difficult place in which to steer a submerged submarine or to control its depth. On one occasion, observed ship to swing to starboard with left full rudder, requiring standard speed to straighten her out."

For Dick Holden and the others who stood diving watches in this area, the sea played hell on depth control. "The currents were very swift and the temperature gradients were very severe," Warder later related. "You could be plowing along there at 2 or 3 knots and hit a cold current and go popping the hell out of the water. Or you would be going along beautifully at 3 knots at periscope depth and have the bottom drop out from under you. It was very bad business."[18]

Seawolf ran at periscope depth, but found no sign of enemy ships. The periscope watch observed an airplane crashing and its aviator bailing out on Nusa Besar during the morning. Warder examined Benoa and Sanur through the day, noting fires on Benoa. The next few days were passed with uneventful cruising off the southern approaches of Lombok and Badung Straits. Radio Tokyo was a source of entertainment in Kelly's Pool Hall for those off duty. The announcer told of the landings on Bali and how "a nest of Allied submarines" had been unable to prevent the landings.[19]

Warder noted an "established absence of shipping [in] Badung Strait and stood out to southeast" on 23 February. During the morning, his periscope watch sighted various single-engined monoplanes in the vicinity of Benoa and Sanur. At 1042, another periscope was sighted off *Seawolf*'s port quarter. Sound picked up a turn count of 78 rpms on this unknown sub's screws, but the periscope and the screw sounds abruptly disappeared. Having cleared Badung Strait, the *Wolf* patrolled the southern approaches through the night.

Early on 24 February, *Seawolf* received an Ultra dispatch that a Japanese convoy had left Makassar the previous morning and was heading south, likely toward *Seawolf*'s vicinity. Warder moved his boat back toward Badung Strait to be prepared to greet this enemy task force upon its arrival. Further clarifying reports of this approaching task force were relayed by *Salmon*, which attacked two cruisers.

At 0500 on 25 February, gunner's mate John Bennett sighted smoke on bearing 074° and *Seawolf* turned to investigate. Bennett had hopes that Lt. Cdr. Warder's promised bottle of whiskey would be his if the *Wolf* could successfully put torpedoes into these ships.

The night was dark and Warder could only scarcely make out the outline of what appeared to be a destroyer, whose smoke had first been spotted. At about the same moment, two more ship outlines could be made out. "Sound reported pinging all over dial on both bows," noted Warder at 0502. He moved forward at 10 knots with *Seawolf* surfaced, but flooded down to keep a low profile. At 0508, she dived to 90 feet and penetrated the convoy's destroyer screen. During the next ten minutes, radiomen Maley and Eckberg reported high speed screws of four different ships and one set of low speed screws. Warder eased up to 60 feet and found one of the destroyers training his searchlight on his periscope "but with shutters closed."

By 0528, Warder could see a transport at about 3,000 yards range. In attempting to set up a forward tube attack, he found a transport bearing down on him at a mere 500 yards. Warder "realized there was not enough range for torpedo to arm, so came left with full rudder and full speed to opposite course." Warder had come so close that *Seawolf* had to dodge the oncoming transport. The convoy included at least three transports, a tanker and destroyer escorts.

"Make ready after tubes!" the skipper called. "Set depths at eight feet." Once they passed by, he would launch his Mark 14s. At 0539, he took a quick setup on another transport. With a range of only 750, the *Wolf* fired tubes 5, 6, and 7 at 0543, aiming at the transport's bow, stern, and forward goal post. Warder "watched one torpedo hit just forward of bridge, throwing up a sheet of water to top of bridge. Felt a distinct thud with the deck moving out from under me."

In the motor room and in John Gibson's after torpedo room, the crew reported hearing two distinct explosions. Nonetheless, Warder complained in his patrol report, "There is not enough of a bang out of these fish to suit me." If not duds, these fish had likely produced only low order detonations. The *Seawolf* sailors, having never fired live torpeodes in prewar training, had never heard their torpedoes explode against a target. In this case, the faulty Mark 14 torpedoes that other submarine skippers were already screaming about had robbed the *Wolf* of at least one more ship in her postwar credits.

Paul Maley reported pinging on the *Wolf* from her port side, so Warder ordered, "Full left rudder. All ahead full. Take her to 90 feet." Gervais obliged and Dick Holden took her down. "Decided to return to northerly course to work on stuff astern of first target, with bow tubes," wrote Warder. *Seawolf* climbed back to 63 feet and 0555 a destroyer could be seen through the periscope at 1500 yards. He made final periscope observations three minutes later and at the point-blank range of 500 yards commenced firing tubes 1, 2, and 3 at 0559. "Saw none of these torpedoes running but believe this due to insufficient periscope exposure."

From sound, Eckberg called out, "Two of these hit something!"

"Down periscope!" he called. Then, to Lt. Holden in the control room, "Take her down! Rig for depth charge attack."

At about the same instant, *Seawolf* was rocked by a series of explosions to starboard which Warder took for depth charges.

"Abandon the conning tower!" he called as his fire control team scurried down the ladder. In his report, Warder noted:

> My last impression of DD was that his stern was coming out of water. I also heard and felt one good bang on the port bow about 2 minutes after firing, as I was leaving the conning tower. There were a number of ships down in this direc-

tion—to right of the DD and to the left of her. (I noticed in the radio broadcasts of the following evening that an Allied plane observing had given us credit for sinking 4 ships including 1 warship.) I think we most probably sank two.

Warder believed he had hit two transports and a destroyer. He could not stick around, however, for other destroyers were racing for *Seawolf*'s position. Whether accurate or not, Japanese records postwar would not give Warder credit for sinking or seriously damaging any shipping this day. Aboard ship, however, the officers and crew were convinced enough to claim two sinkings, and Gunner Bennett was later awarded his bottle for sighting the smoke of these ships.

Seawolf shut down all unnecessary equipment and went to silent running as the destroyers raced in to pound her. "There followed a pretty bad four and one half hours," Freddie Warder wrote. Even after the worst of the depth charging ceased, it would be many hours more before *Seawolf* was fully clear of danger. Radioman Eckberg picked up the sound of three different sets of destroyer screws converging on the *Wolf*. These destroyers kept the crew at stations for hours. Every time the crew tried to start a pump to get the bilge water down, run up the periscope, or start the hydraulic plant to hold out the sound head, the vigilant destroyers would race in.[20]

Warder thought that these tin can sailors were more conservative with their depth charges than those he had faced the previous week but believed their charges "to be heavier and deeper and more accurately dropped." At 200 feet, *Seawolf*'s battery ventilation exhaust began leaking a stream the full diameter of the drain pipe. "It was about three pencil thicknesses before," noted Warder. In Squeaky Langford's forward torpedo room, the gyro setting motor was flooded out by bilge water.

The destroyers kept the *Wolf* down through the morning, allowing the transports to escape. The flat calm sea was not in *Seawolf*'s favor. Warder finally eased up to periscope depth for a look and found the nearest ship some 8,000 yards away. "I watched one of them sight my feather once at 4 miles and turn and head for me. This DD had about 10 lookouts standing on top of the conning tower."

Warder took his boat down while his radiomen tracked her by sonar. The sound of her approaching screws became so loud that everyone aboard could hear the angry swishing of her props as she raced in. The destroyer roared by overhead as *Seawolf* sat silent in the shallow water. This time, the depth charges did not come. Warder later surmised that the destroyer's skipper investigated a lookout's sighting but did not pick up his *Wolf* on his echo-ranging gear. Warder held tight, maintaining absolute silence through his boat.

Temperatures soared above 130° in the maneuvering space. The air in the boat soon became foul with the stench of sweating human bodies. Doc Loaiza, ordered by Lt. Cdr. Warder to distribute saline tablets to the crew, squeaked through the passages in his sandals. The decks were literally sopping wet with perspiration.[21]

Chief Eddie Souza warned that the loss of too much salt for a body could be deadly. Hank Thomson, for one, did not care for the salt tablets. "I thought I was going to die faster taking them, really." One of the mess cooks passed a gallon jug of water for men to take a drink. Men stripped off unnecessary clothing and those not on watch lay in their bunks to conserve oxygen and energy.

"Three times they passed right overhead without cans dropping," recorded Warder. "The record shows only 21 charges were dropped but I think there were at least 30."

Finally, a periscope sweep showed all clear, so *Seawolf* secured from depth charge quarters and stood out of Lombok Strait in order to begin broadcasting her contact report. Warder finally brought his boat to the surface at 1921 and obtained a receipt of his contact report from Surabaya. As the hatch was opened, a small gale ripped through the ship, rustling papers like a fresh breeze. As the diesels roared to life, the air flow reversed as the mighty engines sucked air down through the open hatch. The return of good air brought a new sense of life to the crew.

As their boat raced to the southeast, auxiliarymen and electricians worked to repair the leaking battery ventilation system. They found a leak in the drain line connection to the forward section and repaired it with wicking, white lead, tape and marlin. At 120 feet the next day, Warder noted that "only about a drop per minute" still dripped from the piping.

All told, *Seawolf* had held up pretty well, in spite of running aground twice and taking two Japanese depth charge attacks. The area was active with Japanese invasion forces and Freddie Warder was eager to find more of them to torpedo.

5

"Fearless Freddie"

Fourth Patrol   *26 February–7 April 1942*

The *Wolf* remained submerged on 26 February, sighting only heavy bombers and monoplanes through her periscope in Badung Strait. On the surface that night, *Seawolf* approached what were taken to be two motor torpedo boats, but could not gain favorable attack position. At 0316 on 27 February, the radio gang picked up a contact report from *Salmon* of three cruisers and three destroyers heading for Lombok Strait. Warder increased speed to 18 knots and raced into Lombok Strait to wait. The Japanese failed to show, however, and he commenced patrolling off the southern approaches during the evening hours. Another ship was sighted at 2115 and tracked but it was eventually believed to be the submarine *Spearfish*.

Freddie Warder later found that this was actually Lt. Cdr. Gene McKinney's *Salmon*, making her second war patrol. Warder thought highly of McKinney and later reflected on this close call.

> Gene McKinney damn near shot at me and I damn near shot at him. When we got back to the post, we had to have some heavily increased safety rules. I was up in the northern end of Lombok and he came running down through there on his way south. My guy says, "This looks like fair game." I said, "I don't think so. It could be anybody, but I don't think it's a Jap."[1]
>
> Our intelligence was rather complete. We had records on these guys. He was a day ahead of time. But we made the approach and I was about to shoot.

Fortunately, Warder and Bill Deragon were able to trade peeks through the scope and finally clearly make out the submarine to be American. *Seawolf* continued to monitor and patrol the approaches to Lombok Strait through the next day, but found no enemy ships.

The *Wolf* transited Badung Strait on 1 March and received new orders that directed her to head for an area off Tjilatjap, a Javanese seaport on the south coast of Java used by Allied powers to evacuate personnel from the East Indies to Australia. En route to Tjilatjap [now Cilacap], *Seawolf* sighted the lights of a passing ship during the night of 2 March, but Warder opted not to attack due to unfavorable bright moonlight and his belief that this might be a hospital ship.

The next two days and nights were uneventful as *Seawolf* ran along south of the Java coast. West of Tjilatjap during the night of 4 March, Warder noted "large fires burned all night" ashore from the city. Radioman Mel Eckberg later wrote that, "Bridge lookouts told us the shoreline looked like a carnival of light and fire." The Dutch were reportedly torching everything in sight before the Japanese arrived and capture anything worthwhile.[2]

After four days of watching Tjilatjap burn, *Seawolf* had found no enemy shipping. Lt. Cdr. Warder was thus relieved when a new dispatch was received at 2050 on 8 March. The *Wolf* was redirected to the southern entrance to Sunda Straits, a 60-mile wide stretch of sea between Java and Sumatra. Any Japanese ships bound for the southern coast of Java would likely use the Sunda Straits. Warder and Lt. Deragon pored over their charts, looking for probably shipping lanes the Japanese would use through these deep waters.

Other news spread through the ship by yeoman Sully Sullivan was the promotion of a number of *Seawolf*'s leading petty officers to CPOs, effective 9 March 1942. Sully's list had CRM Eckberg, CMM Hershey, CMM Jenkins, CPhM Loaiza, and CMM Zirkel, pushing the number of CPOs to 14—more than *Seawolf* was allotted.

Seawolf patrolled the southern approaches to Sunda Strait along the coast for three days but found no shipping targets. During the early morning hours of 11 March, Lt(jg) Syverson and one of his lookouts got ashes in their eyes from nearby volcanic activity. "There have been many gas explosions in air and lightning displays during mid and morning watches, in vicinity of Krakatau Island," wrote Lt. Cdr. Warder. By 1100, Warder's men had definitely established the eruptions to be coming from the new island between Verlaten and Lang Islands. "Discharge consists chiefly of white smoke billowing up in great volume."

Fearful of a full volcanic eruption, OOD Jim Mercer turned to duty quartermaster Hank Thomson and said, "If you hear an explosion, surface and steer for the Indian Ocean—even if you are surrounded by Nip warships!" *Seawolf* ran on the surface again that night and found that the volcanic eruptions created an illumination which "serves to blind the lookouts and silhouette the ship" due to the drifting sulphur.

Seawolf continued her uneventful patrol of the southern approaches to Sunda Strait and the Australia-Sunda traffic lane without contacts during the next two days. As its fifth week approached, her fourth war patrol had already become her longest by far.

Radioman Mel Eckberg found that tensions began to rise.

> For the first time, an attack of nerves broke out. Half a dozen men weren't talking to one another. By this time we had been out on the longest sustained run we had made so far—and most of the crew had not seen the sun or been topside all during the patrol. It didn't help any that we were all running short of cigarettes. There were less than half a dozen packs left on the boat. Some of the men had a few cigars, and they nursed these along. Those who smoked pipes weren't in any better fix. Their tobacco was all gone. There were some pretty stretched tempers on the *Wolf*.[3]

Engineman Bob Hanson found that cigarettes were a scarce commodity on the fourth patrol: "I had four packs of cigarettes because my brother Henry didn't smoke. I kept them in my locker in the bunk area and that was the only time I locked my locker."

The black gang had a minor distraction from the mindless patrolling on 13 March when an engine problem developed. The No. 4 main diesel's lubricating oil cooler was found to be leaking salt water to the oil side. After surfacing that night to head for Vlakkehoek, *Seawolf*'s engineers worked on replacing the No. 4 coolers. After submerging again on 14 March, the black gang found that the No. 2

main engine's lubricating oil cooler was also leaking salt water to the oil side, further adding to the repair work.

A radio dispatch that evening told of Axis ships coming in from the south for Tjilatjap or Sunda, so Warder took up station to patrol the southern approaches to Sunda Strait for the next few days. The only enemy contact was a two-masted schooner observed during the mid-day hours of 16 March.

Bill Metz's auxiliary gang received another challenge on the morning of 18 March when salt water was reported in the after battery room. The after battery was immediately pulled clear of its load. Two to three inches of water rose in the forward end of the room before the water could be pumped out. Personnel joined to rip out lockers, clothes, and provisions to find a leak in the riser to the trim line hose connection in the galley store room. Some water leaked in to the battery tank around through the bolt for the hatch's dog, but it only hit the the cell tops. Fortunately, no deadly chlorine gas was emitted. The following day, water was again found to be leaking in a section of the trim line into the clothes locker in the after stateroom but this was easily remedied by a customary soft patch.

Seawolf battled rain and heavy seas off Sunda Straits on 21 March, but found no shipping contact over the ensuing days. Low on food, fuel, and torpedoes, and cigarettes, *Seawolf*'s men found even showers were at a premium. "A lot of times, we would save the condensate water in the engine room off the air conditioning and we'd go rinse off with that," said Little Swede Hanson.

Finally, a dispatch came to Warder from Admiral Wilkes at 2023 on 27 March which ordered him to patrol the Christmas Island area before heading back to port. Christmas Island was British property south and west of Java, valuable for its phosphate. Motormac John Street, known for his endless trivia, claimed that it was once an old pirate hangout. Warder immediately opened up to 14.5 knots and left the luckless Sunda Straits in his wake.[4]

Seawolf approached Christmas Island during the early morning hours of 29 March in excellent moonlight. Diving off the island's northwest point after dawn, Warder ran at periscope depth in order to examine the island for likely landing spots. Bill Deragon and Jughead Casler's charts showed that the only dock facilities on the island were in a small inlet called Flying Fish Cove. The periscope watch at 1006 sighted a sunken ship in Flying Fish Cove with the stern of her grey hull sticking up out of the water where she had slammed into a reef. Warder approached to within 3 miles of her, estimating her size at about 8,000 tons.

By this point in the patrol, most of the men were completely out of cigarettes and tobacco. Ship's cook Gus Wright dried coffee grounds in his oven so that those as desperate as himself could roll the dark grounds in toilet paper to make something to smoke.[5]

Aerial reconnaissance passed the intelligence of an enemy convoy en route to the area and the news quickly spread throughout the *Wolf*. All aboard could only hope that the endless weeks of inactivity were about to come to an end. Bickering and listlessness gave way to renewed energy and new chatter about Kelly's Pool Hall.

After days of reconnoitering Christmas Island, Freddie Warder was convinced that Flying Fish Cove was his best chance for putting a pickle in an enemy vessel.

"I have now decided that cove is only practicable place for landing attempt," he wrote in the late hours of 30 March. "Other possibilities are impracticable due to deep water close to shore, cliffs and rocks, wooded banks, heavy swells and small landing areas." Warder charged his batteries into the early morning hours of 31 March as he prepared to enter Flying Fish Cove. Light rain squalls dotted the area but a good moonlight was available as the *Wolf* made ready.

At 0600, *Seawolf* came to a stop, securing her engines so that the sonar gang could have a good listen. At 0603, sound reported pinging to port. Seven minutes later, a searchlight suddenly fully illuminated *Seawolf* from abaft her starboard beam. Sound reported that the previously reported pinging was not on the *Wolf*'s starboard side. Warder ordered all ahead full and dived the boat, turning with full rudder to close the approaching enemy formation. He quickly realized that the ship which had picked him up was leading the port screen so Warder came left again to close Flying Fish Cove.

Warder's instincts on Flying Fish Cove had paid off, for *Seawolf* was right in the path of an important convoy bound for Christmas Island. She had found the Second Southern Expeditionary Fleet's Occupation Force under Rear Admiral Nishimura Shoji, which had departed Bantam Bay on Java on 29 March. The force was composed of Admiral Shoji's flagship, the 5,195-ton light cruiser *Naka*, and the 5,170-ton light cruisers *Nagara* and *Natori*. They were screened by eight destroyers—*Minegumo*, *Natsugumo*, *Amatsukaze*, *Hatsukaze*, *Satsuki*, *Minazuki*, *Fumitsuki*, and *Nagatsuki*. This heavy array of firepower was escorting the Christmas Island invasion force of the 10,182-ton oiler *Akebono Maru* and two troop transports, the 5,193-ton *Kimishima Maru* and 7,508-ton *Kumagawa Maru*, which were carrying about 850 men of the 21st and 24th Special Base Forces and the 102nd Construction Unit.[6]

By 0623, *Seawolf* had a destroyer at 3,000 yards heading for her. Warder commenced evasion tactics while maintaining his base course heading toward the oncoming convoy. Two depth charges went off at 0627 and he ordered 200 feet using negative tank to evade. The destroyer came over and dropped ten more charges within two minutes. Sound reported screws bearing on various courses as *Seawolf* held at 200 feet, evading oncoming screws. Warder eased back up to 120 feet at 0643 and moved ahead. As the screws faded, he ran her up to 63 feet for a periscope observation.

The important convoy—a submariner's dream—was taking shape above. Warder could soon make out a destroyer at 6,000 yards and two transports at 8,000 yards. By 0703, he had sight of a light cruiser at 7,000 yards to port and soon another cruiser at 8,000 yards to starboard. Warder kept the intercom system open as he took periscope observations and eased his boat into the cove during the early morning hours of 31 March. Unlike most landing areas utilized by the Japanese, the water in Flying Fish Cove was deep. Cruisers, destroyers, and transports were all about, ripe for the picking.

In the conning tower, Rudy Gervais was the battle stations helmsman. Warder looked over the cruisers and began asking for a mark on their bearings. At the plotting table, Jug Casler worked out the angles of the approach on each warship, assisted by communications officer Jim Mercer. In the control room, Doug Syverson fed the information into his TDC, assisted by fire controlman Lefty Leffingwell. Exec Bill Deragon split his time during the setup between the conning tower and assisting Syverson on the TDC's data.

Warder kept up a steady banter of information to his plotting team as he looked over the approaching Japanese warships. He made them out to be light cruisers, each with two turrets forward and one aft. They had catapults for the planes they carried. The

"Fearless Freddie"

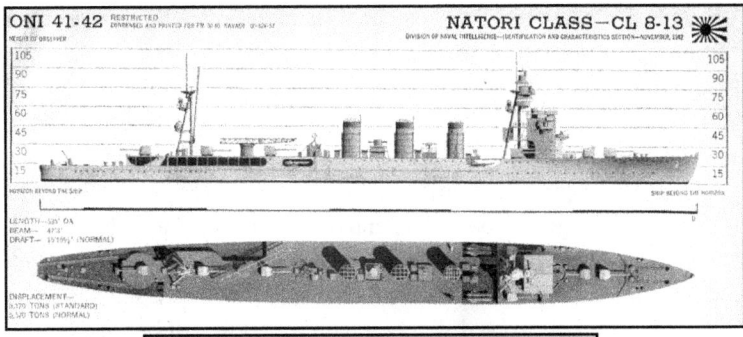

Frederick Warder earned the nickname "Fearless Freddie" after making multiple torpedo attacks against Japanese cruisers during *Seawolf*'s fourth patrol. Wartime naval intelligence book *ONI 41-42* shows one of her targets: the 5,170-ton Japanese light cruiser *Natori*, which was armed with eight torpedo tubes, seven 5.5" guns and three 3" guns, and two dozen depth charges. Lower photo shows *Natori* in 1937 with her observation float plane on its catapult. *ONI 41-42*.

transports appeared to be heading in for Flying Fish Cove. At 0705, he ordered all ahead full to close on one of the cruisers. "These birds are all just milling about except the APs are headed for the cove and are too far to west of me for me to get at them," he noted. Five minutes later, he found himself ahead of the column of cruisers by 4,000 yards. Warder came right at full speed to take a 90° ship track for his stern torpedo tubes.

During the next 20 minutes, he tried jockeying *Seawolf* into position on the convoy as sound continually reported screws on various bearings and ranges. The transports were out of reach, but Warder found the cruisers to be "running back and forth across the harbor entrance." By 0726, he could make out the flagship cruiser *Naka*. She was the first to hoist her flags, kept a searchlight flicking in all directions and appeared to be "lone wolfing" it around the other ships.

Seawolf tried to set up on the flagship but *Naka* turned away at 6,000 yards. Warder then headed for the cove to close the other shipping. *Naka* was conspicuous as she patrolled back and forth across the entrance to Flying Fish Cove at about 15 knots in the moderate seas. At 0752, two seaplanes were spotted flying over the cove and a distant explosion was heard. Warder went to 120 feet to close the cove more rapidly. Making another periscope observation at 0830, he found *Naka* to be at 9,400 yards. The two transports were now in the harbor with one of their destroyers approaching the dock. Two planes were observed southeast of *Seawolf*, heading northeast.

By 0843, Warder found *Naka* to be at only 3,150 yards, speed 16.5 knots. "Mark!" he called.

"Two-two-eight, Captain," replied Casler. The enemy cruiser was now close enough for *Seawolf* to make an attack approach. Warder called for battle stations. From the control room, diving officer Dick Holden quickly reported, "Battle stations are manned, sir."[7]

As the range narrowed, Eckberg continued to call bearings up to the conning tower team when the skipper did not have the scope up for observations. "Make ready the forward tubes," Warder ordered.

"Forward room. Make ready the tubes," repeated Hank Thomson into his headphones.

Assistant approach officer Bill Deragon and Jim Mercer announced that all was set. "Open the outer doors," ordered Warder. Henry Brengelman stood ready at the firing controls. At 0847, Warder made a final observation prior to firing and checked with his plotting team. The range to the cruiser had dropped to 1,400 yards as Warder barked, "Fire one!"

Seawolf spit forth her first warfish as Brengelman hit the firing button. Red Eckberg tracked the torpedo's track as the skipper ordered, "Fire two!" *Seawolf* fired all four forward torpedo tubes, these Mark 14s spread to hit the cruiser's bow, stern, foremast, and mainmast.

Freddie Warder watched intently through the scope as his warfish raced toward the Japanese flagship *Naka* with the range down to only 1,000 yards. Even with destroyers alerted and pinging, Warder coolly watched "men running and shouting on [the] cruiser quarterdeck and measured [the] range as 700 yards." The cruiser was turning toward *Seawolf* as Warder could see the "smoke from torpedo tracks drifting across field of vision."

The enemy destroyers turned and headed down the torpedo tracks for the *Wolf*. "Here they come!" called Warder. "Take her to 120 feet! Right full rudder, all ahead full!"

Gervais, Warder, Deragon, Casler, Thomson, Mercer, and Brengelman all evacuated the conning tower as Paul Maley tracked the incoming sound of angry screws. The hatch to the conning tower was secured and Warder took his spot on a chair in the control room for the pounding that was sure to follow.[8]

Chief of the boat Eddie Souza watched the clock and counted after the torpedoes were fired. At 70 seconds after the first fish was launched, Souza noted one explosion. "It seemed like a year to me," wrote Warder. A number of other men in the control room insisted that they heard two explosions. Warder noted in his report that "the 2nd and 4th torpedoes should have hit the target 163 feet forward of stern and mainmast, respectively; 1st and 3rd passing ahead. There is much contention that there were two explosions. I only heard one." Although this was believed to have been a torpedo hit, Japanese records show that the cruiser *Naka* was able to dodge all four of *Seawolf*'s torpedoes on 31 March. More likely, at least some of these Mark 14s had run deeper than their settings.[9]

Eckberg reported screws coming up the starboard side, fast. The first few depth charges exploded close off *Seawolf*'s starboard side, and the next few came even closer. "It seemed impossible that one wouldn't get us," thought Eckberg. "I thought we were lost. The ship shuddered and rocked. The radio shack was again filled with a blizzard of flying paint and cork."[10]

Locker doors flew open and banged shut. Warder watched the depth gauge and called maneuvering orders as sound reported the incoming destroyer screws. *Seawolf* bounced about as shock waves slammed her. Warder noted "a series of explosions sounding like about 8 depth charges with much water swishing noise, close aboard." He ordered Dick Holden to take her down to 200 feet using the negative tank. The diving officer found that he could not open the flood valve against the intense sea pressure, however. He instead ordered his men to shift to hand power and put a pressure in the tank to get it open and blow the gasket.

"We're going to have one helluva time," Warder wrote. The fact that the close explosions knocked out the *Wolf*'s power made this almost an understatement. Talker Hank Thomson noted that the sudden loss of power came at a poor time. "I was on the phone and could see the pit log. Because of the depth charging, we were doing flank speed submerged, about 13 knots by the pit log in the control room. The loss of power was quickly confirmed by phone."

Seawolf's main motor control went out on the first attack. Chief electrician Edmund Capece was later awarded the Silver Star for his actions this day. He had been stationed on the controllers that drove the propellers when he felt the *Wolf* lose all power. Capece's citation

CEM Edmund Capece was awarded the Silver Star Medal for risking his life to save *Seawolf* during her fourth patrol. Ignoring the risk of electrocution, Capece entered the electrical cubicle and jammed the overload relay with a block of wood to restore power during a depth charge attack. *Courtesy of Pete Lober.*

reads that after *Seawolf* "had been disabled by a serious electrical failure, Capece entered the control cubicle and restored main propulsion by blocking in the main circuit breakers." Lt. Cdr. Warder noted in his log for 31 March, that Capece "blocked in the overload relay with a piece of wood and caught the ship at 270 feet" before she could be crushed by the sea's pressure.

This act was extremely dangerous. In the maneuvering room was the main electrical cubicle, a locked cage with DANGER signs posted all around it. Inside it were the bus bars that carried the ship's electrical power for transfer to the main motors. The breakers were about chest level. No one was allowed in the cage without the skipper's permission, and then only when the circuit breakers were not energized.[11]

Mindless of possible death, Capece unlocked the cage door in the maneuvering room as Clinton Jobe, Mason Poole, and others stood by him. Fully aware that one wrong move would spell electrocution for him, he wedged a 2x4 piece of wood against the overload relay, preventing the breakers from tripping. "I am alive today because of him," shipmate Hank Thomson later stated.

John Bilkey, in the meantime, crawled under the gratings in the control room and worked in cramped, dark quarters while under depth charge attack to help restore power to his boat. Later cited for his heroism, Bilkey would say, "I wasn't trying to save anybody. Hell, I was trying to save my own ass!"

All the while that Capece and the electricians were restoring power, *Seawolf* was still taking a vicious pounding. Throughout such an ordeal, Freddie Warder was a rock. "He was always cool, calm and collected," recalled Little Swede Hanson. "He would walk through and talk with us to see how we were doing."

"This was the worst pounding ship has had but there wasn't much of it," recorded Warder. Another engine lube oil cooler sprung a leak (the ninth such leak of the patrol) and salt water slowly filled the engine sumps due to leaky overhead circulating water valves. The explosions also knocked out two gyro repeaters; the radio transmitter went haywire; and valves flew around the engine rooms. "The engine blower casings did the 'Tokyo Trot,' [and] the depth gauge needles whipped 16 feet," wrote Warder. He also found that the lower conning tower hatch would not hold water due to cable stuffing tubes in the conning tower which were leaking badly.

"That was our worst depth-charging, off Christmas Island," recalled quartermaster Norm Kisver. "We closed off the conning tower and went down into the control room. I kept the quartermaster's notebook and was in charge of counting the depth charges."

Auxiliaryman Paul Zimmerman was standing by the air manifold during this depth charge attack. "This was the most scary depth charging we ever received," he felt. "I braced myself against the periscope well as the depth charges were coming along. There was a lull for a while and then you could hear the screws of the destroyer coming down like they were on a track."

As the first destroyer moved by, sound picked up a second one at 0856 moving at high speed. Two minutes later, another five depth charges exploded well distant. In the radio shack, Paul Maley had a tablet and pencil before him, tallying the ash cans. At 0859, another seven depth charges exploded astern. Eckberg and Maley reported a rapid series of 35 or 40 faint explosions that were not audible without their sonar gear. The destroyers hung on and attacked *Seawolf* for seven and one-half hours this day. The sonarmen listened intently to high speed screws and the skipper conned his boat away from each approaching attack during the ensuing hours.[12]

In addition to the leaking engine oil cooler and conning tower, the starboard stern torpedo tube developed a significant leak. The excess water soon became a problem. Warder found that "every time a pump runs, they come in on us. As we try to come up, negative blows out water (still about 2 tons in it); as we go down it takes more aboard." Warder wished to fill the negative tank completely, but to do so would require getting five tons of ballast out of the ship. His damage controlmen reacted quickly.

They dared not let salt water get into the battery compartment below because deadly chlorine gas would be formed when the saltwater reacted with the hydrochloric acid in the batteries. While Bill

Chief Motor Machinist's Mate Carl Enslin receiving the Silver Star from Vice Admiral Charles Lockwood for his actions on *Seawolf*'s fourth patrol. "Swede" Enslin later served as the *Wolf*'s chief of the boat. *Courtesy of Carl H. Enslin Jr.*

Metz, Red Jenkins, and others worked on plugging and sealing leaks, Chief Swede Enslin led a charge to stop the rising water.

Without pumps to remove the water, Enslin organized a bucket brigade to carry bilge water to the sanitary tanks to keep electrical machinery from being damaged. Water was dipped by hand and the buckets were passed quickly back to the bilges. Motormac Bud McCoy was quick on his feet during the depth charging, and his cool thinking enabled damage control parties to tighten down leaks and control flooding. Enslin later received the Silver Star from Freddie Warder for his quick thinking and McCoy was praised by his skipper for meritorious conduct and advanced in rate to MM1c.

Warder retired his battered *Seawolf* to the north and west. As the fifth hour of cat and mouse with the destroyers above approached, *Seawolf* faced yet another dilemma. At 1310 the trim line in Squeaky Langford's forward torpedo room burst in three places from the stress of the vacuum force it was being subjected to. The torpedo room crews were working to pump water out of the ship to permit the gradual filling of the negative tank, to get the ship under sufficient control to ease back to periscope depth. The crew would pump until the awful racket of the air-binding and high discharge pressure brought another destroyer pinging in their direction again.

Warder and torpedo officer Doug Syverson inaugurated a scheme that used the torpedo room bilges as trimming tanks, pumping water out of them with their drain pumps and drainlines. In order to blow or flood the trim tanks to or from the sea, the torpedo rooms were instructed to use a torpedo tube drain and a torpedo tube vent. This trick worked all right through the depth charging, but would cause a serious problem the next day.

With the negative tank full and the ship in trim, Dick Holden worked *Seawolf* up to periscope depth. Sonar had pinging ahead and astern as Warder took the periscope at 1441 as the *Wolf* reached 63 feet. He observed one cruiser and one destroyer broad on the port quarter on a westerly course at 4,000 yards. He quickly dropped his ship back down to 120 feet before he could be spotted again.

In the forward torpedo room, Pinky Bjerk found the deck was treacherous to walk on as men stood in puddles of their own sweat. One of his companions, Jim Cashero, later noted:

> They kept us down all day long. I think we had 104 depth charges that day. They busted pipelines, waterlines, and everything in the sub. We had leaks all over the boat. We stayed down so darn long we didn't even know if we would have enough auxiliary power to charge our batteries to get the thing running again. We later heard the Japanese claim to have sunk a submarine off Christmas Island, us. But that was just their propaganda.

Warder stayed down until 1618 as his crew worked on the leaks and minor damage. Back at periscope depth, Warder this time found no Japanese ships in sight. There was only "a long flat swell, rain in the west, and a rain squall or smoke over Flying Fish Cove."

Seawolf maintained a deep patrol with hourly observations for the remainder of 31 March. Warder had six torpedoes remaining, but one he was not prepared to fire. He had five torpedoes remaining aft "which are in good shape, we think." The one torpedo remaining in the forward room had an air leak amounting to 500 lbs. per day. This leak had increased progressively until 26 February, when Warder had ordered the flask bled down. "I do not desire to put it in a tube," he wrote. "It would be a sure 'bubbler' with outer doors open."

Seawolf surfaced at 1949 to begin charging her batteries. The first men topside found the deck shelter light to be burning. "Another lesson learned," wrote Warder. The duty electrician immediately defused all topside lighting for the duration of the patrol. *Seawolf* ran away from the moon at 12 knots and sent out a contact report. After the long depth charging, the fresh air that came down through the open conning tower hatch was revitalizing.

In Kelly's Pool Hall, the conversations became lively as the crew discussed torpedoing the Japanese cruiser. *Seawolf* had a bright moon again this night, so Warder kept her about 15 miles offshore as he charged his worn batteries. After the moon went down, he turned the *Wolf* back toward the island again at slow speed. He planned to be at a position about 10 miles east of Christmas Island by 0500 to begin another approach around Flying Fish Cove.

Topside, the watch found the day's depth charging had broken the ship's bell, which was laying on deck. At 2222, a lookout reported two torpedo tracks from abaft the beam. Officer of the deck Jim Mercer turned and dived the ship. The sound watch was unable to pick up any noise source and Ensign Mercer was not sure that these had been actual tracks. Act first, ask questions later was always the safest option in enemy waters. Warder later commented on the phantom "torpedo tracks" seen by his lookout. "I had observed similar phenomena the night before in this general area; definite streaks of refuse material probably washed off the island by the spring tide and drifting to the northwest in the prevailing current."

Seawolf resurfaced and proceeded at 14 knots, jamming juice into her depleted cans. The radio gang received CSAF 98's congratulatory message at 2326. "It helped a lot," wrote Warder. "We were feeling pretty well beaten down, so we put it on the Bulletin Board."

"Bridge to control," called OOD Red Syverson. "Ship spotted bearing two-two-zero. Call the captain."

It was 0151 on 1 April, and Syverson's lookouts had spotted the upper silhouette of an enemy destroyer about 10 miles away as *Seawolf* approached Christmas Island again. Warder, who had just gotten to sleep in his stateroom, raced back to the conning tower.[13]

Seawolf tracked this destroyer with sonar but swung out wide around the island to avoid him. Approaching the island's northeast point at 0300, the *Wolf*'s lookouts spotted a large vessel believed to be one of the Japanese cruisers. Lt. Cdr. Warder worked his boat

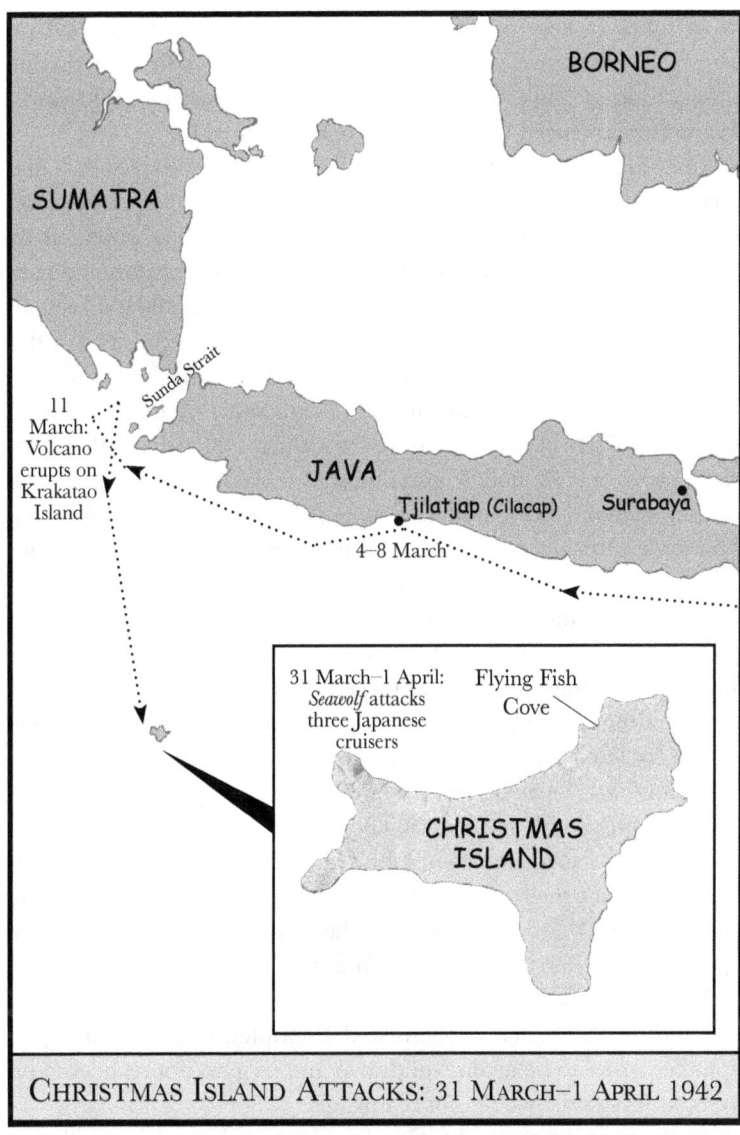

CHRISTMAS ISLAND ATTACKS: 31 MARCH–1 APRIL 1942

toward a firing position to bring his stern tubes to bear. By 0432, his boat had closed enough to positively identify this ship as one of the *Natori*-class cruisers. Japanese records show that this was indeed the namesake ship of this class of light cruisers, *Natori* herself.[14]

Five minutes later, the range to *Natori* had come down to 3,000 yards. Warder came right at full speed to course 345° to bring his stern tubes to bear on a 90° ship track. With *Seawolf* steadied on her approach course, the skipper took one final pre-firing periscope observation at 0440. When the range came down to 1,900 yards, Warder commenced firing tubes 5, 6, and 7 at 0446.

One minute later, Warder logged "one violent explosion" that jarred *Seawolf*. He saw no flames on *Natori* but found observation was becoming more difficult due to the moon being obscured by clouds. Warder called for left full rudder to close his target again, which was turning away. Being submerged, *Seawolf* could not keep up with *Natori*'s much higher speed and she began pulling away. Warder noted her to be "smoking heavily" and felt "positive he is sinking." Red Eckberg heard the convincing "death rattle of the Jap ship in my phones." He clearly picked up "crackling, splintering little explosions" and "a slow, hollow gurgling sound like a man dying" as the cruiser endured her damage. Eckberg felt confident that this ship had sunk. Over the sound system, the captain congratulated his after torpedo room crew on destroying another enemy cruiser.[15]

At 0455, sound reported high speed screws from a destroyer that had obviously been alerted by *Natori* of a submarine's presence. Just over a minute later, sound picked up another set of high speed screws, although neither ship could yet be seen. A minute later, one of the *Natori*-class cruisers was visually sighted on the bearing of the first set of high speed screws. The range was only 1,500 yards and sound estimated his speed at 30 knots.

In the after torpedo room, John Gibson's men completed reloading tube 7 but their other remaining torpedo in tube 8 was giving them trouble. Warder had ordered the torpedo's circular run device removed and—on account of a frozen securing nut—the fish had to be partially withdrawn in order to get a wrench on it. After Bill Reiland and Clarence Kibbons had attempted to engage the gyro spindle, it had to be again withdrawn and its gyro pot rechecked on zero, as its spindle would not re-engage.

In the conning tower, Lt. Cdr. Warder noted that he was "losing an attack waiting on Tube 8, due to my own lack of foresight. Target zigging at high speed. No sense in shooting without spread."

As he continued to wait at periscope depth, *Seawolf* was suddenly jarred by four depth charges, none too close. Warder immediately went back to 150 feet and sound reported screws all around the dial. Returning to periscope depth at 0531, he found the closest ship at 7,000 yards so he came right to head for his bearing and close Christmas Island again.

The crew was secured from battle stations as the *Wolf* drove in toward the island. She eased up toward Flying Fish Cove and Warder called his men back to battle stations at 0825 on 1 April as the task force approached. Once again, Rear Admiral Shoji's invasion force of destroyers and cruisers proved elusive. Warder watched one of the cruisers put up her airplane at 0853. Each time he tried to close a warship to attack, the enemy vessel moved somewhere else. "It was most maddening," the skipper logged.

The flat seas roughened up a little bit as the morning progressed. Warder kept working toward Flying Fish Cove, as this appeared to be the center of where the enemy ships worked out from. Later in the morning, he was frustrated when "the rascals worked over to the northwest into the lee of the island," without having approached any closer than 3,000 yards. In the early afternoon, the flagship cruiser *Naka* and her destroyers moved back toward *Seawolf*. The warships moved about quickly while making signals to the transports in the cove. Just as Warder had expected, the whole force got up steam in the afternoon and prepared to move out of Flying Fish Cove.

Warder began his approach on *Naka* at 1630 from 12,000 yards. His men had already been at their battle stations for seven full hours as the attack approach commenced. *Naka*'s destroyers were taking a rectangular formation ahead of him, well spread out with two to starboard and two to port. All were pinging angrily for the persistent American submarine. The range soon narrowed to 4,000 yards. From the scope, Warder called over to talker Hank Thomson to have the torpedo room prepare their last Mark 14 torpedoes.

Warder waited as the warships passed at 1654 at a distance of 4,500 yards. At just this moment, *Naka* made a timely zig-zag back in *Seawolf*'s direction, thereby putting him in a perfect setup. "This is where he made his mistake," wrote Warder.

At the helm, Rudy Gervais held the wheel steady. The cruiser's range was closing at the rate of 450 yards per minute.

"Angle on the bow, 45 degrees starboard," called Warder as he made his last peek at 1701.

"Course oh-eight-six," chief Jughead Casler read from the scope's dials.

"Bearing three-one-two."

The range was down to 1,300 yards.

This cruiser was flying the flag of a rear admiral. Lt. Cdr. Warder, fully believing that the luckless admiral had already been forced to shift his command flag from one sunken cruiser, quipped, "If we sink him, that Admiral will have to get on one of those destroyers."[16]

"Fire one!" Warder called.

The *Wolf* shuddered as her last two war fish departed from tubes 7 and 8 at a range of about 1,100 yards.

"Right full rudder. All ahead full," Warder called to Gervais.

"Take her to 200 feet," he called down to Dick Holden. "Abandon the conning tower. Rig for depth charges."

The tracking party scrambled down the ladder to the control room and dogged the hatches as Japanese escorts vessels raced to find *Seawolf*. Dropping quickly for depth, the *Wolf* was jarred by a violent explosion 30 seconds after firing her torpedoes. Japanese records would show that this was one of *Seawolf*'s torpedoes exploding in the starboard side of light cruiser *Naka* near her No. 1 boiler.[17]

"I couldn't have missed the son of a bitch," Warder stated.[18]

Warder was correct that *Naka* was a different cruiser from the cruiser *Natori* he had attacked earlier in the morning of 1 April. He noted that *Naka*'s bow was distinctly different from that of *Natori*'s.

From sonar, Mel Eckberg reported that one of the torpedoes he was monitoring was a certain score. "I followed it right to the target," he related. "It was a perfect hit."[19]

Seawolf came under depth charge attack two minutes after firing upon *Naka*. The first three depth charges exploded at 1705. Many men had not been asleep for a day and a half. The boat had been submerged throughout the day. A long depth charge attack at silent running would certainly heat things up. As he later related, Red Eckberg and Paul Maley could hear Japanese destroyers coming.

> The humidity was very high. We waited for the worst, [with] Captain Warder wearing only shorts like the rest of us, sitting in his chair outside the sound shack.
>
> The next hours were hell. At the beginning, I heard the Jap's screws coming toward us.[20]

Eckberg and Maley kept track of the destroyers as they raked *Seawolf* mercilessly for the next nine hours. "After the second hour, the heat, the closeness, the strain, the lack of sleep began to tell," said Eckberg. "We found it difficult to carry out routine orders." Many found it difficult to concentrate as their minds worked sluggishly in the heat and lack of oxygen. "After the fourth hour, a fog of moisture and humidity settled in the compartments throughout the *Wolf*," wrote Eckberg. "We squinted at each other. Some of the men lay sprawled on their bunks, seeking to conserve their strength. Others slumped on stools, their shirts tied about their waists to keep perspiration from running down their naked bodies."[21]

Without circulation, the oil, stale air, and body odors added to the salty, sticky heat that prevailed. Few men needed to urinate because their bodies passed all liquids in the form of sweat. By 1900, the most weary could not even focus to read a book to distract their minds. The refrigerator had been turned off, so even drinking water was warm. Some men drank the warm water anyway and became sick. Doc Loaiza made his way through the boat, passing out saline tablets that had worse effects than the warm water.

Eckberg and Maley continued to track the Japanese destroyers as the hours passed. The last depth charge had exploded at 1601, so Freddie Warder was ready to take his chances three hours later when he believed that he had finally shaken the enemy destroyers. He went to the conning tower to run his boat up to 63 feet for what should have been an uneventful periscope observation. It was nearly his last. The jury-rigged system of blowing and flooding trim tanks via a torpedo tube drain had been created the previous day when the trim line had burst. Warder felt the torpedo tube system worked "all right" until 1 April "when it almost cost us the ship."

Warder hoped to surface his boat and make a run for it. A quick sweep showed a destroyer still close in the vicinity and sound announced that the echo-ranging was continuing. Warder called to take her back down. The main trim line was "hors de combat," so diving officer Dick Holden ordered 1,500 pounds of air blown from the forward trim line to take *Seawolf* back down.

In the forward torpedo room, personnel had opened the master tube drain valve in preparation to blow. When the quick-opening individual tube drain was opened, a hefty 15,000 pounds of water was pushed out of the torpedo tube door. In his report, Warder noted that the "master tube drain valve should be opened last and process controlled with this valve, a small opening, only being used."

The immediate result of the additional 15,000 pounds of water being expelled was to move *Seawolf* from being neutrally buoyant to being positively buoyant. Just like a scuba diver with too much air in his BCD, *Seawolf* popped to the surface like a cork. Her conning tower broached the surface.

"Use negative!" Warder called.

Air hissed inboard from the tanks as *Seawolf* rapidly sucked ocean water into her negative tanks, becoming negatively buoyant quickly. The depth gauge showed her begin to nose back down, and then the needle soon began to move rapidly. *Seawolf*'s descent was fast as the additional water took her down quickly. It was so swift, in fact, that Warder called out orders to check her down angle.[22]

"Blow negative!" Lieutenant Holden shouted.

This command was used during silent running only in extreme emergency. In this case, the *Wolf*'s broached conning tower had already tipped the Japanese to her exact location. The additional noise of the blowers mattered less than the boat continuing her downward plunge. "All back emergency!" Lt. Cdr. Warder also called, putting the screws in reverse to help check the plummet.

"If ever *Seawolf* seemed destined to meet her end, this was the moment," wrote Red Eckberg.[23]

Seawolf reached 80 feet by 1923, but within ten minutes Warder found that the Japanese destroyers were on his submarine and "the party was on again."

Click—wham! The first depth charge was dangerously close this time. *Seawolf* was rocked mightily and men were thrown from their feet. For the next two hours, the destroyers hung on to *Seawolf*, alternately pinging and then dropping depth charges when they felt they had a good fix on her. By 2135, a total of 18 more ashcans had exploded in this latest assault, but *Seawolf*'s depth of 200 feet was enough to keep their effects minimal.

For controlling the boat under such trying circumstances, Warder later recommended diving officer Dick Holden for a Silver Star. "When the *Seawolf* was accidentally blown to the surface and exposed to enemy surface craft, he coolly and skillfully returned his ship to a depth from which she was able to employ maximum defense against further depth-charging by the enemy," Warder detailed.

The bad air in the boat and low battery cells further deteriorated during this second depth charging. "Gravity is now 1150 and hydrogen reading 1.4% in tops of battery rooms; 94°F in control room; 6-1/2 inches air pressure in boat. Ship is most uncomfortable," *Seawolf*'s log noted.

Electrician Bob Parden and another of his gang moved about to check the batteries during silent running, insuring that none had reversed polarity due to the conditions. All other men were to stay put to conserve oxygen.[24]

Seawolf eased back up to 63 feet at 2208 as she continued to pull away from Christmas Island. The pinging continued but no enemy ships were in sight now. The depth charges passed and Lieutenant Deragon made a tour through the boat to check for damages and to check on men's conditions.

Just after midnight on 2 April, the *Wolf*'s sound gear could only pick up the screws of one ship on her starboard quarter which could not be seen. With the air in the boat stagnant and the batteries low, Warder finally elected to bring her up at 0028. After a final sweep indicated that all was clear on the surface, he called for OOD Red Syverson and his best night lookouts. With three blasts of the horn, the *Wolf* rose toward the surface. The skipper wrote that he "went ahead full on everything except the engine which is salted."[25]

At 0110, quartermaster Frank Franz opened the hatch. The fresh air that roared through the boat had a tornadic effect, rustling loose papers and bringing a life-saving chill through with it. The crew was near the end of their rope, both mentally and physically. Many men had been without sleep for 43 straight hours.

Only after pulling further away from Christmas Island did the *Wolf* slow back to 13 knots as she continued to recharge her batteries. Warder set a course east for Australia and had Eckberg send a long dispatch to submarine command recounting the *Wolf*'s latest cruiser attack and subsequent depth charging. Those not on watch sought out their empty bunks and slept heavily.

Warder was certain that he had sunk another cruiser. "Sound did not hear this fellow's propellers again and I feel sure we got him." *Naka* had indeed taken a *Seawolf* torpedo but she would survive. The light cruiser *Natori*, also fired upon by the *Wolf*, took the badly damaged *Naka* under tow and pulled her into Bantam Bay on Java by 3 April. Temporary repairs were made and *Naka* was able to limp out of Bantam Bay under her own power on 6 April. She reached Singapore four days later and went into drydock for repairs.[26]

The *Wolf* dove and ran submerged on 2 April. That night, a radio message came in from command which read, "A wonderful cruise. Your accomplishments rank among the greatest of all time."[27]

Warder had yeoman John Sullivan type up copies of the dispatch and post them about the *Wolf,* including in Kelly's Pool Hall. At the bottom, the skipper added his own postscript to all hands. He offered his "deepest thanks for your ability and your conduct, and above all, your devotion to duty."

Seawolf ran submerged during the day hours of the next two days to avoid enemy air patrols. By the early morning hours of 4 April, Warder decided to run on the surface continually unless forced down, on account of the trim line hazard.

Easter Sunday, 5 April 1942, found *Seawolf* en route to Fremantle, Australia, in peace. Cooks Auston Baker, Gus Wright, and Bill Mallory served up steaks for dinner which had been aboard *Seawolf* since early December from the Philippine cold storage. This was a big morale booster for men who had been without fresh fruits, cigarettes, and other favorites for weeks.

As the *Wolf* approached Australia on 6 April in the bombing restricted area, Warder called down that anyone who wanted to come topside could come up in small groups. The men—most of whom had not seen daylight in well over a month—found the sunlight blinding. Their skin burned in minutes. The scuttlebutt throughout the boat had turned to the prospect of shore leave in Australia and all the mail that should be waiting for them. The phonograph in Kelly's Pool Hall was playing day and night now.[28]

Aboard World War II submarines, it became commonplace to design a war flag to display the boat's "kills." Helmsman Rudy Gervais approached the skipper about decorating the *Wolf*'s conning tower with Japanese flags, but Warder turned down the idea. The sub's salty yet modest skipper considered such a display bragging.[29]

Chief Jug Casler, however, decided that the *Wolf* needed a flag during the fourth patrol. Quartermaster Hank Thomson later recalled that the duty fell naturally upon Casler.

> Flag making and repairs were part of the signalmen and quartermasters' duties, except on larger ships, where a

sailmaker sometimes did the trick. I could do uncomplicated things with the sewing machine but wasn't much good at it. Our flag was made by Casler, as he was our best seamstress.

Casler's original *Seawolf* flag sported a lone wolf riding a torpedo, with one stripe for each ship sunk or damaged by torpedo. "Our first flag was just a torpedo with a wolf riding on it," said Little Swede Hanson. "The first one was always the best. The quartermasters kept the flags during patrol, but we would break out the battle flag and fly it from the mast when we returned from patrol."

Seawolf closed Fremantle after midnight in the early hours of 7 April. The light on Rottnest Island—so named by the first Dutch explorer who discovered the little island—was spotted at 0049. A small boat met *Seawolf* outside of port for a harbor pilot to guide her in through Swan River, the channel and mine fields leading to Fremantle's port. As the boat was being conned in to port, a big mail bag was passed over and all hands instantly had something to occupy their minds. Snapshots and loving words from home were the first the men had seen since the war started.[30]

Seawolf was approaching the main harbor when Lt. Cdr. Lucius Henry Chappell's *Sculpin* came into view. She was returning from her second war patrol, having disabled the Japanese destroyer *Suzukaze*. New to this port, Warder signaled Chappel, "Go ahead. I'll follow you in." Chappel, aware of the *Seawolf*'s fine accomplishments, signaled back his own congratulations and insisted, "After you, sir."[31]

Seawolf docked at 0519 on 7 April in Gage Roads, Fremantle, Australia. After this patrol terminated, Warder's crew nicknamed him "Fearless Freddie." He was credited with sinking three ships for 14,000 tons, including one cruiser, and damaging five ships, including two cruisers, for 30,000 more tons. However, postwar records do not bear out the sinkings. Warder had only damaged the cruiser *Naka* at Christmas Island. She made it back to port and after a year of repairs was returned to the active list. Any other ships damaged or sunk on *Seawolf*'s memorable fourth patrol escaped the official record books. The other issue was that U.S. Mark 14 torpedoes were regularly running deep or failing to explode, which may have allowed at least one of the *Wolf*'s fourth patrol targets to escape.

Regardless of how many ships went down, Warders' "Fearless Freddie" nickname would stick, although the crew was not allowed to use it aboard ship. "We used to call him 'Fearless Freddie, the Fish Slingin' Fool," said torpedoman Jim Cashero. Another crewman, Paul Zimmerman recalled:

> The captain never liked that nickname. He got it after the Christmas Island patrol. Someone coined the "Fearless Freddie" expression after we had pulled into Fremantle.
> Captain Warder always kept his cool during depth chargings. Once we had hit the enemy and their destroyers knew where you were, he would say, "Haul ass and come back to fight another day." It was best to sneak out, pull back, reload the tubes, and then go back in to fight.

Freddie Warder was a little embarrassed by all of the newspaper stories. "All of a sudden all this bullshit starts in the newspapers," he recalled. "'Fearless Freddie is back from Manila.'"[32]

Warder cornered his exec, Bill Deragon, on the source of the story. Deragon admitted that the reporters had caught up with him and some of the crew after a few drinks in the uptown area.

The "Fearless Freddie" tag would hang on Warder for the rest of his career. Although the crew was proud of it, he was not happy about it.

"In war, I was scared to death when I'd shoot torpedoes," he later related. "Fearless? I'm thinking, 'What's gonna happen if I don't hit that son-of-a-bitch?'"

USS *Seawolf* Fourth Patrol Summary

Departure From:	Surabaya, Java
Patrol Area:	Java Sea/Lombok Strait/Christmas Island
Time Period:	15 February—7 April 1942
Total Days on Patrol:	51
Number of Men Aboard:	73: 67 enlisted and 6 officers
Number of Torpedoes Fired:	20
Ships Claimed as Sunk:	3/14,000 tons
JANAC Postwar Credit:	0/0 tons
Damaged:	Cruiser *Naka*
Return To:	Fremantle, Australia

6

The Jinx

Refit and Fifth Patrol   *7 April–2 July 1942*

Once *Seawolf* had reached her Fremantle dock, Lt. Cdr. Warder went out of his way to make sure that fresh milk and cigarettes were available for his men. After he had delivered his patrol report to local command, Warder returned in a launch and had Bill Deragon line his men up on deck. The *Wolf*'s starboard side was peppered with shrapnel from recent depth charge damage and her black paint was chipped and peeling.

Success or no success, what happened on patrol was still classified. Warder told his crew that submarine command was highly pleased with *Seawolf*'s last patrol. The boat would be undergoing a longer than usual overhaul, during which time his crew would get a well-deserved rest. He cautioned his men to keep any stories about the *Wolf* on the boat. "Don't drag her down into the city," he warned.[1]

Part of the crew remained aboard while the first departing group left right away, with pay fresh in hand. A nearby train took men into town about four hours distant. There, they could enjoy their first hot bath or shower in fresh water in two months. Those who remained aboard helped the relief crew in fixing up battle damage from the past patrol. The projector at the end of the No. 2 sound shaft was among the items in need of repair, having been smashed when *Seawolf* ran aground.[2]

The crew put on shoes for the first time in nearly two months, having worn sandals aboard ship. About a dozen crewmen stayed in a Perth hotel the first night, where they followed the local custom of leaving their shoes outside their door at night to be shined.

Eddie Souza had a little fun in the early morning hours, switching out everyone's shoes. Red Eckberg found a pair of size 8 shoes had replaced his size 12s. "We were due back to the boat," he recalled. "We had to get back there. We dashed about, cursing, trying to match our shoes, and finally we all met in the lobby." Chief Sullivan finally stood up on a desk and called off sizes for claiming. Only after returning to the *Wolf* did Chief Souza fess up to the prank.[3]

Back aboard *Seawolf* that morning, the men were ordered into their whites for a presentation ceremony. Commander Submarines Southwest Pacific, Rear Admiral Arthur Schuyler Carpender, came aboard with his staff. The after deck of *Seawolf* was badly chewed up from patrol damage, so Chief Souza formed up his crew forward of the conning tower. He mustered the men into two rows along the port and starboard side with Lt. Cdr. Warder and his officers directly in front of the conning tower.

"We did not have our dress blues to wear, because they were on the *Canopus* when she had been sunk," recalled Hank Thomson. With the *Wolf* crew at attention, Admiral Carpender's staff boarded. His chief of staff was none other than Captain Jimmy Fife, whom *Seawolf* had evacuated from Corregidor months before. Captain Sunshine Murray was another of the officers who came aboard. While the *Wolf* crew stood stiff at attention, Carpender commended them on their recent performances. "You are the envy of every submarine in the fleet," he said. He praised the *Wolf*'s recently completed patrol and told the men that they had set a record "for every other submarine to aim at."[4]

Then, Carpender and his staff proceed to decorate several of the *Seawolf* crew. Navy commendations usually came slower in wartime, but *Seawolf* was making a name for herself and the command wanted to publicly acknowledge her exploits. Warder was awarded the

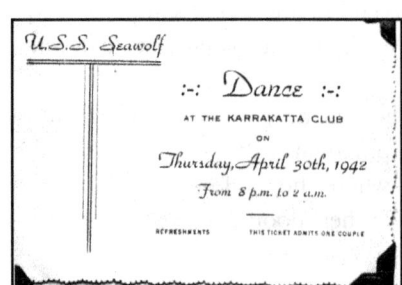

Ticket from a ship's party held after the successful fourth war patrol. *Courtesy of Bob Hanson.*

Navy Cross for this patrol. Admiral Carpender started with Freddie Warder, reading the citation for his award. Warder was praised for entering shallow waters to sink a transport and a destroyer and also for making repeated attacks on heavily screened enemy light cruisers, sinking one and damaging two others. Carpender then pinned the Navy Cross on Warder and shook his hand.

The other awards given during this ceremony were the Silver Star to Lt. Bill Deragon, Lt. Dick Holden, CEM Edmund Capece, and CQM Jug Casler. Lt. Cdr. Warder would also recommend chiefs Red Eckberg and Swede Enslin for Silver Stars. These last two awards were eventually approved and they were decorated in separate ceremonies later. The crew remained at attention forward of the conning tower until after the admiral's staff had departed. Warder spoke to his men, praising them for their contributions. "This cross is as much yours as it is mine, boys," he said.[5]

A little success did not hurt the partying spirit for the *Seawolf* crew. Life ashore was fabulous after so many weeks aboard their ship. Torpedoman Jim Cashero and his buddies seemed to find their share of trouble in bar fights ashore in Australia, as he recalled.

> Some of the guys on the regular ships didn't like submarine men and they used to beat the hell out of us. So, we had to figure out a way to stop that. When we finally got together on it, we'd send one or two guys into a place. When they would start a fight, they'd let the rest of our guys outside know and then we beat the hell out of them.

Auxiliaryman Pete Lober decided that Fremantle was "a great place in the war to be. The people there were very friendly and they made the best beer you ever drank. It was eighteen percent!"

Motormac Casey Mallough later related that he had to get the effects of the war patrol out during his first night ashore on each R&R. He would get smashed the first night, suffer from the effects, and then would avoid hard drinking during the rest of his leave.[6]

The more crafty *Wolf* crewmen knew how to create their own booze. The torpedo alcohol was a source for creating batches of gilley for enjoyment ashore. "We did make some of the gilley juice," admitted motormac Bob Hanson. "We had to run it through a still but just about all of the old-timers knew how to do it. We'd make the batches and save them for R&R time. I never knew of a man who drank it at sea. We had our fun off duty."

Chief electrician Edmund Capece found that his dark complexion concerned local Australians, as shipmate Hank Thomson detailed.

> The Aussies had a very demeaning attitude of the Italian soldier that was fortified by the desert campaigns and the stories they sent home. Capece had a Latin complexion and of course was highly interested in the Aussie girls. Because of the bigot thinking that he heard in the bars, he instructed us to tell anyone who asked that he was of Spanish origin.

Seawolf remained in Australia for more than a month to repair her battle damage and make ready for sea. Before leaving again, there was a big ship's party ashore, attended by Admiral Carpender's staff and various dignitaries from local command.

The most significant personnel issue for *Seawolf* in Australia was the death of her senior machinist's mate, Lloyd Sandridge. The muster rolls show him to have already been transferred off *Seawolf*, but he was with some of his *Seawolf* buddies at the time of his accident.

"He was run over by a taxi in Australia," recalled Little Swede Hanson. "It was late at night and he was going across the road." Sandridge had been drinking when he stepped out into the road. Due to the black-out conditions that were in force in Perth, he did not see the slow-moving vehicle coming. "He was killed by one of those charcoal-burners," said engineman Lucien Rajotte. "They didn't have gasoline, so they used charcoal to propel their cars. The cars couldn't go more than 15 miles per hour."

Lt. Cdr. Warder organized a memorial service in Perth for Chief Sandridge so that his crew could properly mourn his loss. Aside from this tragedy, there were few changes for the *Wolf*'s men to adjust to. Sully Sullivan only had to transfer five men and receive three new men, a low percentage of turnover for the submarine service.

Among those leaving were CTM Langford to *Seal*, CMM Metz to *Holland*, and CEM Jobe to *Searaven*. The unusually high number of chief petty officers did not change with these three departures,

The Jinx 141

for John Bilkey, John Gibson, Pete Lober, and Lefty Leffingwell were promoted to CPOs prior to *Seawolf*'s fifth patrol. With the transfer of Auston Baker, SC1c Bill Mallory became senior cook over Gus Wright and SC2c Henry Wilton, newly received from *Seadragon*.

Electrician's mate Edward Milas was another of the three new hands received aboard the *Wolf* at Fremantle. He joined the main power gang to replace Chief Jobe. Milas would serve in *Seawolf*'s maneuvering room, running the controllers to the generators that were driven by two diesels.

Ed Milas had enlisted in the Navy in April 1939 from Illinois. After first serving on the battleship *Colorado*, he became tired of training on the gun crew. "They asked for volunteers for submarines and my hand went up," he said. After sub school in New London, Milas was assigned to *Skipjack* and was serving on her when the war began. Ashore at Manila on 8 December just like the *Seawolf* crew, Milas had made his way back to *Skipjack*. His boat shepherded the tender *Holland* back to Darwin, Australia, loaded with ammunition and delivered that to Corregidor, just like *Seawolf*. They also transferred some nurses and Army personnel from Corregidor. They were attacked by the enemy, who "beat the living shit out of us" with depth charges. Once aboard the *Wolf*, he found himself one of the Northerners. "*Seawolf* was mostly a rebel ship," said Milas. "All the guys aboard her were Southerners. I think there was only a few of us Yankees aboard her."

Seawolf was underway for war patrol again at 1742 on 12 May 1942. "Fearless Freddie" Warder had secret orders to patrol in the area of Manila in the Philippine Islands. En route to Manila Bay, he was to investigate various ports for Japanese shipping and then patrol in the area off Manila and Subic Bays until about 16 June before heading south. As the *Wolf* passed Rottnest Island, Warder was angry again when the Army searchlights fully illuminated his sub in the darkness, something he had previously complained about.

Seawolf's wardroom had a new face for this patrol, a PCO (prospective commanding officer). In the submarine service, command often assigned an officer who was on the verge of taking his own command to shadow another commander for one patrol. Lt. Cdr.

Lt. Cdr. Edward Stephan, later lost as skipper of the submarine *Grayback*, made his prospecting commanding officer cruise aboard *Seawolf* during her fifth patrol. Some of the superstitious *Wolf* sailors considered the "guest" aboard ship a jinx. *National Archives*.

Edward Clark Stephan reported aboard *Seawolf* at 1400 on 12 May, shortly before OOD Dick Holden had the boat underway. Prior to the war, Ed Stephan—a 1929 Academy graduate—had commanded an old S-boat. With Freddie Warder as a tutor, he would observe the attack procedures of a fleet boat at war. Submariners, by and large, were a superstitious lot, as Mel Eckberg later wrote of *Seawolf*'s new PCO. "He was pleasant, about thirty-five, kept to the wardroom, and was in no one's way. Yet a few of the old-timers grumbled. Some submarine men are convinced that strangers jinx a voyage."[7]

Seawolf ran through rough weather the first two days as she made for the port of Koepang on the coast of Timor in the Philippines. Lt. Cdr. Warder used the time to run his crew through drills and training. On 16 May, he practiced a battle surface with Gunner Bennett's 3-inch deck gun team. With the crew standing ready, Dick Holden blew the boat to the surface and the gun crew swarmed out on deck under the direction of gunnery officer Doug Syverson. Bennett's crew fired one practice shot in 63 seconds from *Seawolf*'s surfacing.

Running on two engines en route, *Seawolf* received an intelligence report on 18 May of five ships sighted at Koepang Harbor the previous day. Warder bent on all four engines at 17 knots to reach Koepang earlier than planned. *Seawolf* dived at 0508 on 19 May and proceeded to examine Koepang Harbor throughout the day. The submarine moved from the 100-fathom curve to water as shallow as 25 fathoms during the day, but found no Japanese shipping.

Warder could only find five small sails in the upper reaches of the bay. He noted the attractive Dutch architecture houses of the town, which seemed unscathed by enemy bombings. The *Wolf* moved out from Koepang and rounded Timor Island during the night en route

The Jinx

Two views of the *Wolf* underway in 1942 from Fremantle, with lookouts atop her conning tower. *Courtesy of Hank Thomson.*

to investigate Dili Harbor on Timor. At 1937 on 20 May, *Seawolf* was making 6 knots on the surface as she recharged her batteries.

"Ship bearing two-five-oh," sang out port after lookout TM2c Bill Reiland. "Distance five miles."

Lt. Holden ordered a course change to head for the contact. Battery charging was secured and Warder raced up on the bridge. He ordered battle stations. The target ship looked to be a two-mast, single-stack, coal-burning steamer of about 5,000 tons and 400 feet in length. With Ed Stephan on hand to observe the attack, the tracking team worked out the angles as the range decreased.

With the enemy ship brilliantly silhouetted by the moon on a smooth sea, Warder ordered his forward torpedo room to open their outer doors. The *Wolf*'s newly painted dark hull was well hidden against the beach behind her.

At 2001, *Seawolf* jarred as a single war fish was fired from tube No. 3. Through the periscope, Warder saw the first torpedo was going to run ahead of his target so he ordered the speed of the target ship to be cut down from an estimated 9 knots to 6 knots. Senior torpedoman John Gibson's forward room was then directed to fire tube No. 4 from a TDC-set range of 670 yards. Warder dived his boat after firing the second torpedo. The enemy ship could not be seen through the periscope in the darkness and Red Eckberg and Paul Maley were unable to pick her up on their sound gear. Quartermaster Frank Franz kept track of the seconds after firing, but the ticking continued without any explosions. *Seawolf* had missed this target.

Warder quickly brought his boat back to the surface at 2012 to pursue this freighter. Ensign Jim Mercer took the deck with his battle stations lookouts and *Seawolf* raced forward to overtake the freighter. The *Wolf* ran along parallel to the ship at 6,000 yards distance while charging her batteries and pulling ahead prior to diving. Warder slowed to 13 knots "as I was overtaking him rapidly and wished to decrease our phosphorescent wake" in the moonlight.

Warder was able to get a firm fix on his target ship's bearing and size. He soon realized that he had misjudged the enemy's course by as much as 45° in this attack and that the range of the ship was at least double what had been estimated and cranked into the TDC. The range set on Doug Syverson's TDC had been 1,000 yards for the first torpedo and 670 yards for the second torpedo.

> These two misses are directly attributable to poor estimation of target angle by myself; about 5 to 10 minutes prior to firing when I persisted in a value of 90° S over three observations, when actually I must have been getting well abaft his beam. This resulted in a generated range of 670 yards for the second torpedo, which should have been more nearly 2,000 yards.

Islands were visible on both sides of the passage as the *Wolf* ran up the Japanese vessel's starboard side. At 2102, the freighter zigged toward *Seawolf*, putting the *Wolf* about 30° off the freighter's starboard bow. "I was expecting him to change course about this time to head for Dili, so attached no especial significance to the zig except to thank him for helping me get up ahead," noted Warder.

The starboard lookout then reported a light to starboard. As Warder turned to look for the source, Mercer saw the maru open fire with her bow gun on *Seawolf*. Warder heard the report of the shot and GM1c John Bennett heard the shell scream by overhead in the shears. "Clear the bridge!" Mercer shouted as he pulled the alarm.

As the men piled through the conning tower hatch, the skipper ordered Dick Holden to take her to 90 feet. The first depth charge exploded at 2110, missing by a long shot to port. Warder prudently ordered 200 feet. Sound reported that the little freighter was milling back and forth in the vicinity above for an hour, searching for the submarine. With no sonar, the little freighter had no chance of anything other than a lucky drop on *Seawolf*. By 2240, sound reported the ship to be retiring toward the direction of Dili Harbor.

Seawolf resurfaced 15 minutes later and headed for Kambing Island. Warder spent the day of 21 May scouring Dili Harbor. The crew waited tensely as Warder moved his boat forward and kept up a commentary on the 1MC of what he saw through the periscope.[8]

"I see two masts up ahead," the skipper narrated. "Looks like a sailing schooner. About 100 tons and 100 feet long. She's got a white hull and looks well kept. Not worth a torpedo, though."

According to press news, the Army had reported intensive bombing of Koepang and Dili on May 18. Through the periscope, however, Warder saw buildings with "clean-cut appearance" and no fires burning. He also commented to the crew that he could make out the white steeple of a church, an airport, and a radio station tower.

Seawolf surfaced at 1816 and spent the next two days running from Dili to Ambon. At 1738 on 23 May, OOD Jim Mercer spotted smoke on the horizon and monitored it until it proved to be a ship. Mercer called the captain as the distance closed to 13,000 yards.

Warder, with Lt. Cdr. Stephan standing by, took over the scope and scanned the approaching vessel as the range narrowed to 5 miles. Warder called out the details of the 7,000-ton, 400 foot long, two-masted freighter to his recognition crew as the ship approached. He estimated her speed at 12 knots and estimated her to be on course for Dili Harbor. Mercer and Jug Casler double-checked all figures before *Seawolf* fired a three-torpedo salvo from her bow tubes at 1815. The firing range set in Red Syverson's TDC was 1,000 yards, but Warder figured the correct values for the torpedoes were between 700 and 570 yards. The skipper kept his face pressed to the scope lens, watching his torpedo wakes as they raced for the target.

"What the hell is this?" the skipper roared as he watched the wake of a single torpedo go under the ship.[9]

As the torpedo passed under the target ship, those in the conning tower could clearly hear a noise that sounded like a detonator exploding. Freddie Warder believed it to be the result of a dud torpedo—another faulty Mark 14!

The freighter turned toward the *Wolf* slowly, but Warder hung on. He had Red Syverson and Chief Leffingwell throw another set-up into the TDC for the final fish in tube No. 2. Aiming 1/4 length ahead of her bow, *Seawolf* fired her last forward tube at a mere 570 yards. Arthur Lamberson, the second leading torpedoman in the forward room, was anxious to prove to Chief Gibson that his torpedoes were properly serviced and not the cause of the poor performance. Pinky Bjerk slammed the firing knob, and yelled excitedly at his fish to go take care of this target.[10]

Red Eckberg heard through his sonar that this torpedo was running true—"hot, straight, and normal." Through the scope, Warder could see that the "maru was firing [his] bow gun at me when 4th torpedo was fired but had not yet manned stern gun."

Seawolf was forced to go deep upon firing. Dick Holden took her to 200 feet as the first three depth charges exploded at 1820. The attack team dropped down out of the conning tower and secured it as the ship came overhead. The ship apparently had no interest in sticking around after unloading its first three charges. Following the depth charging, Warder met with Chief Gibson and Lt.(jg) Syverson, the torpedo officer. Gibson explained that the torpedoes had been serviced the previous night and Syverson had checked them over. Although the fish were felt to be perfect, Warder could not understand how two good attacks had resulted in two misses. The crew shared their skipper's disappointment.

Within 20 minutes, Warder was back at periscope depth eyeing his prey again. The Japanese ship had drawn out to 6,000 yards. *Seawolf* eased up to 55 feet to view the target more readily and continued to close it. The ship was zigging. Warder surfaced and pursued his prey on four main engines, keeping the ship just within sight as he rounded him to the southward. "Decided to pass him during night and throw two more torpedoes at him at dawn," wrote Warder.

Jug Casler's plot established the enemy ship's course as 210° and his speed 9.5 knots. After nearly four hours of running, *Seawolf* slowed to 11 knots, having attained a position off the enemy ship's starboard bow at 8 miles. Lookouts could see the Japanese ship signaling via blinker to an unseen ship on either side of him. The Japanese were obviously very aware of the submarine danger that had almost claimed this ship once. As the night passed into the next morning, *Seawolf* stayed on track and dived at dawn on 24 May.

The Jinx 147

During the course of the pursuit, a dispatch was received with new orders for *Seawolf:* "S197 PROCEED NORTH OF AMBOINA BY DUSK 24TH XX GAR COMING IN THROUGH DJAILOLO." Warder found it "impossible to comply with these orders" due to his pursuit of the maru. He figured his main intent was to stay clear of *Gar*'s territory, so he decided to follow the orders loosely by getting to the desired position via a different route.

The target ship was sighted again at 0605, maintaining its same course about 10,000 yards distant. Intending to be just ahead of the ship, Warder ordered Rudy Gervais to come to course 288 to close the track. Warder hoped to bring his stern tubes to bear at about 1,000 yards. "I had the feeling that my torpedoes may not have armed the day before and I also wanted to get some torpedoes fired from the after room," he reported.

By the time proper depth control was attained, the target ship had reached a range of 1,600 yards. At 0636, the welcome order of "Fire one!" was relayed by ship's talker Hank Thomson. In the after room, Bill Reiland hit the plunger to fire tube No. 5. As luck would have it, the first tube would not fire. After two errant attempts, the after gang fired their second tube to hit the MOT, at her stack. As Warder watched through the periscope, this Mark 14 torpedo ran just forward of the maru's bow but did not explode.

> He did not see it until it had crossed his track. Unfortunately, I did not have another fish ready. The torpedo threw the Japs into great commotion and they ran fore and aft and up and down at a great rate; white uniforms predominating. Water was quite smooth and I was not inside his turning circle as I had been yesterday, so at 0639 ordered 200 feet and rigged for depth charge.

Four minutes later, a depth charge exploded close aboard and overhead. Fifteen seconds later, another ashcan slammed the water. One minute later, a third explosion "stirred the water around in our superstructure," noted Warder. Only depth saved *Seawolf* from damage. While some found the depth charge attack troubling, others were used to it by now. Auxiliaryman Pete Lober: "I wasn't worried too much about the depth charge attacks because our skipper was pretty good. He was one of the best."

Doug Syverson raced back to the after torpedo room to check on the errant torpedo. He was informed by Bill Reiland that the torpedo in tube No. 5 was partially ejected and that the outer torpedo tube door could not be closed. It was 0725 before the after gang succeeded in withdrawing the torpedo and closing tube 5's outer door.

As the target ship faded away into the early morning, frustration set in again. In the crew's messroom, radioman Mel Eckberg found his shipmates down over the three straight misses. Moving to the control room, he found Chief Otis Dishman leaning on the stern planes control, conversing with Fred "Jew" Zirkel and Al Hershey. Eckberg felt Dishman "looked like a mad bull." The men cursed the *Wolf*'s luck, having fired six torpedoes without a hit—five misses or duds and one that hung up in the tube. Hershey argued that *Seawolf* was jinxed by the PCO who had brought bad luck aboard.[11]

The superstitious sailors could find some logic in cursing their poor luck on the PCO jinx but many U.S. submarine skippers were well aware of the true enemy: a cursed Mark 14 torpedo that had serious operational issues.

Seawolf ran on the surface at night to charge her batteries and dived during the day while en route to Kema. Prior to midnight on 26 May, the torpedo gang commenced striking down four new torpedoes to the forward room under bright moonlight. "We carried torpedoes in both torpedo rooms, but we could also carry four more torpedoes in some tubes in the superstructure, just forward of the conning tower," explained torpedoman Jim Cashero. This task was completed in two and a half hours without incident.

During the day, an Ultra was flashed to *Seawolf* about a large convoy bound for Davao from Batavia. *Seawolf* was given an intercept point at the northeast entrance of Macassar Straits. En route on 28 May, Warder investigated Kema's main village via periscope observations. The *Wolf* ran through the Celebes Sea on 29 May and dived in her assigned intercept spot the following morning. Warder constantly had to dodge floating refuse and pesky banca fishing boats to avoid ones that might have a radio to alert the Japanese fleet.

While lying off Stroomen Kaap on 31 May, Jim Mercer made out the tops and stack of a ship at 14,000 yards. *Seawolf* took an approach course and went to battle stations as the range dropped to 10,000 yards. The target was zigging but the *Wolf* narrowed the range to 4,200 yards. Warder ultimately gave up the idea to fire when he could not close the range any further. The extreme distance would have made his steam torpedoes easy to spot in the calm sea. He also did not consider this small ship a worthy target, since firing on her could tip off the important approaching convoy.

Seawolf hung onto her assigned intercept position for two more days without a sight of the reported convoy. Warder then proceeded through the Celebes Sea and Sibutu Passage en route to Tolong. The *Wolf* patrolled north and south of Tolong on 6 June, staying between 2 and 6 miles offshore from the village.

He patrolled along Dalipe Point the following day, moving as close as 1.5 miles to the lighthouse on that point. The only shipping found were pesky native bancas that had to be avoided. On 8 June, *Seawolf* snooped through Looc Bay on Tablas avoiding more bancas.

Seawolf made 13.5 knots during the night en route to Manila amidst a sky dancing with lightning. Visibility was limited due to the heavy weather in the narrow, shallow channel. Lt. Bill Deragon, on the bridge to navigate with Jug Casler, suddenly at 2338 sighted what he judged to be a minesweeper or tugboat sharp off the starboard bow. The ship was turning toward the *Wolf*. Smoke poured from the stack of the little ship, which was a mere 1,500 yards away.

"Clear the bridge! Dive, dive!" Deragon shouted.

The poor weather made it impossible to see the little minesweeper through the periscope. Maley and Eckberg listened intently as the enemy patrol swept the channel, searching diligently. "We were really in a bad spot," wrote Eckberg. The patrol boat passed directly over *Seawolf*, his screws sounding like a train rumbling over a trestle. "We knew he was using his sound gear, and that, coupled with the knowledge that we were practically trapped in this shallow, narrow channel, gave us one of the worst moments of our lives. We were afraid. We were damn afraid."[12]

Fortunately, the patrol boat did not sniff out the *Wolf* and no ash cans tumbled down from above. Gradually, the screws faded and a collective sigh of relief was exhaled. Lt. Cdr. Warder brought *Seawolf* back to the surface just after midnight on 8 June and found the ship to be 1 point off his starboard bow at about 3,000 yards. "Decided to overtake him and take a shot at him at dawn," wrote Warder. The

Wolf continued to run on the surface during the next hours as the electricians charged up the batteries.

The moon came up at 0325, offering a better view of the target, which looked to be a freighter. Warder's tracking team set up an intercept point that would put them ahead of the freighter's track at dawn when he would be about 8 miles off Bongabong Island. The lightning's brilliant flashes temporarily blinded *Seawolf*'s lookouts and probably silhouetted the sub nicely for the freighter's lookouts.

At 0351, the freighter suddenly snapped on a high intensity searchlight and turned towards *Seawolf*. Warder dived his ship and held his course to intercept. With poor light conditions, he "had no confidence in set-up" and had to hold his fire. Once he could definitively make out the target ship in the periscope, the range was 6,000 yards and the Japanese ship was going over the hill.

Seawolf abandoned this approach and proceeded north en route to Batangas. At 1436 on 9 June, Ensign Jim Mercer sighted the tops of a ship through the periscope and turned to head for it. Warder soon found that the contact was actually two ships. *Seawolf* drove in towards the targets but found that their speed was no match for a submerged submarine in the daylight. The skipper reluctantly secured from battle stations as the range quickly advanced.

Seawolf moved through Verde Island's north pass during the early morning hours of 10 June. Warder dived to 110 feet at 0158 when a ship was spotted 8,000 yards away heading toward her. Sound reported pinging on the port bow. Warder eased up to 63 feet and made periscope sweeps as he pushed forward. The ship continued pinging, but it was 0230 before Warder could find him with the control room periscope. He stayed submerged through the night, avoiding detection prior to his planned reconnaissance of Batangas.

Seawolf entered Batangas Bay at 0613 on 10 June and quickly sighted an eastbound ship south of Maricaban Island. Warder turned to cross the enemy ship's bow in order to make an attack with his stern torpedo tubes. This small target, making 5.5 knots by Casler and Mercer's plot, narrowly escaped destruction. At 850 yards, *Seawolf* suddenly hit a tide rip or a fresh water pocket. Holden's planesmen in the control room momentarily lost depth control.

Warder was unable to deliver his attack by the time Holden's men regained depth control. He felt that this was "probably a good thing as I have a hunch there was not enough range for torpedo to arm."

Seawolf abandoned this small maru and proceeded into Batangas Bay. She penetrated to a point 2.5 miles off Batangas City before turning back at 1115. The *Wolf* found a number of sailing craft in the bay, but no worthwhile torpedo targets. *Seawolf* cleared the bay at 1514 and entered Verde Island North Pass.

Seawolf dived at dawn and ran submerged toward her designated station off Subic Bay throughout the day. Rough seas made depth control a challenge this day. *Seawolf* surfaced that night at 1959 with the Capones Island light 6.9 miles distant. Warder took the bridge with Jim Mercer and his lookouts. He kept his sub's decks awash, lying to as she charged her batteries near the expected traffic lane.

At 2015, a lookout reported an object on the port bow whose bearing matched that of Los Frailes Island, which had been in sight upon surfacing. Visibility remained poor and it was the *Wolf*'s listening gear and an alert watch stander which saved the ship this night.

At 2045, Red Snyder—on sound watch in the radio shack—suddenly reported, "I've got a set of screws here on the port bow."[14]

Snyder was one of the two men whom Freddie Warder stood up for against Admiral Hart after their infraction ashore prior to the start of war in Manila. Now, his alert ears had an enemy ship bearing down on *Seawolf*. Snyder's alarm was instantly relayed to the bridge. "On dark tropical nights like this, sonar was our best observer," recalled quartermaster Hank Thomson. "Snyder gave us an early warning and probably saved us from becoming a casualty."

There was nothing to be seen for the bridge watch, but Warder called down to the helmsman to swing the ship to bring the stern tubes to bear. Quartermaster Frank Franz noted that the seas were high and rough this night, with a whistling wind. The skipper and Lt.(jg) Mercer had to yell orders to the lookouts.[15]

Senior radioman Mel Eckberg took the earphones from Red Snyder and confirmed that an enemy ship was approaching in that direction. The starboard quarter lookout Paul Zimmerman made out the enemy ship at 2049 and also shouted a warning.

"I finally picked up a shadow and reported it," said Zimmerman. "The officer of the deck said, 'Are you sure?'"

Jim Mercer couldn't see it and Warder turned his attention in the direction. He wrote that Zimmerman was "endeavoring to show it to me when [a] searchlight fully illuminated us."

Bathed in light, *Seawolf* was suddenly fully exposed.

"Clear the bridge!" shouted Mercer.

Freddie Warder waited while the lookouts dived down the hatch opening into the conning tower, Franz literally riding Doc Loaiza's shoulders as he went down. Zimmerman knocked himself senseless as he cleared the bridge. "I jumped right down the ladder and the binoculars bounced off the rung of the ladder and knocked out two of my front teeth," he related.

Warder was the last to go down the hatch, wanting to make certain that he had all of his lookouts accounted for. He told his men that the Japanese destroyer had him fully spotlighted before he finally leaped down the hatch.[15]

Below, a dazed Paul Zimmerman came up spewing blood from his mouth. "One tooth had pushed out and cut up my lip," he said. "The other one was pushed back in and the doc pulled that one out." Doc Loaiza had stitched more than his fair share of cuts aboard *Seawolf*. "He was great with a needle and thread," said Zimmerman. "He was a good doc and could really sew up some guys."

Seawolf made a crash-dive at full speed and rigged for silent running. Warder ordered full right rudder as the enemy destroyer came over at 2054. Surprisingly, the tin can only dropped one distant depth charge. *Seawolf* remained at 200 feet, slowing to 2/3 speed until a second depth charge exploded further away at 2057.

Warder kept the *Wolf* down for more than another hour, but this destroyer was apparently on an important mission. After dropping only two ashcans on a submarine that she had very nearly charged right down upon, she had kept going without causing any damage.

Seawolf remained in the Manila-Subic area. The morning periscope watch on 12 June sighted a seaplane and the tops of a ship. The first target was not able to be closed but the afternoon watch spotted the tops of another ship at 1357 which was northbound. Warder tracked this ship, which was well in toward the beach. By 1425, the enemy ship's range was closed to 7,000 yards and Warder prepared to attack. He had decided that the target ship was a 2,000-ton minelayer of about 300 feet length. She had a gun forward of her foremast with a central bridge island and stack. At 1449, *Seawolf* fired three forward tubes from 1,250 to 1,300 yards.

"I saw no wakes although I watched target for fully 4 minutes after firing," wrote Warder. "Seas were high and torpedo performance was questionable."

The Jinx

At 1453, the minelayer turned toward *Seawolf*. Warder went to 200 feet and rigged for depth charge attack. At 1501, two depth charges exploded fairly close. Returning to periscope depth a half hour later, Warder found the minelayer was 13,000 yards distant on a northeast course. About ten minutes later, the periscope watch sighted smoke on the horizon from another ship and headed for it. The pesky minelayer was soon spotted to be coming back towards *Seawolf*'s direction. To make matters worse, an airplane was spotted at 1653 zooming the periscope. The pilot must not have had any bombs, for he certainly had *Seawolf* located. He did alert the Japanese minelayer which forced *Seawolf* back to 200 feet. At 1723, one depth charge exploded a good distance away. The sound watch tracked screws for the next half hour until they had faded away.

Warder, realizing he had been detected and attacked twice in this area, moved on. "Decided to run to sea until battery charge is completed and to dive off Corregidor channel at dawn, having found no traffic on this expected steamer lane."

The early hours of 13 June 1942 found *Seawolf* about 10 miles off Corregidor. Ironically, she was patrolling for enemy ships outside of a harbor in which just six months ago she had been at anchor on the day the war started. Now, the Japanese had control of Manila and anything sailing out of Manila Bay past Corregidor was to be attacked. Freddie Warder found a Japanese merchant ship at 0747 making an approach for the familiar old Manila shipping lane. Heavy sea conditions forced the submarine to stay down at 120 feet when not making periscope observations.

This maru came in but met a small station ship and made for the north channel, too distant for *Seawolf* to get in a shot. Undeterred, Warder simply moved up toward the point where this ship came from, reasoning that any other traffic following would be using the same channel this morning. Sure enough, watch officer Red Syverson spotted two more ships at 0913 and the *Wolf*'s crew went to battle stations. Two minutes later, the pot sweetened when a third maru could be seen coming along behind the first two.

Warder conferred with Bill Deragon and Jim Mercer, studying the charts of the area. As the approaching convoy took shape, they soon made out a fourth maru astern of the other three. *Seawolf* crossed the column leader's bow at 5,200 yards. Riding herd with them was a single destroyer coming up ahead of them from their starboard quarter. During the course of the next hour, *Seawolf* worked to get into a favorable position on this convoy as it approached Manila.

The destroyer was as close as 1,500 yards by 0957, so Warder took her to 120 feet to use his sonar to coach his boat into firing position for the stern tubes. Radioman Red Eckberg was ordered to shift his listening efforts from the closest maru to the next maru to the right in order to utilize a target giving a more favorable track and better gyro angle. Eckberg, however, was unable to locate the next freighter for five minutes, allowing the prime firing moment to pass.

Seawolf tried to set up on a smaller freighter but Warder could not attain firing range on a suitable target. Fortunately, the targets were plentiful this day. At 1327, Bill Whitman sighted a small ship running parallel to shore near Luzon Point, apparently the same minelayer from before. This ship outdistanced *Seawolf*, but two more ships were sighted at 1652 standing out of the channel to westward.

The *Wolf* started a normal submerged approach course at 6 knots. This contacted developed into a *Amagiri*-class destroyer leading a freighter out to sea. Warder crossed the destroyer's bow and ordered Bill Reiland's after room to prepare their torpedo tubes.

"Bearing—mark!" called Warder from the scope.

"One-eight-seven," called Hank Thomson as he read the relative bearing from the dials. This was converted to 032° T by Mercer and the data was fed below for Bill Deragon to feed into his TDC. The torpedo officer, Red Syverson, was in the after torpedo room to supervise the firing of these torpedoes. The range had come down to 1,400 yards when Warder at 1722 called, "Fire one!"

Thomson repeated the orders to the forward room as Hank Brengelman, on the firing panel, sent each torpedo on its way. Frank Franz counted the seconds until the torpedoes should hit. From below, Eckberg called, "All hot, straight and normal."

Warder kept his face pressed to the periscope and gave a running commentary of what he saw. He watched the torpedo wakes head toward the enemy. One torpedo exploded with a resounding *ka-boom!* heard aboard *Seawolf* 58 seconds after firing. Lt. Syverson and his torpedomen in the after room all clearly reported hearing the explosion. Warder angrily announced, however, that the torpedo passed under the target and exploded on the other side of her, throwing up a tall plume of water. He wrote that the explosion threw "a column of water up between his stacks to a height well above them."

Paul Maley, keeping tabs on the other ships, announced that he could hear a set of screws approaching. Alerted by the faulty torpedoes, the destroyer swung out of formation and bore down on the spot where the submarine had fired. "Down periscope!" said Warder. "Take her deep."

Warder secured the conning tower and rigged for silent running. The power went to manual, and the planesmen struggled with the big wheels. Strong man Otis Dishman muscled the bow planes, wearing only cut-off shorts and sandals. Gunner Bennett handled the stern planes. Diving officer Dick Holden kept a vigil over his planesman, calling out orders when the inclinometer got off. Swede Enslin stood at the hydraulic manifold, looking at the Christmas Tree, while auxiliary boss Red Jenkins monitored the gauges as the air manifold. PCO Ed Stephan retired to the skipper's wardroom to read a magazine.[16]

Seawolf held her course and went down deep to 200 feet. The destroyer crossed astern and at 1729 dropped the first of eight depth charges. "Some came fairly close," noted Warder. The compensating water line began leaking again after these explosions but no serious threat was posed. The *Wolf* remained down for half an hour until the destroyer's screws faded away and the pinging grew faint. Warder eased back to periscope depth at 1843 and monitored the destroyer as he moved on at high speed. The only evidence of the freighter was a large smoke cloud. Four more explosions were heard an hour later, possibly the destroyer working over a phantom submarine.

The Japanese ships had escaped, so *Seawolf* surfaced that night 21 miles off Corregidor to charge her batteries. She stayed in the vicinity of Manila's busy shipping lane on 14 June and found shipping to be plentiful, sorting through various merchant ships while trying to attain a good firing setup. Lt. Cdr. Warder had made approaches on a small freighter and two escort ships by early evening. The escorts in turn were leading a large freighter out of Manila Bay. In the rain and gathering darkness, the long distance and an untimely zig by the big maru foiled *Seawolf* yet again.

Warder finally secured from battle stations after a long, thoroughly frustrating day off Manila. Enemy ships of all imaginable sizes had been spotted but not one had come into firing range. Like a good fighter, though, Warder would wait out his prey—avoiding being spotted until he could hopefully pick off a good target the next day.

Better fortune smiled on the *Wolf* on 15 June. Coming up to periscope depth at 0612, Jim Mercer spotted a Japanese destroyer a mere 480 yards away. Sound had reported screws, but Lt. Cdr. Warder was "expecting a sub-chaser and not a DD." *Seawolf* went to battle stations, but the tin can was too close for the torpedoes to arm. The destroyer passed only 300 yards away, oblivious to the U.S. submarine taking stock of him.

The destroyer moved off at high speed while Warder waited to see what might be following from Corregidor. Within a half hour, a medium-size maru stood out of the channel making 10 knots. Warder called out the details to his recognition team: estimated length 375 feet, two stick masts, one stack, well decks forward and aft, guns fore and aft.

"Right full rudder," Warder called to helmsman Gervais.

Seawolf fired three forward tubes from 1700 yards at 0700. As the firing was completed, neutral buoyancy was lost and *Seawolf* broached to 50 feet from 63 feet. Warder immediately lowered the periscope as diving officer Dick Holden flooded the negative tanks.

Seawolf plunged rapidly until Holden caught her at 120 feet. As the ship dropped below periscope depth, there was one violent explosion at 0702.40, just over one minute after firing the third and final torpedo. Casler, Mercer, Deragon, Stephan, and several others insisted they heard a second explosion in the wounded enemy ship. This time her torpedoes were good, as *Seawolf* was later credited with sinking a 1,206-ton converted gunboat, *Nampo Maru*, on 15 June.

Two depth charges exploded as Warder evaded to the westward. The enemy's ships screws were clearly heard to stop after the first big explosion. Smaller, rumbling explosions at the site of this ship could be heard over sonar until 0728. "Screws, pingers and depth charges continued until 0920," wrote the skipper. None were "too uncomfortable" at deep submergence. Frank Franz and Norm Kisver counted 12 depth charges, although at least one small explosion was thought to be within the target ship as she was torn apart.

Seawolf eased up to periscope depth at 1009. Warder completed one sweep in low power, finding nothing in sight and the water's surface to be quite rough.

Had just found the top of one ship in line with Corregidor, at least 8 miles away, when we received the first of a string of 6 depth bombs close aboard. They could only have come from a plane. They were not too heavy but he didn't miss us far so I returned to 200 feet.

The plane called in his friends, for high speed screws were picked up on sonar a short time later. This destroyer came up *Seawolf*'s starboard side from above. When he got dangerously close, Warder turned toward him and passed under his stern. On the rapid descent after the torpedo attack, *Seawolf* had lost one gasket from the negative flood tank. Dick Holden did not succeed in adjusting trim until 1503, eight hours after firing.

The auxiliary pump had quit on the day before and was disabled for Red Jenkins' gang to overhaul. The drain pump only delivered nine gallons a minute against the inflow of high-pressure water at 150 feet and this scarcely kept up with the flow of bilge water. Warder's sound watch heard pingers all afternoon on various bearings but his scope sweeps revealed no ships.

At 1953, a half hour before twilight, Bill Whitman was sweeping the horizon with the periscope while at 63 feet. *Seawolf* was suddenly rocked by another four aircraft bombs close aboard from an unseen aircraft. Clearly, she had stirred up the enemy at Manila this day. The *Wolf* went back to 200 feet. By 2023, "pingers and screws were getting quite numerous," wrote Warder. Six depth charges exploded at safe distances over the next two and a half hours.

The most persistent of the pingers stayed with *Seawolf* doggedly. Warder managed to keep him astern as he retired to the north and west. With the air in the boat getting foul, he finally brought her to the surface at 0105 on 16 June. Spotting a ship in the distance, he decided it was worth surfacing and running for it. The lookouts manned the shears and *Seawolf* opened up her speed, leaving the enemy patrol vessel far behind.

The electricians charged the badly depleted batteries prior to dawn. *Seawolf* dived at 0454, some 37 miles off Corregidor to make another go at the plentiful shipping. In light of the aerial bombing the previous day, Warder decided to be cautious, running at 120 feet except for half-hourly periscope observations. At 1101, the watch spotted what was thought to be a destroyer and Warder sounded battle stations. Within ten minutes, Warder realized that his target was only a subchaser "and decided I'd better get out of here."

Warder turned left with full rudder and 2/3 speed to put the subchaser astern but the little Japanese man-of-war was quickly on the *Wolf*'s tail. By 1121, Warder was beginning to think that he had foiled his foe. The subchaser stopped pinging a moment later and Warder decided it was time to take a periscope observation. "Just as I was about to run up the periscope for a look, he started dropping the worst salvo received this patrol," the skipper logged.

Dust and cork rained down from everywhere. In the sound shack, Paul Maley and Mel Eckberg were knocked from their chairs. The effect of the depth charge slammed *Seawolf* down deeper even as Dick Holden was working to take her deep. "It was the closest call the *Wolf* had ever had," wrote Eckberg. "Again and again the Jap dropped his charges. Each one rocked the *Wolf*."[17]

A Silex coffeepot was smashed, a few pipes sprung leaks and one of the emergency lights in the engine room—which was mounted on rubber—was also smashed. A hand was knocked off a clock and the starboard antenna's lead-in was parted.

At 1130, another depth charge exploded fairly close. The sound of the subchaser's screws did not fade away until 1415. For the next hour, Warder had sound track the enemy ship until he felt secure that the escort was not laying back waiting to pounce. *Seawolf* surfaced well after dusk and recharged.

The next day, 17 June, in Kelly's Pool Hall, the men listened to Tokyo Rose while off watch. In between playing Rudy Vallee and Benny Goodman songs, Rose said that several large American submarines had been sunk, a boast which brought laughter.[18]

The long patrol soon began to take its toll as the food stores became thin. Bill Mallory and his cooks were soon down to preparing mainly dehydrated potatoes, rice and bread. There was no canned meat and no butter. Most of the meat taken on in Australia was mutton and Australian hare, both of which were too gamey for most men. What meat remained at this point was Spam. Only the freshly baked bread that could still be made was worthwhile.[19]

Off duty, Warder would retire to his stateroom to read books and magazines or gaze at the picture of his wife and four children on his desk. At night, he often joined Deragon, Mercer, Syverson, and Holden in a game of hearts in the wardroom.[20]

Seawolf swept coves along the west coast of Panay and in the northern Sulu Sea on 18 June. While running submerged just before dusk, the conning tower OOD, Red Syverson, spotted a patch of smoke at 1905 and called the skipper. Warder soon could make out two old coal-burning freighters. He called battle stations and the approach party set to work. It was a race between the *Wolf* and darkness to reach these ships before their smoke was lost in the dark.

Both ships were found to be small and slow. At 2008, *Seawolf* steadied on bearing 270°T and prepared to fire. The moon had not yet appeared, making visibility of the targets extremely difficult. The ships were more easily tracked by sound in the approaching rain. Warder finally "decided to throw 2 fish" at one of the marus, firing at 2020. Warder decided she was a 2,500-ton freighter.

"Fire one!" Torpedoes were launched from forward and the seconds were counted down as they sped toward their mark. The fish were fired from 120-foot depth by night sound attack, utilizing a longitudinal spread of two torpedoes from the forward tubes.

Errors plagued these fish, however. TDC operator Red Syverson forgot to cut in the gyro setters, thus sending out the torpedoes on zero gyro angles. "This point would ordinarily be checked by Assistant Approach Officer [Deragon] or Firing Key Operator [Hank Brengelman] in Conning Tower if firing were done from that station," wrote Warder.

As it was, no check was made this time. Syverson was devoting all his attention to watching the sound bearing input and white light and matching the generated and observed bearing dials. In Syverson's defense, Warder wrote that "I had intended to fire by periscope until I lost the target in a rain squall, so I didn't give the officer operating the computer much of a chance to get 'squared away.'"

This time, there were no hits. The target ship was noted to speed up just over a minute after *Seawolf*'s second torpedo was fired. Warder tailed the two freighters for 20 minutes and then decided to chase after them to get a better look at them.

The *Wolf* surfaced in a rain squall and went ahead on two engines. There was no moon but instead heavy rain and violent lightning to the north. Lookouts could not see 500 yards in the direction of the ships. "When lightning struck it was like daylight but lookouts would be blinded until next black," wrote Warder. He decided to let these little ships go due to the poor weather, the ships' size, the scarcity of his torpedoes, and the rough navigational hazards in the area.

The ships escaped unhit and that night the crew was back to talking about the jinx again. *Seawolf* had fired 19 of her 20 torpedoes and most had been misses or duds. During the next two days, the *Wolf* crossed the Sulu Sea and entered Sibutu Passage. She cleared Sibutu on 21 June and entered the Celebes Sea, hunting along the way. The periscope watch picked up smoke from a ship on 23 June but *Seawolf* was unable to close the distance submerged. During the night, the ship went to battle stations when a ship was sighted, but it proved to be a two-masted sailing schooner unworthy of attack.

As the time approached to make the morning dive on 24 June, OOD Dick Holden sighted a high intensity light shining close aboard on *Seawolf*'s starboard beam. He immediately dived the ship, but periscope observations could not find what must have been a small sailing vessel displaying a light. The *Wolf* transited Macassar Strait and reconnoitered the southern approaches to Makassar City on 25 June. Passing through the Flores Sea en route to Lombok Strait the next day, the only enemy shipping encountered was a small mainsail and jib sloop. *Seawolf* cleared the Lombok–Badung Straits and set her course for Fremantle during the final days of June.

As *Seawolf* headed for port, men took the chance to come topside and get their first taste of sunshine in more than a month. Topside, the ship showed the beating this fifth war patrol had put on her. Her paint was chipped and a large dent was visible aft on the port side. Lieutenant Deragon held field day as the ship approached port and soon the inside of the ship was spic and span.

Outside Fremantle on 2 July, *Seawolf* was met at 0442 by the *USS Isabel* and fell in astern of her to enter the searched channel. She was then brought alongside a tanker and was refueled. Mail was brought aboard and the crew was lost in their own worlds for the next few minutes. Jim Mercer moored the *Wolf* to the starboard side of the tender *Otus* at 0729 and her frustrating fifth patrol came to an end.

The Jinx

Some of the more superstitious enlisted men, feeling that the PCO had brought them nothing but bad luck, were happy to see Lt. Cdr. Ed Stephan depart their ship. Stephan went on to command the submarine *Grayback*, where his luck was not much better. In two patrols, he fired 44 torpedoes for only one sinking that was credited postwar.[21]

Freddie Warder's meeting with Commander Submarines, Southwest Pacific—Rear Admiral Charles Lockwood—was not one that he relished. Lockwood had just recently taken over the position in Fremantle from John Wilkes, who was rotated back to the States. In the endorsement to Lt. Cdr. Warder's fifth patrol report, Charlie Lockwood did not give glowing reports. "In spite of numerous contacts, *Seawolf* patrol reports were disappointing," he wrote. Lockwood suggested that *Seawolf* might have had better results with the numerous targets off Manila Bay had she maintained a more conservative distance from the shipping lane and not drawn the frequent depth charge attacks that she had incurred. Lockwood refused to credit Warder with a sinking, but postwar Japanese records gave Warder one small freighter of 1,200 tons.

This time, there were no Navy Crosses and Silver Stars sprinkled upon the officers and crew. It was not until February 1946 that Warder received a belated Navy Letter of Commendation in acknowledgement of this patrol for sinking and damaging 9,500 tons of enemy shipping. Considering the Mark 14s being loaded in boats during 1942, sinking or damaging any enemy shipping was quite an accomplishment for Warder and his fellow skippers.

Warder took the criticism to heart. He had first met Lockwood in Panama when he was taking *Seawolf* on her shakedown cruise. The admiral had come aboard for a couple of hours in Panama and looked over *Seawolf* stem to stern. Warder later stated that he believed "Lockwood was the greatest submariner we ever had."[22]

In support of this, Admiral Lockwood had gotten to work on the torpedo problem shortly after he arrived at Fremantle. When Lt. Cdr. Jim Coe returned his *Skipjack* from patrol with complaints of duds and deep-running torpedoes, the new ComSubs Southwest Pacific listened. On 20 June 1942—six months into the war—Charlie Lockwood conducted the first true test of the faulty torpedoes. Using a fishing net to verify running depth, Lockwood had *Skipjack* fire a Mark 14 torpedo set at 10 feet. When the net was hauled up, there was visual evidence that the torpedo had punched through at 25 feet—15 feet deeper than the torpedomen had set it. Two more tests in June also showed that the Mark 14 ran far deeper than set.[23]

Submarine skippers like Freddie Warder would forever admire "Uncle Charlie" Lockwood for taking on the Navy brass to question the faulty Mark 14 weapons. The Bureau of Ordnance found that these test were inconclusive, which infuriated Admiral Lockwood and pushed him to conduct further tests off Albany, Australia, during July while *Seawolf* was in port following her fifth run. Lockwood's next round of tests would again show that the torpedoes ran an average of 11 feet deeper than their settings, but more weeks would pass before an official acknowledgment of the poor torpedo was passed.

The morning after reaching Australia, 3 July, the *Seawolf* crew was called together for their skipper to address. As Eddie Souza formed them up topside, some silently worried that skipper Warder would be shipping out to other duties. Standing before his crew, he gave his men congratulations on their latest job well done. Warder advised his crew that although their luck had been bad on this past patrol, his men could "hold up your heads with anyone. I hope to be with you when we set out to sea again."[24]

Admiral Lockwood, of course, knew that Freddie Warder was a fearless skipper and that *Seawolf* was a hot boat. Her fourth patrol had been historic. There was never a question of change, but no doubt the words that Lockwood penned in his patrol report endorsement would certainly motivate Lt. Cdr. Warder on his next foray with the *Wolf*.

USS *Seawolf* Fifth Patrol Summary

Departure From:	Fremantle, Australia
Patrol Area:	Philippines
Time Period:	12 May—2 July 1942
Total Days on Patrol:	51
Number of Men Aboard:	71: 63 enlisted and 8 officers
Number of Torpedoes Fired:	219
Ships Claimed as Sunk:	0/0 tons
JANAC Postwar Credit:	1/1,200 tons
Return To:	Fremantle, Australia

7

"A Marvelous Spectacle"

Refit and Sixth Patrol　　　　　　　　*2 July—15 September 1942*

The *Seawolf* crew wasted no time in returning to their favorite bars in Fremantle. Having spent their previous R&R in this port, they knew their way around. Most men went ashore on 3 July for three weeks' leave as the *Wolf* underwent a lengthy refit through 24 July.

"Fremantle and Perth are twin cities," explained Ed Milas. "We stayed at the King Edwards Hotel. There were no locks and no locking of the doors there." Milas and his buddies found that they did not have the taste for the warm beer that was served in Australia. "The way we got over that was to use fire extinguishers to cool them off," he related. "You had to learn how to do it properly because if you go too fast you break the bottles. It worked good on cans, too."

To get the beer, the men had to journey into town. "You'd have to hire a taxi, one of those charcoal burners," said Milas. "There was no gasoline, so you had to take one of these charcoal burners into town to buy things."

While in Fremantle, gunner's mate Henry "Big Swede" Hanson managed to get himself in trouble with the skipper again. When his section was due back aboard *Seawolf* on 15 July, he returned more than 12 hours late. He was in quite a state when he did return.

"I was off duty below decks with some of the guys," recalled torpedoman Jim Cashero. "All of a sudden, we hear one of our machine guns firing. We raced up on deck and found Swede Hanson."

Hanson, appearing to have perhaps had one too many, proudly announced to his shipmates, "I've stood so many watches on them damn machine guns. I just wanted to see if they'd really fire!"

His brother, Bob Hanson, recalled, "Henry was only a third class gunner's mate for 15 days. When we were in Australia, he came back drunk and tested the .50-caliber out alongside the tender in the harbor." The .50-caliber did fire and Lt. Cdr. Warder was plenty fired up the next morning. Swede Hanson's days as gunner's mate were over. He was busted down from GM3c to S1c and only remained aboard because of how highly thought of his services were on patrol.

By the time *Seawolf* sailed, she had undergone more personnel changes than before any previous patrol. Nine new men were received. Eight enlisted men were transferred and a ninth man, CQM Jim "Jughead" Casler, moved into the wardroom. Freddie Warder believed in promoting from within and wartime afforded him the opportunity to pull strings to get Casler promoted to warrant boatswain and then quickly to ensign. In addition, Doug Syverson had been promoted from Lt.(jg) to full lieutenant. The watch schedule would now benefit from the seventh officer—a far cry from the five who had covered all the watches during the first patrol.

Among the transferees were chiefs Edmund Capece and Fred Zirkel. The biggest change was in the *Wolf*'s radio gang, as both Chief Radioman Red Eckberg and RM2c Joe Ferguson were transferred to the Hollywood Hospital in Perth for treatment. Eckberg had begun to suffer from terrible throat pains while enjoying a beer in Chief Souza's room during the first night ashore. When the pain increased the next day, he visited the doctor on the sub tender and was diagnosed with tonsillitis. Since his condition was not serious, he was put on a waiting list and kept aboard the tender until he could be operated on. Before he could complete his surgery and be discharged, *Seawolf* would sail off on her next war patrol.[1]

In replacement for Eckberg and Ferguson, yeoman John Sullivan picked up two new radiomen: RM2c Ed Hinson from the sub tender *Otus* and RM2c Charles Woodard from the squadron flagship *Sargo*. Woodard was a familiar face aboard *Seawolf*, as he was one of the Corregidor evacuees who had ridden the *Wolf* back to Australia on the second patrol. "Paul Maley and I had known each other before the war, and I had met Eckberg when I was aboard as a passenger," Woodard recalled. "I had been on the beach at Australia for a while, so I went aboard the *Seawolf* as the low man on the totem pole in the radio gang."

Returning from patrol. Lookout seated to right is EM2c Ed Milas. Standing below lookout perch are CEM John Bilkey (left) and Lieutenant Dick Holden. *Courtesy of Edward Milas.*

Paul Maley would serve as senior radioman aboard the *Wolf* for her sixth patrol, assisted on sonar at battle stations by Hinson. Woodard would assist with sound watches but would also oversee the new SD radar that was installed aboard *Seawolf* while she was in overhaul in Fremantle. The SD air search radar was new technology for U.S. submarines, intended to help them detect incoming aircraft. The problem with *Seawolf*'s early set was that it did not come with the more sophisticated PPI plotting screen that later models would have. Woodard and his superiors, Red Syverson and Bill Whitman, would find in the coming weeks that lookouts often could pick up an aircraft visually at a longer distance than could the early SD.

Following her refit, *Seawolf* exercised in preparation to return to war patrol. Lt. Cdr. Warder considered this overhaul "an extensive one for three weeks." When his boat came out, Rear Admiral Lockwood came aboard to check her out. "I took him for a ride from Perth down to Albany, where we had to go to complete part of our refit," Warder recalled.[2]

A large quantity of piping and fittings had been replaced. Warder decided to have a little fun with the admiral aboard to test his boat's depth worthiness. "I told my diving officer, Dick Holden, that we would crash the boat," he related. "We wouldn't bother with any running dive or anything. The order would be 200 feet."

Just as planned, Warder called for a crash dive. The lookouts cleared the bridge and Admiral Lockwood watched admiringly as *Seawolf*'s officers and men scrambled down from the bridge and secured the hatch in seconds flat. Right past the end of Perth, the *Wolf* took a sharp down angle and plunged for 200 feet, as if this sort of emergency dive happened every morning.

"I thought Lockwood might object to it," admitted Warder, "but he didn't. We gave him a good ride."

As the *Wolf* made final preparations to get underway, commissary officer Jim Mercer worked with his cooks to pack aboard all the best provisions. For the sub's upcoming sixth patrol, Doug Syverson's torpedo gang loaded their tubes and torpedo rooms with a truly mixed bag of warheads they found available in Australia. *Seawolf* had one remaining Mark 14 forward from the previous patrol. The torpedomen loaded four British-manufactured Mark 15s with Torpex warheads and 15 battery-powered, U.S.-manufactured electric Mark 16s into the forward and aft rooms. The new electrics ran slower than the old Mark 14 steam torpedoes and had a shorter range, but left no telltale smoke trail from their launching point.

Eddie Souza mustered the crew on deck at 0800 as Sully prepared the final sailing list. Lt.(jg) Mercer had the deck as *Seawolf* got underway at 1036 on 25 July from her berth on the starboard side of the tender *Otus*. From Fremantle, she headed for Lombok Straits, operating under Rear Admiral Lockwood's secret operation order No. 41-42. It called for an offensive against enemy combatant, supply, and transport ships in the vicinity of Sibutu Pass and the western Celebes Sea with a stop of at least one full day off the tip of Mangkalihat on Borneo's coast and another one off Tarakan.

At 1417, Dick Holden dived her to 250 feet for tests, training, and drills. In the afternoon, *Seawolf*'s new radar had its first test when a friendly plane was sighted on the port beam by the lookouts. The radar was started and in about three minutes Charles Woodard had picked up the plane at 4.5 miles. The plane proceeded on an opposite and parallel course and disappeared from the radar screen about five minutes before disappearing from binoculars.

The next morning, the new radar was even more disappointing while in the bombing restriction safety zone. Lookouts sighted an aircraft at 4 miles which paralleled *Seawolf*'s course and then crossed ahead of her. The radar never picked up this plane at all.

Warder drilled his crew en route to Lombok, including two battle surface drills with Gunner Bennett's 3-inch deck crew. On 28 July, they fired three rounds from the big gun and a few rounds of 50-caliber. Entering Lombok Strait on 30 July, *Seawolf* encountered violent rip tides which delayed her submerged passage until evening. After clearing the strait that night, *Seawolf*'s lookouts spotted an object ahead believed to be either a small sail or a submarine. Warder sent his crew to battle stations and headed closer. From the intelligence he had, Warder felt that this was either *Seadragon*, *Searaven*, or *Sailfish*, all in his area. He opened out at high speed to clear the area.

From Lombok, *Seawolf* proceeded to Tanjung Mangkalihat, the far eastern peninsula of Borneo's eastern coast. She passed many sailing vintas on 2 August while observing the area with periscope observations. At 1756, Lt.(jg) Bill Whitman sighted a tanker with a large starboard angle on the bow. Whitman ordered his helmsman to come right with full rudder at full speed and called for the skipper.

By 1811, *Seawolf* had closed the range to 8,500 yards in the setting sun as her attack team plotted the tanker's course. The target was seen to be a 7,000-ton tanker with goal post masts in the middle of about 400 foot length. As the range reached 5,000, Warder's team prepared to fire as best they could with the poor visibility.

In the conning tower, Warder, Deragon, Casler, Mercer, Gervais, Thomson, and Brengelman were at their usual posts. Casler and Mercer had plotted the target ship to be making 7.5 knots, so Warder ordered Chief Gibson's forward room to set their fish on low speed at 24-foot depths. *Seawolf* opened fire at 1837 and shot all four bow tubes—one Mark 15 and three Mark 16 electrics—from a range that was estimated to be as long as 5,100 yards.

At 1840.30, while at 63 feet, there was "an explosion of sufficient violence to kick loose conning tower paint and break [the] diving station barometer." This indicated a torpedo run of only 3,050 yards for the No. 1 torpedo or as little as 2,190 yards for the No. 4 torpedo. "Firing had commenced 27 minutes after sunset and determination of range and target angle had been [a] matter of guess work during entire approach," noted Warder.

Warder, apprehensive of a subchaser that *Seadragon* had warned him about in this area, ordered 200 feet. In further consideration, Warder believed this explosion was from a torpedo. No one had heard the click of a depth charge's detonator, which was almost always heard when an explosion this close was felt. Also, there was no swish of water through *Seawolf*'s superstructure that usually accompanied a depth charge explosion of this violence.

Dick Holden's planesmen caught the ship at 75 feet as the force of the blast shoved her down. Due to the darkness he later admitted: "I didn't have more than the foggiest notion of actual target range and that it could be 3,000 as well as 5,000 yards." *Seawolf* closed the target ship's track but nothing was sighted in the periscope other than the dark land background of Celebes. He believed he had obtained a solid hit on this ship.

One distant explosion was heard at 1849. As *Seawolf* surfaced 13 minutes later, Red Syverson, Bill Whitman, and their lookouts found Borneo's Cape William in sight at 15 miles. The diesels roared to life and *Seawolf* ran ahead at 14 knots in search of her target ship. Warder kept a cautious distance from the beach and decided to run down south, then patrol east and west after the moon came up.

At 2142, Warder had Paul Maley send Admiral Lockwood the following message to alert *Swordfish* (coming north two days astern of *Seawolf*): "UNESCORTED LARGE SOUTHBOUND TANKER GOT ONE HIT FROM *SEAWOLF* ABOUT EIGHT MILES OFF CAPE WILLIAM 1840 H XX STILL LOOKING FOR HIM."

The moon rose at 2302 and lookouts sighted a southbound ship moments later. *Seawolf* closed until a range of 5,000 yards was attained but Warder decided not to attempt a surface attack on account of the impossibility of getting in close undetected. "I decided to open out to westward and go up ahead of target, keeping him in sight and attacking at dawn with periscope," wrote Warder.

Between Cape William and Makassar, the *Wolf* raced ahead of her target ship during the early morning hours of 3 August. Once she was ahead of her target by about 7 miles, *Seawolf* slowed to the target ship's speed of 9.7 knots and took the same course. Warder dived his boat at 0450 and sent his crew to battle stations. By 0528, the target vessel was visible through the periscope at 5,000 yards. Ten minutes later, Warder could see a Japanese destroyer and a small freighter astern of the tanker about 1,500 yards. The DD maneuvered on various courses and speeds but never closed *Seawolf* less than 6,000 yards. The flat and glassy seas made for poor periscope conditions.

Warder's team decided their tanker target was a whale factory ship and the destroyer was a *Shinonome* type, both making 7.5 knots. From a distance of 2,800 yards, Warder began firing three of his forward tubes, using the white light method via periscope to spread his fish. *Seawolf* opened fire at 0551, sending out her only Mark 14 and two Mark 16s, all set for 14 feet. Sound operators Paul Maley and Ed Hinson reported up from the radio shack that all torpedoes running normal at first, but one quickly hooked around.

"I could see wakes plainly in glassy sea," noted Warder. "I saw one torpedo hook to left and I almost went deep to avoid a circular but it straightened out." Maley and torpedo officer Doug Syverson both tracked one torpedo to a perfect interception with the target's screws. Warder watched the steam torpedo's smoke make a perfect run toward her stacks and the torpedo to its right intersect the target's bow. The unreliable fish failed to explode, however.

At 0557, five minutes after the third torpedo was fired, there was an explosion. There were two more explosions within 20 seconds. Warder was watching the destroyer through the periscope and saw no splashes from depth charges, so these were likely end-of-run explosions. He kept his eye on the tin can in case it made a run for him and had his forward room prepare their fourth torpedo. The tanker ran up a red flag up her forward halyard and increased speed, so *Seawolf* dispatched her fourth tube at 0559 from 2,700 yards.

Warder could not find this torpedo's wake in the scope, although Maley reported it running hot, straight, and normal. "I remained at periscope depth, hoping to get a reload completed in time to take a shot at DD." Ed Hinson, keeping his sonar tuned to the escort, reported at 0601 that the destroyer had increased to 120 rpms and was changing course. Warder ran up the scope and observed the DD "turning toward us at good speed."

He ordered Gervais to go left full rudder at 6 knots to present the *Wolf*'s stern to the attacker. At 0604, there was one distant explosion believed to be another end-of-run. Warder ordered 200 feet and rigged for depth charges. Hinson and Maley tracked the destroyer over the next ten minutes, but no ashcans fell. Like a cat with a mouse, the Japanese DD skipper worked the seas in search of his prey. Finally, at 0618, he dropped his first depth charge at *Seawolf*.

The first blast rocked *Seawolf* off her starboard quarter. A minute later, a second charge erupted astern. Two minutes later a third blast ruffled the submarine. The crew stood silent as the destroyer's screws went up the port quarter. Warder maneuvered to put his enemy astern, but the DD soon went over *Seawolf*'s starboard quarter.

Ed Milas was on duty in the maneuvering room, working with EM3c Jesse Jobe to respond to any changes in speed for *Seawolf*. "On watch, I'd have two diesels and the other guy had two diesels to work the generators. Submerged, we ran off the batteries and answered a series of commands to get the turn counts on the screws we needed." At silent running, he found that the temperature was quite extreme. "It was not like you see in the movies," Milas said. "All we had on was sandals and a pair of shorts. We were really sweating it out."

During the next hour, *Seawolf* managed to slip away from the circling destroyer without any further depth charging. Warder eased up to periscope depth at 0800 and found no shipping in sight on the glassy seas. He then proceeded on toward his assigned station at Mangkalihat, avoiding various small craft throughout the day. Warder searched along the coastlines for his damaged tanker to see if she had beached herself, but found nothing.

The periscope watch sighted white smoke at 1020 on 5 August. *Seawolf* raced at full speed to overtake the contact, sending the crew to battle stations. After chasing the contact for more than an hour, however, the sub was unable to narrow the distance any more than 10,000 yards. Warder then completed his assigned reconnaissance off Mangkalihat, on Borneo's easternmost tip.

At 0432 on 7 August, lookouts sighted another ship and OOD Red Syverson headed for it. "Looks like submarine," Warder noted in his log at 0450. "Sounds like submarine." Deciding this might be *Spearfish* heading southbound, Warder turned at 15 knots to clear the area. After scouring Mangkalihat during the day, *Seawolf* moved northward up Borneo's coast toward the little island of Tarakan, located just off the coastline of Borneo.

Seawolf was off Tarakan late on 8 August and eased in to investigate during the morning hours. Doug Syverson picked up a ship at 0422 on 9 August while approaching the Tarakan light. The ship was on *Seawolf*'s port quarter, so Syverson turned to head for it. Visibility was poor, but Warder called for battle stations as two ships were made out. *Seawolf* flooded down to reduce her topside visibility as she approached the coast and the enemy ships. Warder believed the lead ship to be a destroyer. "Intended to give him one and then swing on to ship astern with three," he wrote.

The lookouts reported that the lead ship began signalling with red yard arm blinkers. Warder thought at first he might be signaling to his ship astern but quickly realized that he was challenging *Seawolf*. "Before I could catch him, I realized he was filling up my binocular field and was probably inside firing run," wrote Warder. The escort, believed to be a torpedo boat, turned and raced for *Seawolf*.

Warder submerged at 0437 and continued to monitor the situation. He was so close that he was actually inside the escort ship's turning circle. He planned to cross ahead of the second ship. The second ship crossed ahead of *Seawolf*, however, and Warder realized his only hope was to head for Tarakan at high speed in hopes that the tanker would wait for an escort ship or would mill around while the torpedo boat searched for *Seawolf*.

At 0452, one depth charge from the A/S vessel "missed badly," and Warder ordered 150 feet to avoid the torpedo boat. The next depth charge was further to the southeast. Three minutes later, a pair of depth charges exploded a good distance off. Sound tracked the torpedo boat pinging astern but he did not make another run.

Seawolf eased back to periscope depth at 0512 and found that the tanker—estimated to be 4,000 tons—had headed toward her position, now only 4,000 yards away. "The Gods have smiled," wrote Warder. "The A/S vessel not in sight." He took a convergent course to that of the tanker and worked across her bow at high speed, slowing to bring his stern tubes into action.

Seawolf fired tubes 5, 6, and 7 aft at 0545, setting the depth at two feet on high speed. The torpedo run was 1,300 yards. Paul Maley reported all fish running hot, straight, and normal again. Warder watched through the periscope but could not find the smoke of the torpedoes due to the smoke screen they were laying out. At 0547, Maley reported a "plop" of a dud hit. No explosion was heard or felt aboard ship. "Tanker remained very healthy, turning away and firing at us with his stern gun. I watched four salvos," wrote Warder. "From flash, could judge caliber to be 3" fire at about 5 second intervals."

Freddie Warder was furious about the torpedo performance his ship was enduring. He blamed the poor results on a complete lack of the sub skippers to test fire torpedoes during the months leading up to the war. "We should have been permitted to fire torpedoes at live targets to see if the goddamn things would explode," he later related. "This business cost us a hell of a lot of lives and a hell of a lot of unnecessary hazard. So, I'm shooting these things and finding out what's wrong with them!"³

The tanker then headed for Tarakan at good speed. Sound reported an end of run explosion at 0551. The *Tidori* torpedo boat was sighted again at 0556, smoking heavily and working his searchlight on the tanker. Warder remained at periscope depth "with idea of not picking a fight with this ship but in firing at him in case he attacked us at favorable range." Four small explosions were heard at 0607, but no splashes could be seen. Two minutes later, the torpedo boat dropped five more depth charges. Warder was able to spot one of these splashes astern of the angry little escort ship.

Warder exercised his fire control party on this torpedo boat until 0800, "at which time he stood in towards Tarakan and we knocked off for breakfast." The escort's pinging was audible until 0955 and his smoke remained in sight until 1114. Warder then reversed course towards Tarakan in order to have favorable attack position by dusk.

The afternoon periscope watch picked up another torpedo boat, possibly the same one from the morning. This boat was moving fast and pinging, but did not come within torpedo range. *Seawolf* surfaced at 1911 and stood out towards Makassar to send out her radio report of the day's attack. "From conversation with Commander [Sunshine] Murray I judged that there had been no actual knowledge of enemy use of Tarakan and very little knowledge of activity in Makassar, of which we had observed considerable," wrote Warder. He therefore had his radiomen send Murray and his staff an intelligence report of his day's activity with the tanker and torpedo boat.

Seawolf remained off Tarakan on 10 August, playing cat and mouse with another of the persistent torpedo boats during the morning. By afternoon, the *Tidori* boat had departed south, leaving Warder to deduce that his tanker charge had departed also. Having spent enough time, he headed *Seawolf* toward Sibutu Passage.

By 13 August, *Seawolf* had entered the northern approaches to Sibutu and she moved into the western Celebes Sea during the early hours of 14 August. The boat was almost immediately rewarded with a contact. Lt.(jg) Jim Mercer sighted smoke on the horizon through the periscope at 1008 and headed for it. The recognition team soon decided this was a passenger-cargo ship, which they pegged as the 3,113-ton ex-British *SS Wenchow*. She was later positively identified as the 3,113-ton passenger-cargo ship *Hachigen Maru*.

The *Wolf* went to battle stations at 1018 as Warder made on a collision course straight for his prey. Her speed checked at 8 to 10 knots and her course varied. On the TDC, Syverson and fire controlman Lefty Leffingwell soon had everything in check. "Doug Syverson was a wonderful officer and the most outstanding TDC operator I ever worked with," recalled Leffingwell. *Seawolf*'s older TDC was located in the control room, where the data was fed to the firing team in the conning tower.

Seawolf crossed *Hachigen Maru*'s bow and commenced firing at 1128. Tubes 8 and 5 were fired from 950 to 1,000 yards range. The torpedoes—one Mark 15 and one Mark 16— were set at 6 feet on high speed. Sound reported both running hot, straight, and normal, although Warder could not see their wakes.

Sixty-four seconds after firing the first torpedo, *Seawolf* obtained one hit which "blew hell out of shaft alleys, No. 3 hold, steering engine room, after steering station, mainmast rigging, and after gun platform. I am not sure a gun was mounted. If there was, it went overboard," Warder logged. *Hachigen*'s deck broke through midway between her mainmast and counter stern.

Warder maneuvered to within several hundred yards of her submerged and studied her through the periscope. By 1205, Warder could see that the crew had completely abandoned this ship. Its flooding, however, seemed to check itself after the crew abandoned.

Warder decided to finish off his cripple, firing another torpedo at 1228 from tube 7 at a distance of 1,500 yards. The Mark 16 electric torpedo was set at 6 feet but Warder watched it "run under target at after corner of passenger quarters, about 50 feet aft of point of aim." Clearly the torpedoes were running too deep, because Warder estimated that *Hachigen Maru* had at least a 17 foot draft, based on the draft marks visible in Roman numerals through the periscope.

Upset, Warder decided to go ahead and have Bill Mallory, Gus Wright, and Henry Wilton feed his crew while watching the ship before continuing his attack. He also wanted his torpedomen to thoroughly inspect this last torpedo. He would then fire another torpedo at zero feet "if the target still floated." Warder found that the crew of about 40 men were evenly divided in two whaleboats, lying to about 1,000 yards off their stricken ship. "They row poorly. There are two wounded in one boat. They signal smartly by semaphore," he logged. "Uniforms are nondescript dungarees, khaki, whites, sweaters. Some caps, some hats."

During this time, Warder gave his crew the unique chance to take a look at one of their torpedo victims. Auxiliaryman Paul Zimmerman was one of those who jumped at the chance.

> My station was right next to the conning tower on the air manifold. Captain Warder asked if anybody wanted to take a look. There were no other targets and no other ships in sight at the time. I took a quick look. That was the first time and the only time I ever looked through the periscope. I could see debris and the survivors in the water. I never wanted to take another look.

At 1427, the crews of the little whaleboats raised their masts to attempt to sail for land. Warder, having fed his crew, now moved in to fire another torpedo at zero depth. He observed the target ship "to be down one mere foot by stern with row of ports in steering engine room now submerged." He could see a ragged hole of about 18 inch diameter in her side plating on the port side below where the deck was buckled, about one foot out of the water.

At 1512, Warder instead decided to use this badly damaged ship as a decoy during the afternoon and sink her by gunfire during twilight if she remained afloat. Gunner Bennett's deck gun crew was robbed of the chance to shoot this ship, as she went into her death throes an hour later. *Hachigen Maru* sank at 1645 with much steam, smoke, explosions, and flying debris as her bulkheads collapsed.

Seawolf ran out of the western Celebes Sea that evening, sending a contact report to ComSubs of her "torpedo experiment," as Warder put it. She returned to Sibutu Passage for the next two days and patrolled the northern approaches. All was quiet until a maru

"A Marvelous Spectacle" 175

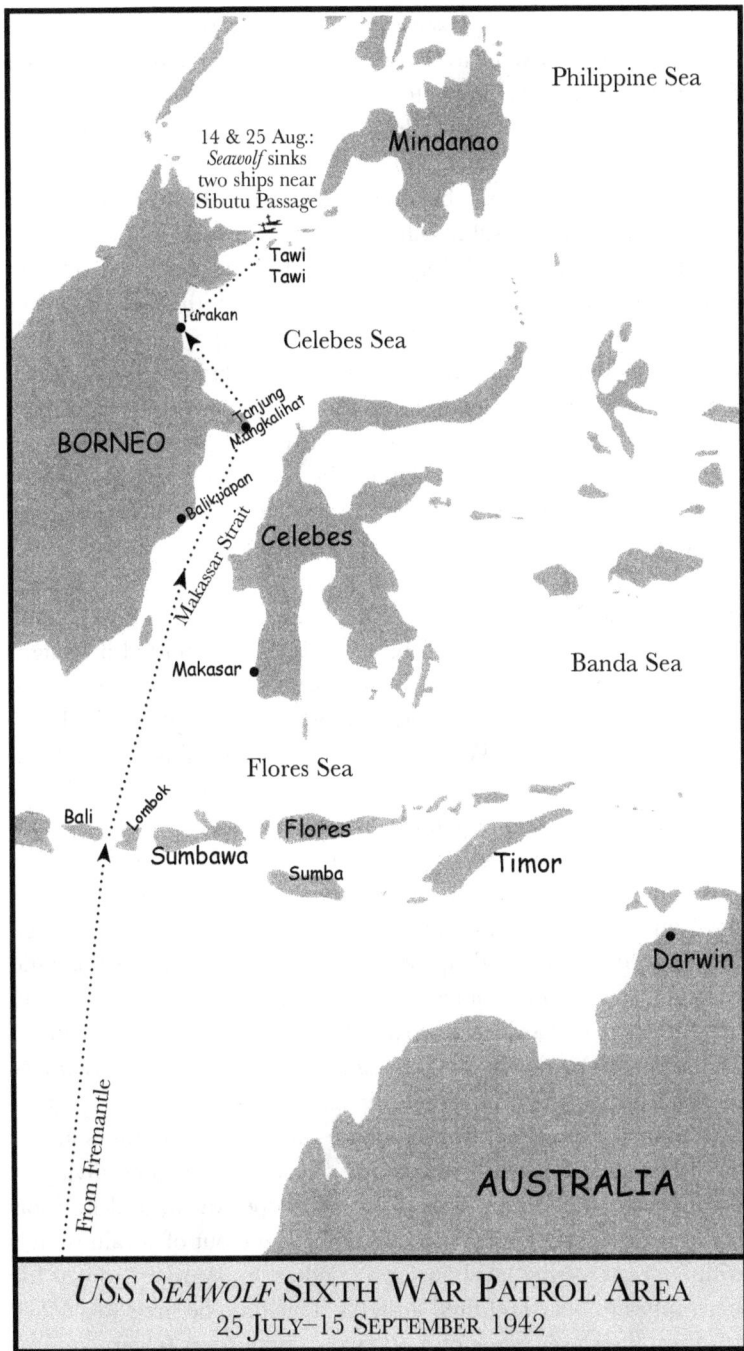

USS *Seawolf* Sixth War Patrol Area
25 July–15 September 1942

trailing a destroyer was picked up in the late morning of 18 August. Although the *Wolf* went to battle stations and tracked this pair, her fire control team was unable to reach a firing point.

Officer of the deck Dick Holden spotted the bright barrel of a periscope that afternoon 3,500 yards away. *Seawolf* avoided it but could detect no noise via her sound gear. Lt. Cdr. Warder decided this was probably the submarine *Grampus*, also working the area.

Seawolf continued her lonely vigil over the northern approaches to Sibutu Pass for the next few days. By 22 August, the lack of targets was frustrating Warder, who decided to begin operating in the pass during daylight and back and forth in the moonlight on the northern approaches during the night. Two more days passed with this new strategy without effect. Shipping was clearly not plentiful in Sibutu.

Lt. (jg) Jim Mercer finally sighted the smoke of a ship at 2108 on 24 August off the northern approaches. He called for the skipper and approached to get ahead for a moonlight attack. This ship was believed to be a 5,000-ton tanker making 8.5 knots speed. *Seawolf* fired a single Mark 15 torpedo from tube No. 1 at a distance of 3,300 yards, aiming for the MOT. Paul Maley reported a straight and normal run from sound. Warder kept his periscope cross wire on the target at the time the torpedo should have exploded by Hank Thomson's stopwatch. Warder was unhappy with this miss as "real pain [was] taken to insure correctness of torpedo course." He felt the range had been too great to warrant a good chance of a hit and vowed that "such type of attack will not be attempted again."

Warder pursued his target, but secured from battle stations at 0005 on 25 August as the ship went out of view. A short time later, he surfaced and opened up on four engines to overtake this little freighter's last known course. The ship was assumed to be heading for Tarakan, so Warder maneuvered to gain position 7 miles ahead of the target by dawn. Passing through very heavy rain, *Seawolf*'s lookouts did not spot the target ship again during the night.

By 0602, *Seawolf* was at the point estimated to be 7 miles ahead of the maru and she reversed course to look for her prey, diving at 0605. Lt.(jg) Bill Whitman had the periscope watch, and he soon had the little maru in sight by 0635. She came out of a rain squall, with *Seawolf* crossing her bow for a good setup. Warder brought his stern tubes to bear and fired at 0705. This time he fired two Mark 16s from the closer range of 1,300 yards, both set at 6 feet.

This time his torpedoes worked as planned. The first torpedo exploded 59 seconds after firing, hitting the 1,349-ton freighter *Showa Maru* between her bow and middle of target. Warder felt the "explosion did not sound violent but the ship blew apart, the forward third bursting into flames and then plunging downwards." *Showa Maru*'s propellers lifted into the air and the ship disappeared at 0707, less than two minutes after the first torpedo was fired and one minute after being hit. "It was a marvelous spectacle," wrote Warder.

Freddie Warder moved in close to the scene to investigate and called for his deck crews to prepare. Dick Holden brought her up quick and as her decks were still awash, the conning tower hatch popped open. Gunner Bennett, Swede Larson, Big Swede Hanson, and their gun crews took their spots. On the bridge, Chief Yeoman Sullivan and the other battle lookouts climbed the shears. Sullivan noted about six Japanese sailors hanging onto floating boards and one man in a life preserver.

Electrician Bob Parden found war a sobering reality. With orders not to let the Japanese get aboard *Seawolf* unless Warder found a prisoner who spoke English, Parden noted that the men not manning the guns were detailed to insure that no one clambered aboard. Their small submarine could not handle all of the survivors from the standpoint of either feeding them or safety.[4]

There was much wooden wreckage about, floating empty boxes, and canvas tarpaulins. One of the lookouts reported a vinta to westward during this time. Lt. Cdr. Warder spotted a life ring floating in the water. Hoping that it would contain the name of the ship, Eddie Souza dived overboard and swam out to retrieve it. Unfortunately, the ring did not contain the ship's name. Bill Deragon managed the *Wolf* crew on deck, insuring that no one came aboard. Souza tried calling out for any survivors who spoke English, but found none. At this point in the war, bringing back survivors was not a priority.[5]

Torpedoman Jim Cashero was topside to assist in case a prisoner was brought aboard. As he went up, he argued with another shipmate who was eager to go topside in hopes of getting to shoot at some of the Japanese sailors. "You, bastard!" Cashero snapped. "You're brave, with them being there in the water. But if they had guns to shoot back, you wouldn't want to be going up there."

Chief Souza and Lt. Deragon threw over six *Seawolf* life preservers in exchange for the one they took aboard. "After they recover from shock, they will be able to make a raft of wreckage," reasoned Warder as he submerged and moved toward Sibutu Passage on the little freighter's former route.

Following this successful attack, *Seawolf* prowled along Sibutu's northern entrance for another week, running periscope patrols of the lanes during the day. Patrol orders called for *Seawolf* to depart this area on 31 August, but Warder hung around one additional day due to his being delayed a day earlier in the patrol.

For this patrol, *Seawolf* had not carried the extra torpedoes in her superstructure. Instead, she had loaded with 20 fish: eight aft and a dozen forward. "We used those extra tubes to store potatoes this time," recalled Jim Cashero. "When they sent us up one night later in the patrol to get them, there was a problem. We found that the potatoes were all gashed up and rotten from being in the tubes!"

When no enemy shipping appeared, Warder headed south through Sibutu Pass en route toward Tarakan. Bill Whitman again sighted smoke on his watch at 1741 on 2 September off Tarakan. *Seawolf* surfaced and commenced tracking this target, which was hull down on the horizon. The little ship was hugging the coastline. The coral reefs in this unfamiliar area and a moonless night made close pursuit in shallow waters hazardous. Warder opted to proceed to the middle of the channel entrance ahead of this small ship "rather than close him and the shore line in the darkness."

Seawolf moved ahead and dived off Tarakan before dawn on 3 September. Mercer, Casler, and the plotting party had the enemy pegged to arrive at their location around 0500 if she maintained her 10-knot speed. By sunrise, the enemy ship had not appeared and *Seawolf* had nosed out from the channel entrance to cover other approaches to Tarakan. At 0640, Red Syverson reported the top masts of a northbound ship and dived to begin an approach.

By 0746, Lt. Cdr. Warder found that this target was actually two northbound vintas, sailing close-hauled. He stayed submerged close to land to cover Japanese shipping routes. By late afternoon, it was obvious that the ship had changed its course. *Seawolf* proceeded on in order to spend the next day submerged off Cape William. The pesky vintas and numerous schooners of about 100 tons were sighted during the night and throughout the next day.

The *Wolf* moved along Cape William and then toward Makassar City once again on 5 September. A small ship was sighted at 1746 but it was a small pilot vessel or station ship not worthy of attack. *Seawolf* moved along the western approaches of Makassar but found no shipping. After two more fruitless days spent in this area, Warder finally turned for base. He cleared Lombok Straits on 10 September and made a course for Fremantle, Australia. En route, Warder put his boat through some extremes to look for things that needed fixing during overhaul. He had his electricians run the batteries down at the three-hour rate, meaning making full speed submerged to drain the batteries in short order. He also had Dick Holden taken the boat down below test depth to 265 feet to check for stress and leaks.

Seawolf exchanged calls with her escort vessel at 0345 on 15 September and proceeded into Fremantle's harbor after 52 days on patrol, one day longer than her previous two patrols. In his patrol report, Warder noted that Doc Loaiza had to deal with five cases of catarrhal fever on this patrol, including two officers, and one case of rheumatic fever. In addition, one chief electrician's mate "will have to be transferred on account of nervous condition. The patrol was in no way arduous. In fact, it was the easiest patrol conducted to date." As for *Seawolf*'s newly installed SD air-search radar, Warder simply noted that it was "without value to this ship."

Freddie Warder had made six attacks on his sixth patrol and was credited with sinking two ships, both confirmed in postwar records, and he had damaged another ship. Warrant officer Jug Casler broke out his sewing kit and added two more stripes to the *Wolf*'s battle flag for the two latest kills. *Seawolf*'s wartime credit of two ships for 8,100 tons was reduced postwar by JANAC to two ships for 4,462 tons.

Both Submarine Squadron Two commander Jimmy Fife on *Sargo* and Admiral Lockwood were again critical of his performance, however. "Two hits out of seventeen torpedoes fired is far below the expected standard of performance," Fife complained on 15 September. He felt *Seawolf*'s fire control party erred in target speed estimates and some attacks were made from excessive range. Fife did credit Warder with making a good attack on the 3,113-ton maru on 14 August, including the illustration of only one of the three torpedoes firing. "The experience of firing third torpedo at ship dead in water and abandoned and having it pass under without exploding emphasizes need of modern testing equipment for Mark 6-1 exploders in all submarine tenders. *Holland* and *Otus* have been handicapped by absence of such equipment since beginning of war."

In the second endorsement to *Seawolf*'s sixth patrol report, Admiral Charlie Lockwood was even tougher.

> The failure of *Seawolf* to inflict greater damage on the enemy can be attributed principally to improper solution of the fire-control problem. Decisions to fire were made when excessive ranges existed and when lack of reliable data indicated that only a bare possibility of hitting would result. *Tidori* class destroyer in sight for several hours on August 9 should have been attacked.

In his report, Warder detailed each torpedo's performance. He had heard or seen one torpedo hit the target that was a dud. Three others were heard or seen to explode under the target. Eight others were heard to explode at end of run after missing or underrunning their targets. Of 12 fired during daylight, "at least four and perhaps six smoked heavily." Warder further criticized the "erratic warhead behavior, non-firers, prematures, poor exploder design, [and] uncertain depth performance." The torpedoes had "too great a distance required for arming," tending to make approach officers take greater firing range than they should.

In response to his fire control party's solutions not being good, Warder later responded, "The goddamned torpedoes were no damned good. That was the problem."[6]

USS *Seawolf* Sixth Patrol Summary

Patrol Area:	Borneo/Celebes Sea
Time Period:	25 July—15 September 1942
Number of Men Aboard:	74: 67 enlisted and 7 officers
Total Days on Patrol:	52
Total Hours Submerged:	589.6
Total Hours Surfaced:	673.2
Fuel Burned:	78,774 gallons
Total Miles Steamed:	9,410.7
Number of Torpedoes Fired:	17
Ships Claimed as Sunk:	2/8,100 tons
JANAC Postwar Credit:	2/4,462 tons
Shipping Damage Claimed:	1/7,000 tons

8

Sagami Maru

Refit and Seventh Patrol   *7 October–3 November 1942*

Red Eckberg was at work on the tender *Holland* when he caught sight of something that stopped him in his tracks. It was *Seawolf* making her way into the harbor at Fremantle on 15 September 1942. "I'd recognize her in a thousand," he thought. Since being hospitalized before her sixth patrol, he had since received orders to report to another submarine, *Skipjack*, but Eckberg had other plans. He met Lt. Cdr. Warder walking up the dock after the *Wolf* was moored. Warder confidently told his senior radioman to come aboard again and that he would work out the details.[1]

The crew enjoyed liberty in the twin cities of Fremantle and Perth once again. She was in refit for three weeks as minor repairs were made by *Holland*, including work on the new SD radar which the skipper had no respect for. During the overhaul period, Freddie Warder promoted two more of his chief petty officer into wardroom country. *Seawolf*'s newest "mustangs" were CEMs Clinton Jobe, a plankowner, and Larry Crane. Warder transferred Jobe on to other duty but kept Warrant Electrician Crane, bringing the *Wolf*'s wardroom up to eight officers.

Eight crewmen were transferred and ten new men were received. The new hands were sized up by their officers as they reported aboard. Jim Cashero recalled that engineering officer Dick Holden was the duty officer part of the time during the R&R while other men were ashore. A few *Seawolf* crewmen would come back to check on the progress of the relief crew. Cashero found that Holden was constantly checking out what new men or reliefers were doing.

Freddie Warder believed in promoting chief petty officers into the wardroom. Among those he promoted during 1942 were (left to right): CMoMM Dave Butler, CEM Larry Crane, and CTM John Gibson. *Courtesy of Hank Thomson.*

Our engineering officer was Lieutenant Holden, a great big guy. When new guys got aboard ship, he would ask them if they drank or did this or did that. If they said, "No," Holden would say, "Well, he probably wouldn't be worth a damn anyway."

We would have some of the guys in the engine rooms who wanted to tear the engines apart just to be doing something. He would ask, "Are they running?"

They would say, "Yes, sir."

And Holden would say, "Well, leave them alone then!"

Warder's new patrol orders called for *Seawolf* to proceed to Pearl Harbor following his patrol to become a part of Submarines, Pacific Fleet. In a near repeat of his sixth patrol, Warder was to proceed to Exmouth Gulf for refueling. He was then to head to Davao, Makassar City, Balikpapan, Mangalihat, and Tarakan, "conducting offensive patrols en route and on stations for as many days as practicable. Operate at least two weeks in Makassar-Tarakan area, including four days off Tarakan." *Seawolf* was also to reconnoiter the Palau islands, an island group no submarine had yet investigated and where the codebreakers reported carrier activity. Finally, *Seawolf* was to then proceed to Yap, reconnoiter that island until 16 November and then head toward Pearl Harbor—arriving there about 30 November. Warder's orders called for him to "destroy enemy forces encountered, including merchant shipping."[2]

Seawolf buddies TM2c Keith Bjerk, TM1c Bill Reiland, and MoMM2c Edward Chapman pose for the camera in 1942. Reiland would later serve as chief of the boat and became an officer. Bjerk made the first seven runs of the *Wolf*. Reiland and Chapman are on eternal patrol with *Seawolf*.
Courtesy of Hank Thomson.

On 6 October, final loading from *Holland* was completed in preparation for patrol. Doug Syverson's torpedo gang loaded a dozen Mark 14 steam torpedoes in the forward room and eight Mark 9, Model 1 torpedoes in the after room. John Gibson, Art Lamberson, Bill Reiland, Clarence Kibbons, Pinky Bjerk and their torpedomen gingerly moved the 3,000-pound fish with the precision of surgeons. "We loaded the torpedoes using booms and hoists to swing them over and then we would slide them down the skids below decks," recalled Jim Cashero. Once again, the four extra tubes in the superstructure were not used.

The Mark 9 torpedo was 21 inches in diameter but was shorter in length than the Mark 14. Designed for use in battleships during World War I, the old Mark 9s had been given to some of the older R-boats. Their 210-pound warheads packed a smaller punch, and the torpedo could only travel as fast as 27 knots—compared with 46 by the Mark 14. The most troubling fact about these Mark 9s was the fact that they were more than 20 years old!

Seawolf's diesels rumbled to life during the morning of 7 October. By 0834, OOD Jim Mercer had the ship underway and he conned her out of Fremantle Harbor for her seventh patrol. At 1048, she made a quick dive and then commenced the first of three practice approaches on an Australian vessel.

En route to Exmouth Gulf, *Seawolf* dived for battle problems and drills for her new hands. She entered the Exmouth Gulf area on 10 October. As she approached, Warder made a stationary dive in 7 fathoms of water to observe leakage of fuel oil from his compensating water system when using the drain pump on bilges. The engineers found only a slight amount of oil noticeable, but wanted it fixed during the brief fuel stop this day.

The *Wolf* moored to a fuel barge at 1115 so that oil king Pete Lober and his gang could take on 8,000 additional gallons of diesel fuel. By 1420, *Seawolf* was underway for Lombok Strait. PBY contacts were made in the afternoon and the tuned-up SD radar was notably better in actually picking up the aircraft.

On this patrol, Freddie Warder was fully prepared to capture his sinkings on film to back up his claims of destruction and poor torpedo performance. "Captain Warder and other skippers had been complaining that they knew the torpedoes were hitting the targets but were not exploding," recalled torpedoman Jim Cashero. "Later on, they also found that they were shooting underneath the targets because the torpedoes were running too deep." Lieutenant Jim Mercer had been experimenting, taking photographs through the camera with a 35mm camera. Chief Red Jenkins helped him clip it to the eyepiece of the periscope in the conning tower so that he could snap pictures of sinking ships. Having lost the chance to sink or seriously damage as many as eight additional ships due to faulty fish, Warder intended to prove his case.

The fact that Warder meant business was evidenced by a notice Chief Souza posted on the bulletin board. In the event of *Seawolf*'s capture, the notice reminded the men what information they could give out: name, rate, service number, and home address.[3]

Warder drilled his crew during the first two days out of Exmouth, conducting a battle surface for the benefit of Gunner Bennett's deck gun crew. For the gun crew, it was an experience they did not have often. Warder was not a proponent of deck gun actions against surface ships. He would use the gun only in an emergency. Third loader Pete Lober later described, "There was four of us in the loading line. We took the shells out of the container from the conning tower. We passed them to the last guy in the line, who would shove them into the barrel." Bennett's men fired six rounds of 3-inch shells and some .30-caliber and .50-caliber rounds. Lt. Syverson proudly noted the time from surfacing until firing the first round was 70 seconds.

Passing through Lombok Strait on 13 October, Jim Mercer sighted a southbound ship close aboard on the port bow at 2310. He called for battle stations and ordered emergency speed with right full rudder to avoid. Lookouts identified this southbound ship as either *Sargo* or *Seadragon* about 600 yards westward of *Seawolf*. The other sub also turned away and rapidly disappeared.

The *Wolf* patrolled from Surabaya to Ambon and along the shipping route from Makassar to Lombok or Madura Straits. Her lookouts sighted another friendly sub at 0149 on 16 October and turned away. "From time, location, and message traffic decided that ship was probably *Sargo*, bound for Lombok," wrote Lt. Cdr. Warder.

Seawolf approached Makassar City early on 17 October. Red Syverson had the watch at 0902 when he sighted smoke inside the reef off Pulu Lanyukang. Syverson went to normal approach course at full speed and called for the skipper. The target, judged to be a 4,000-ton transport, appeared to be heading out of port for Balikpapan—which had been seized from the Dutch by the Japanese early in the war—on the southwest coast of Borneo at 14 knots. The tracking party was called and *Seawolf* made her approach.

Two forward torpedoes were fired from 1,500 yards at 1005 and Warder watched them through his periscope. Conditions were excellent and these torpedoes were observed to run well. They missed ahead, however, and Warder decided that his target had slowed its speed at the time of firing. By this time, forward torpedoman Jim Cashero recalled that *Seawolf* was "shooting to hit the target." On earlier runs, Warder and other skippers had been under orders to use the Mark 14's faulty magnetic exploder. For Cashero and his fellow torpedo gang, they merely followed the orders given.

> With the magnetic detonators, we would shoot them under the ship so that they would go off and break her keel versus just shooting them to knock a hole in her side. When it came time to fire them, though, we just adjusted the torpedo settings to what was called down from the captain. They didn't always tell us whether they hit or not, so we didn't think too much about it. We were always busy reloading the tubes after firing. It was all just part of our job.

The spread applied was not sufficient to cover such a speed error. The target was not seen to maneuver to avoid. The ship—having spotted the torpedo wakes—swung at 1008 and headed right toward the position of *Seawolf*, her range rapidly decreasing to 1,200 yards. Warder thought briefly of giving the transport a torpedo right down the throat, but the enemy ship steadied on a course which gave him a small starboard angle. "Abandoned the idea of giving him one with 0° track and went deep for depth charge attack," Warder wrote.

Red Eckberg and Paul Maley called out the bearings to Warder as he prepared to conn his ship out from under the falling ash cans. Thankfully, no charges were dropped. The only two explosions, between 1010 and 1011, were believed to be *Seawolf*'s torpedoes at their end of runs. Eckberg finally announced at 1021 that the maru had turned and was fading away from sound. Warder returned to periscope depth and quickly sighted his target ship again just inside the reef. The ship had entered the channel just south of Pulu Lanyukang at 15 knots. With such a high speed, Warder let the ship run toward Makassar City while he took his boat toward the northwestern approaches of Borneo for Kapoposang harbor.

In the early afternoon, *Seawolf* had a close call while running at 120 feet. As the boat climbed past 100 feet for a periscope observation, the crew was surprised by a loud explosion. "I believe it was a plane or planes making drops on flotsam," Warder theorized. He went to 200 feet and cleared the area. Eight more explosions were heard during the next hour, none too close. Clearly, Japanese aircraft were responding to *Seawolf*'s morning attack.

The *Wolf* ran submerged through Makassar Strait en route to Balikpapan on 18 October. After surfacing that night, lookouts spotted a large vinta. Shortly after passing, the vinta made sail and a light was seen aboard her. Warder told Bill Deragon to pass the word that he planned to investigate her. If it was a radio-transmitting spy sampan, he planned to sink it with his deck gun. *Seawolf* surfaced quietly and approached. Gunner Bennett and his 3-inch crew opened the gun locker and broke out the machine guns and small arms. The bridge was silent as the *Wolf* approached the vessel.[4]

Warder eased his boat close to the sampan and looked her over, but did not find any radio antenna or gun mounts. All he found were two men, a woman, and a boy—all in native attire. Warder judged them to be harmless and moved on. Bennett's gunners stowed their weapons again, still hoping for their first gun action of the war.

Seawolf approached Balikpapan and nosed into the harbor's shallow waters in the early hours of 21 October, avoiding another large sailing vinta. Shortly after noon, OOD Dick Holden spotted a Japanese sub chaser at 4,000 yards on a northeasterly course, moving at full speed.

Holden called for battle stations and brought *Seawolf* on a collision course at full speed to approach the speeding sub chaser. The approach team read the bearings and angles over the next few minutes. "I listened to the screws, and the thumping of my heart was so loud and so strong it seemed to shake me from head to toe," recalled Chief Eckberg. By 1229, Warder figured his best opportunity would be to make a snap shot at the range of 2,400 yards. "In view of short length, light draft, inadequacy of data," and Admiral Lockwood's urging "to avoid this kind of shooting, decided to let him pass," Warder noted.[5]

Seawolf secured from battle stations and continued to patrol the 21-fathom-depth waters off Balikpapan. *Seawolf* investigated the same harbor the next day, moving in very close to shore at times. Through his headphones, Red Eckberg "heard surf breaking on the beach, the water crashing and clashing over shoals and reefs." Warder brought his boat into dangerously shallow water with battle stations helmsman Rudy Gervais at his station. He let Dick Holden take his turn at the scope, looking for enemy shipping. At 1312, Lt. Holden spotted a tanker and an escort vessel that looked similar to the previous day's sub chaser—both bound for Balikpapan. By 1404, the closest range was 3,200 yards on a poor angle, so Warder was eventually forced to abandon the approach.[6]

The *Wolf* continued along this trade route in hopes of picking up a similar target, running at 100 feet with periscope observations every 15 minutes. On the evening of 24 October, a coal-burning tug was sighted by Dick Holden and Jug Casler's watch team. The tug appeared to be bound for Tarakan. Warder briefly pursued this little ship in hopes of it leading to larger game, but to no avail.

The next days were equally void of shipping. *Seawolf* carried out her patrol orders by staking out the shipping around Tarakan for five days. Freddie Warder set course for Davao on 31 October. The radio gang picked up an Ultra at 1840 which ordered *Seawolf* to intercept a convoy off San Augustin about 1600 on 1 November.

Seawolf bent on all main engines at full speed—80% power or 17.9 knots—and changed course to take the most direct route to the San Augustin entrance to Davao Gulf. Deragon and Casler took a star fix at 0240. They noted that *Seawolf* was fighting an adverse current of 1.8 knots. Warder continued to run hard toward the Ultra-directed contact area, opting to remain on the surface past sunrise on 1 November. The current drift increased to 2.8 knots in the morning hours, further working against his chances of reaching the area. "Realized I was just throwing away precious fuel for no good reason except possible airplane detection," he wrote at 1034 as he finally decided to dive 42 miles off Balut Island.

Seawolf approached the southern entrance to Davao Gulf during the overnight hours, fighting the strong westerly current. Early on 2 November, she was continuing to move toward the San Augustin traffic area. Finally, at 1145, Lt.(jg) Jim Mercer called, "Smoke on the horizon. Bearing three-four-eight, True. Call the skipper."

Seawolf exposed 12 feet of periscope but could find neither ship nor smoke. At 1324, Doug Syverson also reported smoke on the horizon in the same general direction and he was able to find the ship's topmasts. *Seawolf* went to battle stations and full speed to close the San Augustin traffic lane. Warder ordered two forward bow tubes opened as he approached, preparing to fire Mark 14s. At 1433, however, he found himself dead ahead of his prey by 2,700 yards so opted to fire two stern tubes with Mark 9 torpedoes. During the approach, the sea had progressed from calm to rainy to rough.

Seawolf opened fire at 1441 with tubes 5, 6, and 7 aimed at the target ship's stern, MOT, and 1/4 length ahead of her. The Mark 9s, set at four feet, were fired with only a 1,100 yard run. This time her torpedoes performed as desired. The first torpedo struck with a rocking explosion felt aboard the *Wolf*. The ship was the 2,933-ton freighter *Gifu Maru*. The resulting explosion was one of the loudest felt aboard the *Wolf*. "Many people insist on two hits," wrote Warder. "I only saw and heard one." Over the 1MC, Warder gave a play-by-play of the sinking ship. He described how the explosion ripped lifeboats right off the freighter and collapsed the smokestacks. Japanese sailors began jumping overboard, followed by a second explosion.

In the control room, Chief Pete Lober listened to the skipper's commentary. "He described how it rolled over and the crew was climbing on the bottom of the ship," said Lober. "They were close to shore and could swim to the shore." Recalling his plan to photograph a sinking ship for evidence, Warder called, "Jim, hurry up if you want a picture of this. Only the stern is showing now."[7]

Sagami Maru

Gifu Maru, a 2,933-ton Japanese freighter, was sunk by *Seawolf* on 2 November 1942. This dramatic photo by Lt. Mercer shows *Gifu Maru* taking her final plunge with only her stern sticking out of the water. *U.S. Navy photo, courtesy of Mike McCoy.*

Lieutenant Mercer slipped his camera into the special periscope mount that Chief Jenkins had fashioned. He lined up the sinking *Gifu Maru* and snapped several pictures of her death throes. Mercer passed the periscope back to the skipper, who continued his loudspeaker commentary as the freighter finally slipped beneath the waves. She was gone 30 seconds after Mercer snapped his pictures.

Gifu Maru sank in full view of Cape San Augustin, some 5.5 miles away. Bill Deragon poured over the ONI-208J book to identify the enemy ship. In his report, Freddie Warder described the men in the water after *Gifu Maru* disappeared.

> Counted 41 men in water, about 16 of whom were holding on to wreckage. Starboard whaleboat could not be manned [on] account heavy list to starboard after lowering part way. Believe it was pulled under with ship. Port whaleboat forward falls carried away and spilled boat about 5 minutes before ship sank. Believe it also broke up when ship sank. Most men came off collision mat or cargo net rigged down

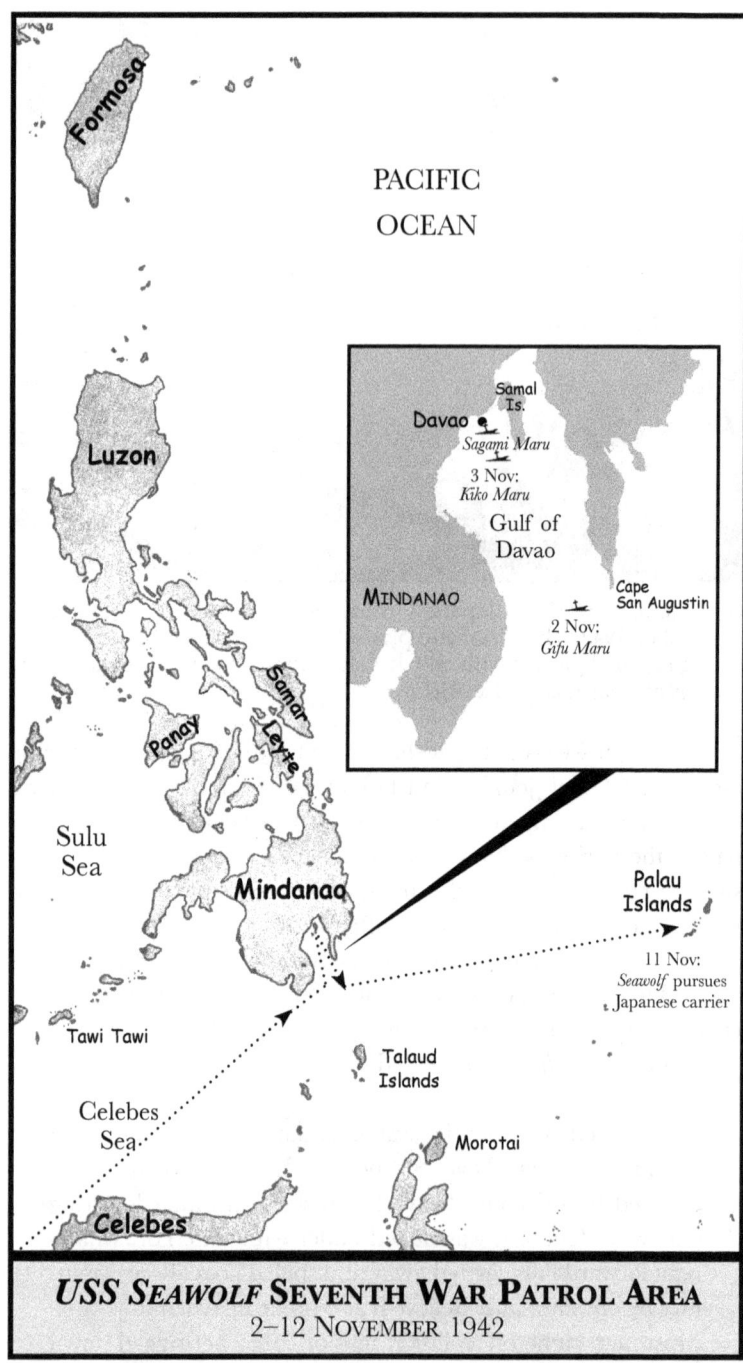

port side. Remained vicinity until nightfall as I expected DD or other rescue ship to show up. No such ship appeared and it is believed sinking was not observed on beach. With prevailing weather and strong southwesterly current it is believed that there will be few, if any, survivors.

Following *Gifu Maru*'s sinking, *Seawolf* surfaced that evening and stood in toward Davao Gulf. At daybreak on 2 November, she dived 12 miles south of Davao Harbor's light and moved northward outside the 300 fathom curve in order to try and establish the existence of a swept channel before going into shallower water. A motor boat flying a Japanese ensign was spotted at 0629 headed south 2 miles away. Warder spotted two small freighters in the harbor, anchored off. Beyond them lay some smaller shipping near the shoreline which were later determined to be "of anti-sub character." One larger ship was anchored in Talomo Bay. "Latter was definitely more important and easier to attack, so laid course accordingly," wrote Warder.

The wind was from the north and a surface chop on the water helped disguise *Seawolf*'s periscope. Warder called for battle stations at 0921 as he approached what his tracking party deemed to be a 9,300-ton motor vessel. Warder crept his boat closer, while calling out his periscope observations over the 1MC for the crew.

"What a camouflage job!" he said. "She's a beauty."[8]

The large transport—anchored in 30 fathoms of water—was painted in wartime camouflage. She appeared to be tied to a wharf or anchored right off one. Workmen moved about her, loading some cargo. The plotting team checked the enemy ship's position against the prevailing current and depths, lining up the *Wolf* for a zero angle shot and an escape course. This ship was the 7,189-ton transport *Sagami Maru*. Ironically, *Seawolf* had already unsuccessfully attacked *Sagami* and her sister ship on 2 February during her fourth patrol.

Warder worked *Seawolf* in through the shallow water, ever mindful of the possibility of mines. He had talker Hank Thomson call John Gibson's forward crew to make ready the bow tubes. Rudy Gervais had the helm as he moved her in slowly. Warder announced that sailors aboard *Sagami Maru* appeared to be loading hemp onto her. With all in check from a zero angle, he finally called for the firing of two torpedoes. Standing by, Jim Mercer and his camera would document the firing of these torpedoes from point blank range on a stopped target. If there were torpedo failures this time, he intended to bring back visual evidence.

Warder fired his first torpedo at 1050. *Seawolf* was making 2.4 knots at the time, so the 1,100-yard range from the first firing would decrease. The first torpedo—a Mark 14 aimed at *Sagami*'s forward goal post—ran under the point of aim and exploded on the beach 45 seconds after firing. Warder ordered Gibson's crew to change the depth setting for the No. 2 fish from 18 feet to 8 feet. Chief Eckberg reported this fish running "hot and straight" toward the target.

Boom! The resulting explosion from the second fish brought a rush of excitement through the boat. The first torpedo hit *Sagami Maru*'s machinery spaces at 1052.30 right at the point of aim just under her stack. "Much flying debris," Warder happily noted. The ship listed 30° to starboard toward *Seawolf* and settled five feet in the water.

Warder ordered Chief Gibson to set his fish even shallower to four feet. At 1053.05, three minutes after firing tube No. 1, he fired tube No. 3. The range was now down to 950 yards. Warder observed this torpedo's wake in toward the target and saw it inexplicably run under the point of aim, 10 yards abaft the stack. A torpedo set at four feet had thus passed under a stopped ship whose draft was clearly four times that of the setting! This torpedo did not explode but instead hit the beach and stuck. Eckberg continued to hear the torpedo "running" after the time of explosion had passed.

Frustrated, Warder had John Gibson's men fire their fourth forward tube, which was loaded with another Mark 14, also set at four feet. It was a repeat performance of the previous firing. *Seawolf* had closed to the point blank range of 880 yards when she fired this fish at 1054.20, aimed halfway between *Sagami*'s stack and forward goal post. This one also ran under the target and hit the beach, although Eckberg and Maley could again hear the propellers continuing to turn. It became obvious that the faulty Mark 14s were running at least a dozen feet deeper than their settings.

Warder kept his eye glued to the periscope, breaking only to let Jim Mercer snap photos. Warder noted that gunners had manned the heavy guns mounted on her forward and after decks and were shooting wildly in all directions. *Sagami*'s damage controlmen apparently dealt with the torpedo damage by counter-flooding other compartments for the freighter soon righted her starboard list.[9]

"Reload the tubes," Warder called as he ordered Rudy Gervais to swing out for a moment to reload. He had come point blank to the anchored freighter and now needed to move back out to give his torpedoes time to properly arm.

Japanese transport ship *Sagami Maru* viewed through *Seawolf*'s periscope before being torpedoed in the Philippines. Note her camouflage paint scheme and the small boat astern of *Sagami*. This photo was taken by Jim Mercer on 3 November 1942 during the seventh patrol as part of Freddie Warder's efforts to document torpedo failures that *Seawolf* and other U.S. subs were encountering. *Official U.S. Navy photo, courtesy of Bob Hanson and Mike McCoy.*

Seawolf's naval intelligence recognition books offer this aerial view of *Sagami Maru* (above) and an artist's rendering of the 476-foot, 7,100 ship. *ONI-208-J.*

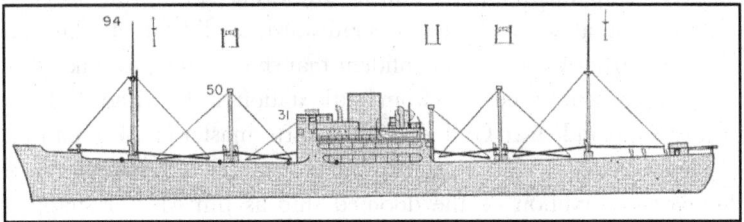

Once he began his second approach, he found *Sagami*'s gunners were firing all the while. "His shells make much bigger splash than do ours and make a bigger racket on exploding than any I have yet observed," wrote Warder. "On second approach the after gun was firing about 500 yards to port of us and the forward gun about 500 yards to starboard."

Seawolf made her second torpedo attack at 1131 from the port side. This time, he fired a Mark 9, not trusting the Mark 14s that had run under the ship deeper than their settings. He closed to 800 yards before firing, again aiming at her stack. This torpedo hit in the after part of the ship. When the smoke cleared, the after gun platform and the entire topside was clear of people. "Forward gun still manned but not firing," Warder noted. "Deck is lined with people."

Warder had Gervais swing *Seawolf* around to bring her stern tubes to bear. He intended to deliver another Mark 9 torpedo "as it is evident that ship can take a lot of punishment." *Sagami Maru* was now down by the stern about ten feet but not sinking as she should be after two torpedo hits. The port anchor chain was still up and down, clearly indicating zero speed. Jim Mercer snapped more pictures for documentation of the attack as men abandoned their ship.

By 1139, much of *Sagami Maru* was deserted. Warder could only see "one or two people on forecastle head, two Jacob's ladders rigged on port side." Two or three people were seen on the flying bridge wearing khakis. *Sagami Maru* was now flying a "Victor" flag from her midships halyard. Japanese ensigns were flying at the forecastle and the main truck and the flag staff were down. Warder reckoned these two halyards were probably carried away by the force of the explosions. Five boats were seen carrying people to the dock.

At 1143, *Seawolf* delivered a third attack to finish the job, firing one Mark 9 torpedo from tube 8 from 1,100 yards, set at a four-foot depth. This torpedo exploded in one of *Sagami*'s forward cargo spaces, causing her to flood forward and a fire to break out. Satisfied that this ship was doomed, Warder withdrew. "Fire grew progressively worse as we withdrew to southward, as did target trim and draft," Warder wrote. "Am confident that ship is going to sink."

Warder secured the crew from battle stations as he withdrew from Talomo Bay in Davao Gulf. At 1200, quartermaster Hank Thomson logged a heavy explosion, source unknown. At 1205, *Seawolf* made her last observation of the doomed ship as Jim Mercer snapped

Sagami Maru burning and smoking in Talomo Bay after enduring a torpedo hit from *Seawolf*. *Official U.S. Navy photo, courtesy of Bob Hanson and Mike McCoy.*

more photos from his periscope camera. *Sagami Maru* was settling by the bow, with the forward half of the ship smoking heavily.

At the same instant, another ship was spotted on a southerly course, making high speed. Three minutes later, Warder spotted three aircraft on his port quarter, evidently radioed in by the sinking ship. Two flew away but the third zoomed straight in on the *Wolf* with her guns spraying bullets.

"Take her to 120 feet!" Warder called to Dick Holden as he and his men scrambled out of the conning tower.

The planes apparently scoured the area and called in surface ships. Maley and Eckberg reported fast screws approaching. The first of five depth charges exploded at 1220. Holden took the boat on down to 200 feet and silent running as she stood out for deeper water. The two enemy ships continued to sweep the seas for *Seawolf* until 1405, when their screws finally faded out to the southeast.

Warder headed his boat toward the gulf's entrance, but the exit would take many hours submerged. The skipper retired to his bunk for some rest, leaving Bill Deragon in charge. In the forward torpedo room, Red Eckberg tried to rest but found that his "nerves were still tingling" from the recent shallow water attack. He watched torpedoman Art Lamberson play solitaire until he finally drifted off.[10]

Freddie Warder could gripe about torpedo performance when he returned to Pearl Harbor. Armed with Jim Mercer's photographs, he had clear evidence of torpedoes running much deeper than aimed. It had taken six torpedoes fired from point-blank range at a motionless, deep-draft cargo ship to obtain three hits. "The failures of the first attack are typical and merely add weight to the previous complaints of other COs and myself as to the erratic performance of Mark XIV torpedo and its warhead attachments," he wrote in his seventh patrol report. "All six of these fish smoked. None of these tubes had been made ready until about 2 min. before firing."

Warder requested that the officers reading his seventh patrol report also reference seven specific dates on his first, fourth, fifth, and sixth patrols in which he had experienced duds, erratics, deep runners, and prematurely exploding torpedoes. The two-decade old Mark 9 torpedoes which *Seawolf* had carried on her seventh patrol performed far better than the new Mark 14s. Torpedo officer Doug Syverson, in fact, was praised in a commendation for "making possible effective discharge of battleship torpedoes which had not been fired for twenty-three years."

The *Wolf* had been potentially robbed of as many as eight sinkings due to faulty fish in the first year of war—a fact which would make Freddie Warder want to boil over even 40 years later.

Seawolf's original battle flag was sewn by CQM Jug Casler during Freddie Warder's command. A stripe was added for each torpedo victim claimed by the ship. *Courtesy of Bob Hanson.*

9

"Circular Run!"

Seventh Patrol   *3 November—1 December 1942*

Freddie Warder did not have long to stew over the number of fish he had wasted on *Sagami Maru*. An hour after the escort vessels had moved on, *Seawolf* eased back to periscope depth and a quick sweep showed all clear. Ten minutes after securing the crew from depth charge quarters, OOD Dick Holden made a new contact. The previous torpedo victim was 8-9 miles away and no longer visible, but he spotted the smoke of a new target.

This ship was about 7 miles distant with a large starboard angle. This ship was standing in closer so Warder called for battle stations yet again at 1542 on 3 November. He began an approach and found another ship well over on this ship's starboard side. "Unidentified smoke is on his port side well inshore," he noted. Her decks were loaded with what appeared to be invasion barges. Exec Bill Deragon stood ready in the approach party and Hank Brengelman was at the captain's right shoulder with his hands on the solenoid controls, eager to once again push the firing buttons.[1]

Warder conducted an approach with the intention of firing two bow torpedoes but at 1558 found himself too close to the target's track. He had helmsman Rudy Gervais swing the boat around to bring the stern tubes to bear. With a flat surface above, he made only brief periscope observations on his target and did not search for other ships.

In spite of the presence of two anti-sub vessels, Warder had his soundmen use ping-ranging to get the final distance to target. From a range of 1,910 yards, he told Hank Thomson, "Tell the after room to set their fish at four feet."

Thomson relayed the word to Bill Reiland's gang, who quickly adjusted the depth of their Mark 9s. The target vessel was judged to be a 5,000-ton freighter-transport ship. Warder lined up his first shot one-quarter length ahead of the oncoming maru.

"Fire one!" he snapped at 1612.

The *Wolf* recoiled as she belched forth another warhead.

Moving his aim to the MOT, Warder called, "Fire two!"

Brengelman hit the firing button as Reiland in the after room manually fired the second fish, just to be sure. Three seconds later, *Seawolf* launched her third aft Mark 9, spread to run one-quarter length behind the ship.

Seconds after the third fish was fired, the captain shouted a hair-raising alarm: "Circular run! Take her to 150 feet!"

Mel Eckberg, tracking the torpedoes in toward the target, agreed with the skipper that the number two torpedo had hooked to the right and was circling back toward *Seawolf*.

Lieutenant Holden's planesmen pointed *Seawolf*'s nose downward as he flooded negative for an emergency descent. The after room's second torpedo had gone erratic—as seen by Warder—and was hooking back around toward its point of origination. At least two U.S. submarines sunk themselves during World War II with circular run torpedoes. *Seawolf* dived deep enough quickly enough for her own warhead to churn by safely overhead but it may have been the silent prayers of many that helped save her hull this day, as well.

"We had a lot of duds and some erratics," said Jim Cashero, one of the forward torpedomen. "A circular run is also considered an erratic. The only thing you can do when a torpedo makes an erratic run is to get down safely below its depth until that thing either straightens out or runs out of compressed air."

Eckberg kept a ping ranging on the target ship's position and when a solid explosion was heard and felt at 1614.10, it timed perfectly with the third fish having made a 1,950 yard run. Thirty seconds later there was another explosion, which Warder believed "to have been a secondary explosion in the target."

Eckberg reported screws coming up the starboard quarter and Warder ordered depth charge stations. The enemy ship came right on over *Seawolf* and the first depth charges rocked the waters nearby at 1615. The breaking up noises of the target ship were heard in *Seawolf* "in contradistinction to the depth charging we were getting," wrote Warder.

For the next 45 minutes, two ships made runs on the *Wolf* and dropped a total of 22 depth charges. "Some were close," Warder noted. Some minor electrical damage was sustained and one leaking main motor air cooler was discovered.

The last charges exploded at 1701 and the enemy ships gradually departed the area. Henry Wilton, Gus Wright, and Bill Mallory made sandwiches for the exhausted crew while stewards Bugawisan and Tamayo ran snacks and coffee to the officers. Two attacks during the day had made for a long day. As fate would have it, there would be no rest for the weary. *Seawolf* eased up to periscope depth at 1815 and at 1853, smoke was sighted on the horizon.

Seawolf surfaced at 1902 and posted her battle lookouts. Two minutes later, Chief Pete Lober, the port after lookout, sang out, "Ship on the port quarter." Warder quickly made out "a ship with one stack smoking heavily and heading in our direction with a bone in his teeth."

The range was only about 1,500 yards, so Warder ordered his men to clear the bridge. The ship dived to 100 feet as the destroyer came up near her stern but did not attack. The Japanese ship was apparently content to clear the area, for his screws were gone from sound in ten minutes. Warder kept his boat down another two hours to let the area cool off a bit. He stood out on two engines on the surface, enjoying a fresh breeze from the northeast, rain squalls, a moderate sea and limited visibility—"all in our favor."

Seawolf withdrew to the southern approaches of Davao Gulf in order to rest the crew and the captain. Chief Jenkins' auxiliary gang went to work repairing the bow plane tilting panel, trim and drain pumps, leaking section drain piping, a leaky main motor air cooler, and one air conditioning plant that had all suffered from the day's depth charge attacks. Doc Loaiza found that headaches were prevalent, likely on account of excessive pressure that had been created on the first brief surfacing. Many other griped about prickly heat from the excessive heat endured during the prolonged silent running.

Seawolf lurked off the southern approaches of Davao Gulf the next day. That night, Warder sent a detailed report of his boat's successful attacks. During the early morning hours of 5 November, the crew enjoyed the propaganda of Tokyo Rose that her radiomen piped in over the loudspeakers. Radio Tokyo proudly boasted of the sinking of an American submarine in this area. "I hope they meant us," Warder logged triumphantly.

Seawolf continued to patrol the traffic routes to Davao for another two days. Finally, at 1126 on 7 November—after finding no Japanese shipping—Freddie Warder decided to go inside Davao Gulf again. The only problem this time was that the sea was like glass. Red Syverson made out smoke at 1257 and by 1400, *Seawolf* had sighted a minesweeper 5 miles away, heading towards her. Deciding that a plane may have spotted her scope in the calm water and sent this ship, Warder went to 150 feet and reversed course. The sweeper went south of *Seawolf*, close aboard. Thankfully the 150-rpm screws faded out by 1450 without leaving any calling cards.

An hour later, Warder had closed on the area where he had attacked *Sagami Maru* on 3 November in Talomo Bay. Exposing a full 12 feet of scope, he was unable to find any trace of her. He moved in close enough along Davao Harbor to notice several houses along the island's bluffs.

Seawolf surfaced that evening in Davao Gulf to recharge batteries. By 0222 on 8 November, she was 17 miles off Talicud Island when Ensign Jug Casler sighted a minesweeper on the port quarter. The enemy ship was heading for the *Wolf*, about 3,000 yards away. Casler immediately cleared the bridge, dived to 100 feet and steered clear of the danger.

Warder opted to reverse course and operate in the Davao–San Augustin shipping lane, remaining submerged. At 0837, Lt. Syverson noted a small power boat, alternately stopping and starting again as if using a listening device. This boat gradually worked in toward Davao, seemingly aware of a submarine in the area.

The periscope watch picked up smoke and the tops of a coal-burning freighter a little after 1000. The target appeared to be the same 5,000-ton freighter-passenger ship observed in Davao Harbor on 3 November. The ship was escorted by one minesweeper similar to those observed previously. The freighter zigged radically, never maintaining a steady base course. "Four minutes was the most time he was observed to be on the same course," noted Warder.

The maru's zigs made the plot quite challenging for Casler, Deragon, and Mercer. By 1103, *Seawolf* was in a good position, hop-

These photos by Jim Mercer shows the demise of the 2,929-ton gunboat *Keiko Maru* on 8 November in Davao Gulf. In the upper photo, *Keiko* is burning from two *Seawolf* torpedo hits. Freddie Warder monitored *Keiko Maru* closely. In the photo below, *Keiko Maru* has her fires extinguished and sits low in the water. Shortly after this photo was taken from close range, *Seawolf* finished off her cripple. *Courtesy of Bob Hanson.*

ing for a left zig. Instead, the dancing maru zigged toward *Seawolf*. With the range down to 925 yards, Warder ducked down to 120 feet and ordered standard speed, hoping to open the track for a stern tube attack. The target ship instead turned again, this time coming up the *Wolf*'s port side and crossing her bow very close aboard.

"He was a very tough ship to lose," noted Warder as the maru's screws roared by overhead. Back at periscope depth at 1123, Warder noted another inbound ship at 14,000 yards.

After approaching this target for a half hour, *Seawolf* fired two forward torpedoes at 1201, both aimed for MOT with 4 foot depth settings from distances of only 800 and 775 yards. Warder used what he called the "Voge Speed Spread Diagram" to aim his torpedoes, a method proposed by Captain Dick Voge early in the war. The first torpedo slammed into or under her boiler room. The starboard whaleboat after falls carried away in the explosion and the ship appeared to bend upwards in the middle. Steam streamed out of her stack and uptakes at the base of her stack. The ship had a large number of people on board, all dressed in white and apparently "well disciplined." Warder believed his target was a 4,000-ton freighter, although she was later identified as the 2,929-ton gunboat *Keiko Maru*. The second torpedo hit was under her bow. *Keiko*'s port whaleboat pulled away from the ship and the force of the second explosion blew her bow gun overboard.

Minutes later, as Jim Mercer took pictures through one periscope, Warder noted the damage control efforts aboard *Keiko Maru*.

> The after gun went right to work on us, and so did an AA gun on flying bridge (about 20mm). Fires were extinguished in boilers; and ship lying to. People on board did not panic. A number went to work in the forward well deck clearing two large power boats from gripes. Stern gun fired case ammunition and believe it was shrapnel as they littered the water with many short splashes. Shell explosions were noisy. A number of men in the waist of the ship were engaged in throwing overboard floating material. Ship settled to the level of the anchor hawse pipes.

Seawolf had one torpedo left in her after room "which did not promise to be any good to me by itself." Warder thus decided to fire this fish into his stricken ship to insure a good sinking while he was able to witness it. In the meantime, a minesweeper was detected in the distance. She was racing at high speed from the direction of the shore, obviously responding to *Keiko Maru*'s distress calls.

At 1218, *Seawolf* fired again from 825 yards with tube No. 8. The target speed by this time was 0 and the depth setting was again 4 feet, aimed MOT. This third torpedo ran hot and straight. The damage controlmen aboard *Keiko* saw it coming for them and they ran for the upper works of their ship. It hit right in the maru's midsection, filling the air with water and flying debris to at least twice the height of the vessel. "When the air cleared, there was no more ship," Warder wrote, "only much refuse and one troop barge in water."

One minute after this last torpedo hit home, there was another very loud, violent, and long-sustained explosion. This lead Warder to believe that *Keiko Maru* had been carrying some high explosives aboard. By this time, the minesweeper was 4 miles away.

Dick Holden took her down to 120 feet and the boat was rigged for silent running. The minesweeper made his first run on *Seawolf*'s port quarter at 1249, missing with his first two depth charges. Holden eased down to 150 feet. The minesweeper then made two more runs on the *Wolf*, dropping several more depth charges during the next hour. The ship stayed close to *Seawolf* until 1515, when her screws could no longer be heard. One final depth charge was heard at 1528, far astern. Warder surfaced 14 miles off Cape San Augustin that night and radioed attack reports to Admiral Lockwood. *Seawolf* then headed east from Davao toward the Palaus to complete her final patrol assignment. En route the next day, Lieutenant Holden sighted a westbound Japanese flying boat at 4 miles off the port bow. The SD radar failed to pick up this aircraft, which caused Holden to make a precautionary dive to 120 feet.

The passage to Palau was uneventful. By 0615 on 11 November, the Palau Islands were in sight and *Seawolf* dived to begin her initial sweep through this virgin territory. By 1100, OOD Jim Mercer had sighted smoke bearing 067° and turned to approach it. This ship appeared to be on an opposite course moving away and *Seawolf* was unable to close it submerged. The area seemed ripe, however, and by 1428 Dick Holden had sighted another ship 15,000 yards away.

This target was a small patrol vessel, so Warder quickly secured from battle stations. This ship patrolled on various courses, passing within 1,200 yards of *Seawolf*. Warder continued to steer toward the channel entrance, approaching to within 2.7 miles of it. The lighthouse at the entrance was a tempting target for Lt. Mercer's camera, but every time he prepared to take a picture his efforts were foiled by another zig of the patrol vessel or a rain shower.

The patrol vessel remained in the area as the rain increased during the afternoon. Sound reported screws at 1718, and one minute later a single depth charge exploded while the *Wolf* was at periscope depth. Warder spotted the patrol vessel two minutes later, 4,000 yards off and headed northeast. As Warder continued to monitor this patrol vessel, he suddenly picked up two destroyers charging out of Palau. He sent *Seawolf* to battle stations and commenced an approach at full speed.

One of the destroyers was 8,000 yards off at 1729. The heavy rain made visibility limited, but Warder suspected that the destroyers were moving out in front of a prime target. The other destroyer was closed to 2,900 yards over the next ten minutes. Warder changed his course to try and hang on to the course of the speeding destroyers while he maintained his best submerged speed.

Seawolf steadied on the course of this destroyer, as Warder continued to swing his periscope about. At 1746, he suddenly found a large aircraft carrier in the rain and twilight. She was at 3,800 yards distance, so he ordered full speed to try and close the range. This was most likely the 24,140-ton *Junyo*, which had departed from Truk on 11 November. Committed toward one of the destroyers, Warder cursed his luck for being out of position to fire on the carrier.[2]

Having decided to approach the nearest destroyer, Warder later wrote that he had been "sucked in to the north and east," thus finding himself out of position when the carrier came out of the channel in the rain and stood west "with no opposition from us."

Seawolf's electricians ran her at full submerged speed, but *Junyo*'s range quickly increased to 6,000 yards. Warder ordered Gervais to come to 340° T at full speed in order to hang on. "My assumption at this time was that CV was going to head for homeland," he wrote. By 1758, *Junyo* had opened her range to 9,000 yards with a large starboard angle as she moved toward the southwest.

Seawolf continued to push her batteries at a high submerged speed. "The cables were running hot to the batteries, so we could not make enough speed to catch her," recalled motormac Bob Hanson.

The carrier easily inched ahead, reaching 9,750 yards distance by 1802 and 9,900 yards by 1805.

Moments later, *Junyo* was completely out of sight. Warder brought *Seawolf* to the surface at 1822, putting his battle lookouts on the bridge. The boat went ahead full at 17.5 knots on all four diesels. In the control room the speed indicator "vibrated all the way up to a point that we hadn't seen in eight months," recalled Red Eckberg. A destroyer was soon in sight again, well distant. During the course of this high speed pursuit, Jug Casler and Bill Deragon could not give their captain certainty as to their boat's position in relation to a reef known to jut out in this area. On account of the rain and darkness, there was neither land nor any other navigational marks in sight.[3]

Seawolf chased the carrier task group on four engines for the next hour without spotting *Junyo* again. At 1925, Warder ordered the engineers to take two engines off propulsion for a quick charge. Should the boat reach a point where she could make a submerged attack, the depleted batteries would need some juice. The radio gang worked on sending out carrier contact reports to Pearl Harbor in the meantime.

Seawolf continued in pursuit of the Japanese carrier south of Palau into the early morning hours of 12 November. At 0059, a lookout spotted a white light on the horizon. OOD Dick Holden thought the light was getting closer so he held his course. By 0104, the bridge personnel could see that the light was drawing aft rapidly so Warder slowed his boat to 1/3 speed to give this ship a chance to catch up. Uncertain as to the source of this light, Lt. Cdr. Warder decided to at least turn to investigate it.

In the midst of this approach, Chief Pop Mocarsky suddenly called for the captain. "We've got a serious problem, sir," he reported from the maneuvering room.

Mocarsky alerted the skipper that the readings on main generators No. 1 and 2 each showed dangerously low readings of 10,000 ohms each. The No. 4 generator showed a level of 100,000 ohms, which in itself was less then ten per cent of desired strength. Mocarsky and his electrical gang found that the cables to the main motor generator cables had become dangerously hot and fire was feared.[4]

Back in August, some of *Seawolf*'s generator cables had failed. Sister submarine *Searaven* had suffered a bad generator fire the previous May which was also vivid in Warder's mind. With the possibility

Photo taken by *Seawolf's* Jim Mercer on 14 November 1942, showing a Japanese schooner under sail off Yap Island. *U.S. Navy photo, courtesy of Mike McCoy.*

of starting a dangerous electrical fire if he overheated these cables any further, Warder opted to break off his pursuit. "I decided to call everything off and head for Pearl," he wrote.

Mocarsky, John Bilkey and their electricians set to work on trying to determine the problems with the generator cables. By 0156, *Seawolf* was proceeding at 10 knots on the No. 3 and No. 4 engines at 50 per cent load each.

Eckberg's radiomen finally heard NPM transmit their earlier carrier contact report to *Seal* at 0328, seven hours and 35 minutes after *Seawolf*'s original transmission. Warder sent out another contact report on the carrier, describing her course and his pursuit. Eventually Pearl Harbor received the report. *Seawolf*'s carrier find was a major event in 1942. Only about 23 sight contacts were made by U.S. submarines on major Japanese war vessels that year. Five contacts were made on battleships and 18 on aircraft carriers. *Seawolf* and seven other boats made only visual contact with their carrier. The other ten 1942 Japanese carrier sightings developed into actual attacks in 1942, which resulted in slight damage to three carriers.[5]

Freddie Warder departed the Palau area with great disappointment at having missed the chance of a lifetime to fire at an enemy carrier. Pop Mocarsky's electrical gang worked through the night on their overheated cables. With little use and blower cooling on them, the main generator cables improved appreciably by midday on 13 November. Warder decided he could continue with his assigned reconnaissance off Yap Island on the 14th through 16th. He did not bother to alert Pearl Harbor that he had changed his mind about coming straight back home with his electrical problem.

The following morning, he commenced a patrol back and forth about 2 miles off Yap's Tomil Harbor entrance, keeping outside the 300 fathom curve. Jim Mercer put his camera to use snapping pictures of the Japanese-held island, which held a small ship in the harbor. On closer inspection, this proved to be a small schooner unworthy of a torpedo. The following day, Warder could still see the schooner sitting in Tomil Harbor but no other shipping appeared. *Seawolf* remained off the harbor and made a detailed reconnaissance of the shore installations, noting radio masts, gun emplacements ashore, and enemy fortifications on various hilltops.

Yap was void of shipping so the *Wolf* surfaced that evening and proceeded to pass south of Fais Island to get on the great circle route for Pearl Harbor.

Doug Syverson sighted two masts on 18 November at 0714 off the starboard bow. It was a large freighter on a southerly course. *Seawolf* came to normal course and bent on all engines. The electricians kept a close watch on the main generators No. 1 and 2, which showed 800,000 and 600,000 ohms respectively. The enemy ship was closed as it came in and out of the prevailing rain squalls.

The electrical readings dropped, but did not reach dangerous levels before a Japanese plane was spotted 10 miles away. *Seawolf* dived and continued tracking the ship, but found that the aircraft remained over the maru providing aerial cover. The rainy seas and the presence of the patrol aircraft were enough to prevent the *Wolf* from reaching a favorable attack position this day and she eventually lost this target.

Seawolf had two more sightings en route to Pearl on the morning of 19 November. Warder hoped to gain position on one of these vessels by running ahead on the surface. He decided the best target to be a large tanker, which the plotting team tracked to be making 15 knots. *Seawolf* ran at 80% power at 15.5 knots, but soon had to cut her pursuit short on this ship. Pop Mocarsky reported that the ground detector voltmeter was climbing steadily and had reached

100 volts. *Seawolf* broke off on the tanker approach and shut down her No. 1 and No. 2 diesels. The generator armature circuits had dropped as follows: No. 1 from 1.5 megs to 100,000 ohms and No. 2 from 0.7 megs to 40,000 ohms, after 2 hours of operation.

With smoke still visible on the horizon, *Seawolf* dived and headed back for her first target freighter. It developed to be a coal burning maru with eight motor fishing sampans about her. "She spent her time going from one side to the other of the latter and from back astern to up ahead of them," wrote Warder. "It's a miracle we didn't get tangled up in the trolling lines or nets. We were in a spectacular school of fish."

Red Eckberg made a ping range on the target ship, having been unable to previously do so due to the pesky sampan hulls in his way. With a successful ping, Warder knew his target ship's true range. *Seawolf* fired two Mark 14 torpedoes from her forward tubes at 1407. The run was only 1,750 yards, but they missed.

Through the periscope, Warder watched the Japanese maru break out a flag hoist and then wheel around to the northward. The enemy skipper had apparently slowed his speed about the time the torpedoes were fired, to which Warder attributed the misses. He ordered Holden to dive to 150 feet. *Seawolf* endured four depth charges from the escort ship during the next 40 minutes, but none was close.

Returning to periscope depth at 1540, Warder monitored the sampan fleet as it worked to the southeast. He found the weather too rough to play with these vessels on the surface, so he continued his course for Pearl Harbor. "Fuel, as well as generator cables, has become a problem," he noted.

Seawolf exchanged radio messages with Pearl Harbor that evening and she continued to clear Yap over the next two days. The periscope lookout sighted another submarine on the surface at 0843 on 22 November. The other sub was on a parallel course and was believed to be *Seal*. *Seawolf* increased speed to 11 knots and the ship gradually drew aft and was out of sight by noon.

The next week was uneventful as *Seawolf* ran toward the Hawaiian Islands. The quartermaster gang updated the *Seawolf* battle flag. Motormac Edward "Chief" Chapman entertained himself during his off hours by designing faux medals. He would then tack them to the stateroom of various officers, for some particular act of noteworthy valor accomplished on war patrol.

Seawolf approached Pearl Harbor on 1 December 1942, the anniversary of her third birthday. She had completed seven war patrols and was well known throughout the fleet. The men became anxious for Pearl Harbor and the prospect of returning home for liberty.

Her escort ship, *USS Litchfield*, was sighted at 0556. After exchanging signals, *Seawolf* fell in astern and followed *Litchfield* in to Oahu. The channel watch was stationed as *Seawolf* made her return to Pearl Harbor, the first she had been there since before 7 December 1941. Warder allowed those men in proper uniform to come topside as the *Wolf* moved through the channel. Bomb damage and shipping wreckage were reminders of the surprise attack of one year prior.

Torpedoman Jim Cashero was thrilled. "During my first year of war, I don't think I'd seen the sun more than 30 times total," he said. "We would dive before the sun came up and on most every day we wouldn't come back up until the sun was down." Cashero fortunately had stood his share of lookout watches. Some of those in the engine rooms had seen a lot less of the sun than he had.

Auxiliaryman Paul Zimmerman was elated to be in Hawaii. "It had been a year since we had gone out to fight and a year since the attack on Pearl Harbor. We finally got to really see the damage that the Japanese had inflicted on us."

As *Seawolf* tied up, there was a crowd on the dock to welcome her. This was Warder's first meeting with Admiral Bob English, who had put together a special greeting party for *Seawolf*'s crew. "He was just tremendous," Warder later stated. "He had John Brown and the whole damn band there" to greet *Seawolf*.[6]

As soon as the ship was properly secured, mail bags and fresh fruit came over to be enjoyed by the crew. Admiral English came aboard to greet Lt. Cdr. Warder and to discuss his patrol results. When he reviewed the seventh patrol report, English was more pleased than Lockwood had been with the sixth patrol. He credited *Seawolf* with a 9,310-ton passenger-freighter (silk liner) sunk in Toloma Bay, a 3,500-ton freighter and a 4,000-ton freighter transport. He gave Warder 16,810 tons sunk and one 5,000-ton freighter damaged. As these kills were achieved prior to *Seawolf* joining Commander Submarine Force, Pacific Fleet, English credited them to Commander Submarines, Southwest Pacific Force, Admiral Lockwood.

The only down note English gave Warder was in his decision to continue the patrol to Yap. He had reported trouble with his main generator cables and that he was heading for Pearl. When the cables improved, he instead continued his mission to Yap. "While displaying commendable initiative," Warder had no way of knowing whether English had deployed other subs to this area.

English offered Warder and his crew a "well done" on this outstanding patrol. His remarks were written on 9 December. He said that *Seawolf*'s seventh "was conducted with intelligence, thoroughness and aggressiveness." He noted that the sinking of *Sagami Maru* anchored off Talomo Wharf was "especially worthy of commendation in that it is the first instance noted by this command that one of our submarines has operated for any appreciable period in waters as confining as Davao Gulf."

For his seventh patrol, Freddie Warder would be awarded his second Navy Cross, allowing him to select other members of his crew for commendations. Silvers Stars would go to Jim Mercer, Doug Syverson, and Dave Butler. Bronze Stars were awarded to TM1c Rudy Gervais, TM1c Bill Reiland, CMoMM Bud McCoy, and CTM Eddie Souza.

After being secured, Lieutenant Deragon announced to the crew that those not on duty could enjoy free beer at the base swimming pool, courtesy of the executive officer of the submarine base. After addressing the crew on deck, Deragon turned it over to Lt. Cdr. Warder, who was dressed in a new khaki uniform.[7]

Warder gave his crew a speech of how proud he was of them and the record their boat had achieved. He then went ashore and released his crew to the beer party. Once his official visit was complete, he would go to the pool to enjoy a beer with his men.

Warder met with Admiral English ashore for a few drinks to continue talking about his patrol.

"You're going to be my first selection for division command," English told him bluntly.[8]

"Sir, I'd be honored," said Warder.

"It means you're going to be a long time in the Pacific," warned English.

"That's fine," Warder said. "I want to serve under you. But Admiral, I'm not sure I'm old enough for this command."

English looked at him and said, "I think you're old enough."

Warder and English finished their dinner and Warder left with the satisfaction he had been picked to a division command position over others with more seniority than him. His wife and kids were still living in the Baltimore area, but he felt certain that they would support his next big career move. "They knew what I had to do whenever I could do it," he later stated.[9]

A few of the men enjoyed themselves too much. Torpedoman Jim Cashero and electrician Jim Hughes were among these. "I used to get in a lot of trouble," Cashero admitted. "In Honolulu, Hughes and I stayed out too late and ended up being reported AWOL." Cashero and Hughes ended up in the brig, to be bailed out by Eddie Souza—who was less than pleased with how they had chosen to mark their return to the Hawaiian Islands.

Chief Pete Lober for one enjoyed the brief liberty in Honolulu with his buddies the Hanson brothers. "We were only there a short while as they made minor repairs on the ship, but we got to go ashore and stay in a hotel right along the ocean." The Royal Hawaiian was a treat. For Lober, he found an even bigger treat. He received word of his promotion to warrant officer and pending transfer to new construction of the new fleet submarine *Rasher*. Although a big loss as *Seawolf*'s fuel king, Lober was the sixth chief petty officer to be promoted to officer by Freddie Warder, following on the heels of Jug Casler, John Sullivan, Larry Crane, Clinton Jobe, and Dave Butler.

Deservedly proud to be making officer status, Lober was nonetheless remorseful in the approaching prospects of leaving his buddies behind.

USS *Seawolf* Seventh Patrol Summary

Patrol Area:	Dutch Indies
Time Period:	7 Oct.—1 Dec. 1942
Number of Men Aboard:	74: 66 enlisted and 8 officers
Total Days on Patrol:	55
Number of Torpedoes Fired:	19
Ships Claimed as Sunk:	3/16,810 tons
JANAC Postwar Credit:	3/13,000 tons
Shipping Damage Claimed:	1/5,000 tons
Return to:	Pearl Harbor/California

Our crew was just like family. The skipper had to go bail some of the guys out of the brig a couple of times. Swede Hanson was always getting into trouble when we went ashore. I asked him one time later, "Did you ever win one of your fights?"

He told me, "No. Never!"

Seawolf's stay at Pearl Harbor was brief as she took on fuel and supplies and made minor voyage repairs. During her second afternoon in Hawaii, Chief Souza mustered his crew on deck for the good news from skipper Warder. Admiral English sent his blessing to the crew and the best gift of all: *Seawolf* was approved to return to the States for a badly needed overhaul and modernization at the Mare Island Naval Shipyard in California.

Hours later, the veteran *Wolf* cast off her mooring lines, steamed out of Pearl Harbor, and pointed her bow toward San Francisco. *Seawolf* had survived one year of war, had become the subject of many a newspaper and magazine article, and had seen more action than most other 1942 U.S. submarines.

But to her seasoned crew there was only one inviting prospect at this point: they would be home in time to celebrate Christmas with their families.

10

"Hope This Old Man Knows His Stuff"

Refit and Eighth Patrol   *December 1943–3 May 1944*

En route to Mare Island, the mood on the boat was festive. "Card games were in full swing in Kelly's Pool Room, and bull sessions went on at all hours," said Red Eckberg. He found motormac John Street tallying up *Seawolf*'s kills before chiefs Otis Dishman and Swede Enslin, proudly proclaiming that the *Wolf* had sunk over a dozen ships and damaged perhaps half a dozen.[1]

Officially, Freddie Warder would be credited with six ships sunk for 18,719 tons in JANAC's postwar analysis. Wartime credit was much more favorable toward *Seawolf*. Submarine command had thus far given her credit for 16 ships sunk or damaged, totaling 85,400 tons. This included eight ships sunk for 38,900 tons and another eight ships damaged for an additional 46,500 tons. Actual and wartime credits would be nearly the same had *Seawolf* gone to war with a more dependable torpedo than the faulty Mark 14—which certainly had allowed at least eight ships to escape sinking or heavy damage.

As *Seawolf* approached California, the weather grew increasingly colder. Men unaccustomed to cool weather broke out their heavy gear as December in the United States began to grip the boat. When she finally came within sight of the Golden Gate Bridge at dawn of 10 December, there was a rush of excitement. When Lt. Cdr. Warder passed the word down from the bridge that the *Wolf* was approaching the famed bridge, men scurried to ask for permission to come topside.

Seawolf met her escort prior to the Golden Gate Bridge and passed under the big gate. On the other side, Warder allowed the deck hatches to be opened and heavily bearded men pulled on their heavy clothes to come topside. They could see the bridge behind them and in the distance the white line of surf on the mainland of America.[2]

A pilot boat approached and the pilot clambered aboard. *Seawolf* steamed into San Francisco Bay and made her way to Mare Island's navy yard, which had greatly expanded since her last visit to the West Coast. The proud builders at Mare Island were turning out new submarines and overhauling older ones with record efficiency. Established in 1854 in the San Francisco Bay area, the shipyard at Mare Island had first been commanded by Cdr. David Farragut. Of the 17 fleet-type submarines built at Mare during the war, eight ended up as top 20 producers in terms of enemy shipping sunk.

After *Seawolf* moored to the port side of another sub, Lt. Cdr. Deragon called the crew to muster at 0800. Deragon announced that the *Wolf* would be in for at least two months for overhaul. The best part was that he was allowing the crew to take 30-day leaves back home in shifts. Half the crew would leave first, while the other half stayed behind to work the ship. Those who were leaving first were frantic to get in touch with loved ones whom they had not spoken with in more than two months.

All ammunition was transferred to a naval ammunition depot during the morning and *Seawolf* was prepared for her extensive overhaul. By the next morning, leave requests had been approved. At 0930, half of the wardroom—Red Syverson, Jug Casler, and newly promoted officers John Sullivan, Larry Crane, and Dave Butler—departed on leave. Bill Deragon, Jim Mercer, Dick Holden, Pete Lober, Bill Whitman, and Lt. Cdr. Warder remained behind for the second leave with half of their crew. Lt.(jg) Whitman was transferred to temporary duty aboard *Sargo* (SS-188) to fill a void in her wardroom as commissary officer, although he would return to *Seawolf* in February 1943 following this patrol.

Big Swede and Little Swede Hanson, the only brothers aboard ship, took leave together to return home to Eau Clair, Wisconsin. "I flew as far as Kansas City and then caught a train from there," recalled Bob Hanson. He married his sweetheart Marilyn, whom he had dated since before the war. He had last seen her in August 1940 when *Seawolf* shipped out. Reunited with her in February 1943, he wasted no time in getting married.

Others of the *Seawolf* crew were married during their 30-day leaves. John Bilkey had the advantage of living close by. He was able

During her overhaul in California following the seventh war patrol, *Seawolf* received a new logo. Walt Disney artists created this new image of a wolf biting a torpedo for the famous submarine. *Courtesy of Bob Hanson.*

to regularly return home to see his wife and young son near Mare Island. Motormac Paul Zimmerman was in the first group to leave, along with his buddy Bill Reiland. "I went back to New York and married my girlfriend on December 19, just over a year after Pearl Harbor," he said. "Our guys were given a lot of priority for flights and trips across country. I was able to hop a freight flight on a Navy plane that was going to Washington. From there, I got A-1 priority to get another flight on to New York."

Mel Eckberg was among those staying aboard ship for the first 30 days. He wired his wife and child to catch a train to San Francisco, where he would be working with the ship during overhaul.[3]

Seawolf's overhaul commenced on 10 December 1942, and would extend through 24 February 1943. During these weeks, quite a few changes took place on the old *Wolf*. The yard workers replaced the main motor armatures with skewed-slot types. The troublesome main generator motor cables that had threatened to cut short the seventh patrol were renewed. The ship's trim, drain, and compensating water lines were renewed also. An SJ surface search radar set was installed, along with a proper radar antenna. The *Wolf*'s 3-inch/50-caliber deck gun was moved from its after deck position to forward of the conning tower superstructure. Modified air conditioning coils and main motor and generator coolers were also installed.

Seawolf also took on a new battle flag during her Mare Island overhaul. She had become famous and the Walt Disney artists drew up a new wolf character. "Our second flag was a wolf with a torpedo in its teeth, a Disney design," recalled Bob Hanson.

Aside from a battle flag and the many engineering changes made at Mare Island, the personnel changes in California were extensive.

The biggest—and not unexpected—one was the change of command for SS-197. Freddie Warder, promoted to a full commander on 15 December, returned home to his wife and four children in Baltimore for Christmas and a three-week break. Word got around that Cdr. Warder had new orders. One of his final acts was to work on good assignments for his veteran men who were transferred on.

On 18 January 1943, the Mare Island drydock that held *Seawolf* commenced flooding at 0900 and the *Wolf* was soon waterborne again. She shifted to a nearby shipyard dock after lunch and at 1330 a relatively simple change-of-command ceremony was held aboard her. Some of her older crew members had already been transferred to new assignments, and more than half were still on leave.

Chief of the boat Eddie Souza assembled the rest of the crew for the change-of-command ceremony. Souza had put together a committee of crewmen to get their skipper a going-away gift. They bought a wristwatch and some luggage. Freddie Warder—dressed in uniform, complete with medals—gave a farewell speech to his men.

It was a tough moment. The crew that had coined the nickname "Fearless Freddie" had known no other skipper since the *Wolf* had been commissioned in 1939. They were immensely proud of him and steadfastly loyal. Warder told his men that he had been fortunate to have "the best submarine crew ever gathered together." He explained that he knew his successor, Captain Gross, and that those who were remaining aboard should give him the same unswerving loyalty that they had shown Warder. "He is a good man and, he knows his submarines," said Warder. The skipper then made his way among his crew, shaking hands and talking with each man personally. "We saw tears rolling down his cheeks, and some of us were beginning to sniffle, too," related Red Eckberg.[4]

Bill Deragon escorted his former skipper to the gangplank, where they shook hands. Warder slapped his longtime Exec on the shoulder, turned to wave to his men and then walked across the *Seawolf*'s gangplank toward his next command.[5]

Lt. Cdr. Royce Lawrence Gross officially relieved Commander Warder of command on 18 January. Warder was already familiar with "Googy" Gross. "He was a terrific guy," Warder thought. "My first experience with him, he was the diving and engineering officer of the *S-36*." Warder had to go on *S-36* before the war with senior commanders to make a simulated torpedo approach and attack.[6]

Lt. Cdr. Roy Gross, *Seawolf*'s second skipper, receives the Navy Cross from Admiral Nimitz in early 1943. *Courtesy of Submarine Force Library, Groton, Connecticut.*

"It was a complicated practice with four destroyers running around like crazy. I got inside the box in a good attack position," Warder later related. Once in a good position, Warder called for "all stop" in order to keep *S-36* in the set-up he had chosen for attack. He relied heavily upon the diving officer he did not know to keep his boat neutrally buoyant and undetected.

"Gross was just as cool as a cucumber," said Warder. "If I had been diving officer, I couldn't even have begun to handle it like he did." He let the ships sweep by to the firing point and then fired his practice torpedo for a called hit.

Gross had hoped to be a pilot before submarine school. "I failed in flight training at Pensacola, then went to submarines," he would admit. "I fancied the small ship concept, the camaraderie. I would have stayed in submarines even if the extra pay was cancelled."[7]

Gross had picked up the nickname "Googy" at the Naval Academy. It stuck with him when he met up with his contemporaries, although to his family he was always "Roy" or "Royce." Gross later related that Googy was "given to me plebe year by Johnny Corbus, classmate, based on the funny sheet character of that time named Googenheim. Why, I'll never know."[8]

When Pearl Harbor was attacked on 7 December 1941, Googy Gross was in New London, a lieutenant in charge of the de-mothballed *R-9* training boat. During the war, he attended PCO school and made his PCO run aboard *Silversides* with Cdr. Creed Burlingame and his Exec, Roy Davenport, during their fifth patrol in 1942.

Freddie Warder was promoted to division commander of SubSquad 122. His first and only meeting with Admiral Bob English had been a good one upon his return to Pearl Harbor after *Seawolf*'s seventh patrol. When Warder took his boat back to the States, English "sent in a good letter in the meantime that he wanted me to be division commander."

Just three days after Warder left *Seawolf*, the radio and newspapers were full of the tragic news that Admiral English and members of his staff were missing on a flight from Hawaii to San Francisco. All aboard perished when their plane crashed into mountains north of San Francisco during bad weather. Command of the Pacific submarine force soon passed to Rear Admiral Charlie Lockwood.[9]

Many of the *Seawolf* crewmen were thrilled to find the February 8, 1943, issue of *Life* magazine which came out while they were on leave. It contained a special two-page spread entitled "U.S. Subs at Work." Although the sub's identity was not revealed, all of the photos were taken by Lt. Jim Mercer and each ship was a victim of *Seawolf*'s handiwork, including the stubborn *Sagami Maru*.

The Mare Island overhaul was extensive, but it was the first yard overhaul *Seawolf* had since being commissioned on 1 December 1939. The Japanese had spoiled her previously planned overhaul at

Manila Bay in December 1941. Upon completion, *Seawolf* underwent a ten-day readiness for sea period at San Francisco. It was important for Lt. Cdr. Gross to shake down his new hands, as well as check the seaworthiness of the yard's repairs. His wardroom alone had its share of changes. When *Seawolf* returned to war, she would carry eight officers. Gross' Exec was Lt. Cdr. Bill Deragon, who had been on the *Wolf* during her previous seven war patrols. His third senior officer was Lt. Doug "Red" Syverson, who had also been aboard since the start of the war. Syverson moved into the role of engineering and diving officer to replace Dick Holden, who had received orders placing him in command of the training boat *R-14* in Key West, Florida. Roy Gross considered Deragon and Syverson to be "outstanding, aggressive, cool officers."[10]

Fourth officer Lt. Jim Mercer stayed on as communications officer and would continue to serve on the conning tower attack team in plotting. He would be assisted by Ensign Jug Casler, whose duties for the upcoming patrol would be that of assistant navigator.

When Bill Whitman returned from his brief assignment on *Sargo*, he became torpedo and gunnery officer and as such would operate the TDC during torpedo attacks. To the five veteran *Seawolf* officers Roy Gross had retained, he received two new ones: Ensigns John J. Kennelly and Hubert Eugene Gluski. A reservist whose wife lived in Detroit, Gluski took over the commissary job, while Kennelly--who hailed from Chicago—became the first lieutenant and Red Syverson's assistant engineering officer. In addition to Dick Holden, four other officers had been sent on to new assignments: Pete Lober, John Sullivan, Larry Crane, and Dave Butler.

In the months between returning returning to the war zone, the *Wolf*'s enlisted men went through equally drastic changes. More than one-third of her crew received transfers prior to her eighth patrol. Among those departing for an administrative position with ComSubPac was Chief Eddie Souza, whose billet as chief of the boat would be filled by Swede Enslin, one of the plankowners. "Swede Enslin, the chief motor machinist, became our chief of the boat," recalled Paul Zimmerman. "That was a rare thing because most of the time the chief torpedoman was the chief of the boat."

In addition to Souza, 14 others departed before the eighth patrol who had been with *Seawolf* since the war started: TM2c Keith Bjerk, OS1c Mariano Bugawisan, Ens. Dave Butler, TM2c Wilbur

Chubbuck, MoMM1c Orval Cross, CRM Red Eckberg, SM1c Frank Franz, EM1c Robert Hutchison, CMoMM Red Jenkins, QM3c Norm Kisver, Ens. Pete Lober, CEM Pop Mocarsky, EM1c Lee Bob Parden, and MoMM1c John Street. Of this veteran group, Butler, Eckberg, Franz, Jenkins, Lober, Parden, and Street were also original 1939 *Seawolf* plankowners.

For the old-timers who stayed on, there were many advancements in ratings. CTM Bill Reiland, CPhM Frank Loaiza, CMoMM Sandy Randazzo, CMoMM Otis Dishman, CGM John Bennett, CEM Hank Brengelman, and CMoMM Lucien Rajotte were all new to the CPOs' quarters since the last patrol. They would help direct the 31 new men who reported aboard *Seawolf*.

Among the new hands was 19-year-old S2c Delbert Mar, a native Californian who had joined the Navy right out of high school. "They say I was one of the first Chinese Americans to join the submarine service," he recalled. "My mother and father were born in China and came over to the United States before World War I." Mar's younger brother would also join the Navy and serve in World War II. Having gone to school to learn fire control for battleships, Mar would serve as a striker under fire controlman Lefty Leffingwell.

Another of the newest young sailors was S1c Wilson Mills, a 21-year old from Hope, Michigan, who had worked for Dow Chemical after high school before enlisting in the Navy. Mills had put in for the Air Force but got tired of waiting around. When the call came for submarine volunteers, he was quick to head to New London for sub school. He held the class record for holding his breath in the diving bell. A brown-eyed youngster with red hair, Mills became known to his buddies as "Red." He raised hell during his leaves ashore, but on board ship, Red Mills was all work.

Just prior to departing California, *Seawolf* picked up a few more new hands on 6 March. Among them was RM2c Joseph Hale Strong from *Saury*. His boat had completed five runs and was also Stateside for an overhaul. He was not thrilled with his unexpected new orders. "I came over to the *Seawolf* at California from the *Saury* as a replacement for one patrol," said Strong. "I was so goddamned mad to get transferred to her because I was due to go on 30-day leave Stateside and I got screwed out of that."

Her refit complete and her crew reorganized, *Seawolf* departed California for the week-long voyage to Hawaii. She nosed into the sub base at Pearl Harbor on 13 March 1943, three months after departing the war zone. In order to shake out the bugs and work out the newest crewmen, the boat went through training exercises for two weeks. Roy Gross initially held a six-day training period, firing eight exercise torpedoes. This was followed by an eight-day period in the Submarine Base floating dock for repairs to the QC sound head and to the propeller shaft bearings to eliminate a pronounced shaft squeal that could prove to be fatal in the war zone.

The firing of test torpedoes was important. Lt. Cdr. Gross was uniquely qualified to criticize faulty American submarine torpedoes. The Navy was working to produce a new electric torpedo but the modified Mark 14 was still the principal torpedo in early 1943. Gross later wrote of his assistance in 1942 in testing the errant Mark 14s.

> The torpedoes, contact exploders, and magnetic exploders were terribly unreliable. I blame this on the unrealistic test procedures of the Newport Torpedo Station. As far as I know, they were never fired through nets, nor were the exploders actually impacted at speed against a solid object. Not until Lockwood did this in the graving dock at Pearl Harbor, and with live torpedoes against the cliffs of Oahu. I helped with the 90-foot drops with sand-loaded warheads in the drydock.[11]

As *Seawolf*'s training maneuvers came to a close, the men faced the sobering reality that they were heading back into the war zone to face the wrath of enemy ships, planes, and depth charges for the first time in four months. Many got in last-minute partying ashore, while others took time to write home to their new spouses. Sadly, some of the marriages that happened during the war for the submariners were not sincere. They would learn that there was a small percentage of women eager to marry just to get a GI's paycheck. Some were later found to be married to multiple men under aliases.

Seawolf torpedoman Jim Cashero was caught in this ploy.

> I got married during the war, but I also got divorced too. The gal I married, she married several different ones. When

we got back to 'Frisco, was when I got married. Come to find out, she was married to several guys. So we had it annulled during the war. She was just marrying them for them to send her their money.

Motormac Bud McCoy was another *Seawolf* veteran unwittingly caught up in this scam. Since he would not survive the war, it is unknown whether he ever learned the sad truth. Late in the Mare Island refit process, he had written home to his parents:

> I have been working pretty hard these last four weeks trying to get the ship ready for sea, but we are leaving here this Saturday or Sunday. Where we are going, I don't know. I am going to get married this week but am not sure which day.[12]

McCoy related to his parents that he was not marrying his longtime sweetheart, but instead had met another young lady named Penny, whom he had fallen in love with. Penny claimed not to know her mother and that her father had been lost in action in the Pacific. "I am going to send her back there to stay while I am at sea," McCoy wrote, promising also to continue sending money for his new wife's expenses. He did marry Penny before leaving California and began sending money back home to his parents to help her out. While at Pearl Harbor during *Seawolf*'s sea preparations, he sent his parents an update on his new wife on 29 March. McCoy hinted that he was heading out on war patrol again shortly.

> I haven't been ashore but once since we left the coast. I go on the base and have a few beers once in awhile, but it's not worth a damn—about half water. That allotment I made out in February will start in April, so you should get it in the last of April.
>
> If you don't hear from me for two or three months, don't worry. I may be at sea.[13]

Bud McCoy would continue to support his new wife for many months and would make leave again to visit her in 1944. Months later, however, his parents were paid a visit by the FBI and found that she had married other servicemen.[14]

CMoMM Walter Glen "Bud" McCoy (right) was aboard *Seawolf* when the war started and was the most senior enlisted man aboard when she was lost in 1944. McCoy kept a secret diary while on some of the *Wolf*'s patrols. *Courtesy of Harold and Jeff McCoy.*

Repairs were completed by 1 April and *Seawolf* began preparing for war patrol. Hubert Gluski and his senior cooks, Bill Mallory and Gus Wright, filled the ship's stores with fresh meats, produce, vegetables, and special treats. Bill Whitman's torpedo gangs oversaw the loading of a mixed batch of Mark 14-3A torpedoes, some loaded with TNT and some with Torpex. John Gibson's forward room took on 12 fish, while Chief Bill Reiland's after room loaded eight.

Two days later, the *Wolf* was ready for sea and departed on her eighth war patrol on 3 April 1943. *Seawolf* was underway at 1300, escorted by *PC-596*. She made a trim dive off Barber's Point, released her escort at dark, and set course for Midway. Roy Gross held training dives, battle problems for his TDC team, and battle surface drills for the gun crews. As the new engineering officer, Doug Syverson made daily trim and training dives to gain experience.

The engine rooms of *Seawolf* were presided over by veterans, to which had been added a certain number of new recruits during the overhaul. Lucien Rajotte, who had been recently promoted to CMoMM, was in charge of the small gang of auxiliarymen that he and Chief Al Hershey had at their disposal. Otis Dishman had the forward room and Sandy Randazzo, also a new CPO, had the after room. One of Dishman's most seasoned men was MoMM1c Bud McCoy, the newlywed who had been aboard the *Wolf* for more than two years and was a veteran of all previous patrols.

Yeoman John Burruss made several last-minute changes of personnel before *Seawolf* departed on 3 April. Received aboard this day was GM2c Rex LaVere Mickey, who had quit high school a half-year early to join the Navy in February 1942. "They put me through gunner's school and I made third class gunner's mate," Mickey related. "I got on the destroyer *Moffett* and third class was supposed to know something, so I got transferred to submarines!" He caught *Seawolf* fresh out of sub school at New London.

Gunner Mickey was assigned a bunk in the forward torpedo room, where he became buddies with two torpedomen strikers, seamen Dallas Leroy Malone from Michigan and John Colby Sadler from Kentucky. A farm boy from Oregon, Mickey had followed his two older brothers into service—one of whom served aboard a sub tender and the other being in the Army in Italy.

It was forbidden by the Navy for anyone aboard ship to keep any sort of journal or diary. Engineman Bud McCoy, however, had acquired a little diary while on leave and he began scribbling notes each day. He and his buddies had long fought the great Winton diesels to keep them in operation. As the following excerpt from his diary suggests, McCoy and his fellow motormacs knew that the Navy Yard overhaul had not freed them from future engine issues.

> April 3, 1943: Left P.H. today on way to [Midway] Island; usual sea routine, morning on surface.
>
> April 4: Still on surface, made trim dive. Had training. Everyone in pretty good humor.
>
> April 4: Still headed the same way. Made a trim and a training dive. Got to give these new fellows some training. Don't like the idea of training new men in this point of the ocean. I don't feel so hot today for some reason. Guess not enough exercise...Wonder if my wife has gone home yet.
>
> April 5: Still traveling some place. Still more drills and dives. Getting tiresome.
>
> April 6: Same routine. None of the engines have quit yet, but the month is a pup yet.[15]

En route on 5 April, Roy Gross continued to drill his gun teams. At 1715, Gunner Bennett's 3" crew fired five practice rounds of target-detonating and five rounds point-detonating 3" practice shells.

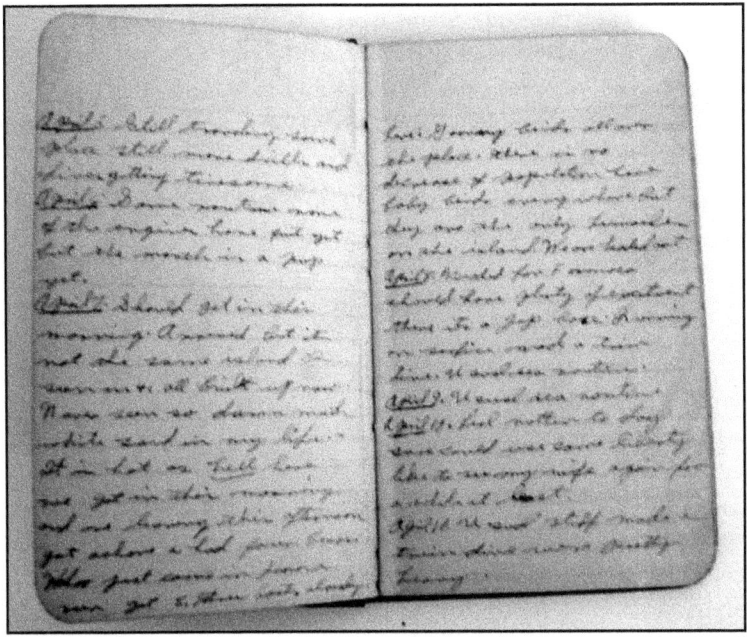

CMoMM Bud McCoy's personal diary: this photo shows his handwritten entries from April 1943 while aboard *Seawolf*. Although forbidden by Navy regulation, diaries like this were occasionally kept by submariners to help pass the time. *Image courtesy of Harold and Jeff McCoy.*

Seawolf's gun crews also practiced firing both 20mm guns. After dark, Gross opted to test some new flashless gun shells his boat had acquired. Bennett's team fashioned a target made of oil-soaked rags burning in a tin can and set them adrift. Sitting in the seats on either side of the gun, Lucien Rajotte and Big Swede Hanson handled the duties of pointer and trainer for the weapon. The 3-incher fired five rounds of point-detonating flashless shells at this target. "Flashless qualities appeared excellent," wrote Gross in his patrol report.

Seawolf moored alongside one of Midway's docks at 0858 on 7 April. Five other submarines were at Midway, including Cdr. Mush Morton's famous *Wahoo*, which had returned from her fourth patrol––during which she was officially credited with sinking an incredible *nine* Japanese ships. At 1204, there was an air raid alarm at Midway and *Seawolf* got underway. The alarm soon proved to be a false one and the *Wolf* returned to Midway's dock to finish refueling.

Some of the crew not involved with the refueling took the chance to get ashore. Bud McCoy wrote in his new diary:

It's not the same island I seen in '41; all built up now. Never seen so damn much white sand in my life. It is hot as hell here. We got in this morning and are leaving this afternoon. Got ashore; had four beers. Gooney birds all over the place. This is no decrease of population, having baby birds everywhere; but they are the only females on the island.[16]

Seawolf completed taking on water and fuel and was underway from Midway by 1630. Once underway, Lt. Cdr. Gross let chief of the boat Swede Enslin spread the word to the crew that they were headed for Formosa, where plenty of action was expected. The *Wolf* crossed the 180th meridian at 0309 on 8 April and advanced her calendar to 9 April.

Exec Bill Deragon continued to lead the ship in training dives and battle problems daily en route to station. Chief Rajotte's auxiliary gang had to work on the gyro when it crapped out on the second day out from Midway due to low vacuum on the north rotor. All gaskets were found to be rubber and these were replaced with brass gaskets from the spares. The gyro performed satisfactorily thereafter.

The crew passed their off duty time reading books and working on puzzles in Kelly's Pool Hall. Ship's cooks Bill Mallory and Gus Wright were old pros at whipping up special dishes to help pass the monotonous days to station. Newcomer Al Barboni, a ship's cook third class who had just joined the *Wolf* at California, was still considered a "bum cook" by some of the crew. To top off things, the beloved washing machine "Baby" was broken by an unfortunate crewman who soon endured the wrath of his shipmates—forced to wash their clothes the old-fashioned way, in a bucket again.[17]

Bud McCoy was looking forward to action, but wondered how the greenies would handle counter-attacks. "Don't care to hear any depth charges this early," he wrote. "Wonder what these new fellows will do when they get bad. I think I know."[18]

On 13 April, one of the lookouts sighted a Japanese two engine land bomber at 6 miles distance. The older SD radar had not made contact with this plane. Jim Mercer cleared the bridge and the *Wolf* made a crash dive. This plane was probably out of Marcus Island, which was 220 miles distant. No bombs fell, so *Seawolf* had apparently escaped the bomber's attention. She returned to the surface and resumed her course.

"I stood my share of lookout watches, being a seaman, second class," recalled Delbert Mar. "In most cases, we saw them before they saw us. We had added radar equipment in 1942, but us lookouts were still pretty good at picking up planes."

At 0927, Gross received an Ultra from Pearl Harbor which reported a freighter near Guam. *Seawolf* changed course to the southwest and increased speed to three engines, making 16.5 knots. He submerged his ship at the assigned position at 0849 on 14 April and lay in wait all through the day and the following night. The following day, 15 April, the freighter *Kaihei Maru* appeared right on schedule. At 1614, Seaman Red Mills called out, "Mast coming into view at about 8 miles."

Mills had spotted the upper works of a large freighter which was escorted by a small, coal-burning trawler escort. Lt. Mercer called for the skipper and tracking commenced at once. The ships were using radical zigs, making 10.5 knots through the water and 7 knots good along their base track, general course north-northeast. They were heading for Marcus Island, with about 240 miles to go, giving *Seawolf* two nights and one day to work with. The large maru was at first estimated as a 10,000 ton motor freighter.

The time waiting for the ships to approach close enough to attack seemed an eternity. In his diary, engineman McCoy scribbled, "About dark now. We can hear their screws and see the ships. Hope this old man knows his stuff."[19]

Roy Gross was more than ready to prove his skill. He worked ahead of *Kaihei Maru* on his port flank and got in good position ahead of the maru just as darkness closed in. There was clear, bright moonlight and a calm sea to contend with. Visibility was 7,000 yards or better. *Seawolf* dove at 7,000 yards range, using her SJ radar at 40 feet in to 4,200 yards. Thereafter, bearings and angle on the bow were used only, the enemy's speed being constant at 10.5 knots. *Seawolf* remained at periscope depth—63 feet—for attack.

Quartermaster Hank Thomson was now one of the old hands in the conning tower. "Roy Gross' voice became softer and softer during an attack to the point we had to tell him to speak up," Thomson later recalled. "Captain Warder's was just the opposite."

Thomson passed the word to the after torpedo room to flood their tubes. Senior torpedoman Bill Reiland opened the air pressure to flood the torpedo tubes. The air was vented back into the room to prevent telltale air bubbles from escaping to the surface. Gyro settings came down from the TDC and Thomson repeated the orders to set the tin fish to run at 16 feet depth.

Roy Gross found the setup to be under "ideal conditions." He could see his target almost as well as if in daytime, due to his periscope being treated for night vision. With the range down to 1,950 yards, he called out, "Fire one!"

The tubes were fired both electrically from the conning tower and manually by Reiland's men. With a slight jar and a whooshing sound, the first Torpex-loaded war fish sped on its way at 1925. Fourteen seconds after firing the first torpedo, *Seawolf* was rocked by an explosion as the warhead detonated prematurely. Roy Gross saw spray blown in a line ahead of the target and the resulting plume obscured the target as the remaining fish were being fired. The remaining three torpedoes were fired in a spread eight seconds apart. "Points of aim in doubt due to target being completely obscured by premature," Gross noted.

Reiland's tubes had been flooded only five minutes before firing and the outer doors were cranked open only two minutes before firing. Gross indicates how the premature explosion affected his remaining war fish.

> Should have held up other torpedoes, but the Commanding Officer was so amazed at having the first war shot he has ever fired be a premature—particularly after having requested that the magnetic feature be left operative—that he forgot to hold up the others, and they left at their assigned 8-second intervals. Gyro angles were small, tracks were favorable, range good, speed certain. Depth settings 16 feet, sea flat.

By the time the last torpedo had been fired, the enemy ships were fully alerted by the premature torpedo explosion. Both *Kaihei Maru* and her trawler escort fired their guns in *Seawolf*'s general direction. Through the scope, Roy Gross noted their muzzles blasting orange, circular flashes in his direction. The escort immediately turned to head for *Seawolf*.

"Take her deep!" Gross called at 1954.

As *Seawolf* was going under, sound operators Paul Maley and Ed Hinson, after torpedo room and other personnel all heard an explosion that was clearly not a depth charge. This came 90 seconds after the first torpedo had been fired, which timed as a hit for the second torpedo with about a 2,000-yard run. Bill Whitman's TDC range had been 1,900 yards for the second fish. "Other two missed due to maneuver of target, as he had over a minute after the premature to evade," wrote Gross.

Diving officer Red Syverson took the *Wolf* deep to 250 feet. The last sight of the trawler showed her smoking heavily and racing toward *Seawolf* with a bone in her teeth. BOOM! BOOM! Two depth charges exploded at 1957, not too close. One minute later, a third exploded safely distant. The trawler circled around and made another run five minutes later. A fourth deadly ashcan exploded at 2004, also safely distant. Hinson and Maley continued to track both the escort and target ship's screws during the next hour.

By 2105, the screws were fading out and Lt. Cdr. Gross brought his boat to periscope depth. After a quick sweep, he ordered her to the surface to chase these ships toward Guam. Bill Whitman and his battle lookouts took their stations as *Seawolf* raced in pursuit. By 2137, the SJ radar had the ships and lookouts could see the maru via binoculars at 15,000 yards. RT2c Ben Rogers was credited with picking up the enemy ships on radar. Gross started around the enemy's port flank again for another attack. His electricians charged the batteries with one engine during the end around. After a 90-minute run, *Seawolf* was in good position ahead of *Kaihei Maru* and dived, with radar range showing to be 10,000 yards.

Everything checked out on the TDC as Gross called out the angles on the approaching ships. At 2318, he eased up to 40 feet to let Rogers take his last radar range of 4,200 yards. The new radar beat the old system of taking only sonar pings of targets to confirm ranges. Thereafter, tracking officers Jim Mercer and Jug Casler worked on their plotting table with called bearings and angles on the bow reported by Gross. By 2322, the target ship was filling half a field of Gross' periscope window. He estimated the target's length at 550 feet and speed to be 10 knots.

"Open forward tubes," he called and Thomson repeated the order forward.

This time, *Seawolf* would fire Torpex-loaded torpedoes set to explode on impact. John Gibson's forward gang set them to run at 16 feet again. The tracking party figured a 100° starboard track—a virtual straight shot—with the range dropping down to 1,650 yards by the time of firing at 2329.

"Fire one!"

Gross shifted his point of aim from her midsection to the bow. Eight seconds passed and he called, "Fire two!"

Eight more seconds, and a shift of the target line to *Kaihei Maru*'s stern and he whispered, "Fire three!"

Seawolf shuddered as each torpedo was belched forth toward this large transport ship.

"All three running hot, straight and normal," called Paul Maley from sound. His assistant, RM2c Ed Hinson, kept track of the escort vessel while Maley monitored *Kaihei Maru*.

Two explosions were felt at about one minute ten seconds and another five seconds later. Gross witnessed only one torpedo explode but black smoke immediately obscured the ship. "Believe two hits, but do not know which one of the three missed," he wrote. He saw black smoke but no flames on the target after the explosions.

At 2333, Gross watched the escort trawler drop one depth charge at a phantom submarine. He kept track of the escort for a few moments until he wheeled around and headed directly for *Seawolf*. Gross took her deep as Maley continued to listen to the target vessel. He reported that her screws had stopped, indicating clearly that she had suffered severe damage at the least. The *Wolf* was rigged for depth charge and the men did not have long to wait. The first exploded at 2337, not too close. The angry trawler worked the area for 12 minutes before unleashing her next explosive.

Roy Gross kept his boat down another half hour, expecting a more severe depth charging. When he eased toward the surface shortly after midnight on 16 April, he could see *Kaihei Maru* was dead in the water with the angry escort circling her at full speed.

"Decided to fire two torpedoes at escort to allow a free hand in finishing off the target, even though he had the lines of an old shoe," wrote Gross.

Seawolf commenced her third approach on these two ships, with Gross planning to use his bow fish again. He eased in to 1,600 yards from the circling escort and lined him up for a straight shot with a port 90 track. Estimating his draft to be 10 feet, Gross had the torpedoes set for eight feet and fired two bow tubes at 0119, spreading them to hit bow and stern. There were no hits, so Gross decided one of the fish must have been a dud. More likely, they both passed too deep. The escort must have spotted the fish passing under, for he opened fire with his bow gun. Two distant end of run explosions were heard about 3.30 and 5.30 minutes after firing.

"Escort appeared baffled and we withdrew at periscope depth," logged Gross. "By this time, we had fired nine of our 20 torpedoes, with not much to show for it except one big ship stopped." At 0143, he decided to abandon further torpedo fire at the escort. He would instead put two more good shots into the main target, *Kaihei Maru*. With only 45 minutes of moonlight left, he started maneuvering for position once again. More than nine hours had passed since this transport had first been sighted.

Seawolf set up to once again fire her bow tubes. Her target speed was 0, making Bill Whitman's TDC work easier. From 1,400 yards, the fish were fired at *Kaihei*'s stack with no spread at 0152 on 16 April. Watching through the scope, the skipper clearly saw "one big explosion at base of stack" just one minute after firing. "Sound lost both torpedoes in the explosion, but had reported both running dead ahead," wrote Gross. "Assumption is that one torpedo either ran deep and under without detonating, or it hit the side without exploding." *Kaihei Maru*'s bow immediately dipped under all the way to the midsection stack as the ship broke in half.

The escort ship raced toward *Seawolf* once again and Gross took her deep to rig for depth charge attack. The Japanese must have been low on charges, for only one depth charge exploded as the *Wolf* retired at deep submergence.

Returning to periscope depth at 0250, he found nothing in sight. *Seawolf* surfaced. Ben Rogers' SJ radar quickly picked up two pips, one large and one small, almost immediately at about 7,500 yards, where the target and her escort had been left. Gross was distressed. "Decided to withdraw to 10,000 yards and await dawn to see what manner of ship this was. The expenditure of eleven torpedoes without sinking anything, yet getting four hits, gave cause for concern."

As the *Wolf* continued to stalk her prey with dawn's approach, many exhausted crewmen grabbed a quick meal. By 0438, it was serious business again as the boat dived and began an approach on the scene of the night attack. By 0752, he was able to get a good look at the *Kaihei Maru*. "Bow broken off at bridge, which was only slightly less than half her length, port side badly torn up in way of stack and a little aft of stack, down by the head 10 degrees, with a 15 degree port list, no signs of life," the skipper wrote in his report. "Escort passed within 300 yards of us while studying picture."

Gross decided to patrol submerged all day in the vicinity for the possibility of either the escort leaving—allowing deck gun practice on the derelict—or maybe a fast destroyer coming out of Marcus Island, a ten hour distance for a warship making 20 knots. "Neither happened, but a 2-engined land bomber circled the spot most of the morning," Gross wrote.

Seawolf surfaced at 1929 and Gross opted to move on. "Decided that neither derelict nor escort worth more torpedoes since only nine remaining on board, and that gunfire would be foolhardy with the escort still hovering around," he wrote.

In his diary, motormac Bud McCoy agreed with the skipper's logic. "Escort still there. Sure glad the old man didn't try and battle surface on the escort boats. They got bigger deck guns than we."[20]

Roy Gross found it hard to leave this battered cripple behind that ComSubPac had found important enough to direct him to via an Ultra. *Kaihei Maru*, a 4,575-ton armed transport ship, had suffered mortal wounds from *Seawolf*. Soon after the *Wolf*'s departure, *Kaihei* slipped under the waves and headed for the bottom, her demise later confirmed postwar.

Lt. Cdr. Gross set course for his patrol area at 11.5 knots. His fuel was already lower than his patrol plan called for due to the end arounds chasing *Kaihei Maru*, which had added some 800 miles to the distance to his designated patrol area. The *Wolf* dove after dawn and moved toward the Bonin Islands. "Running on the surface again in middle of Jap country," wrote McCoy. "May see plane or ships any time. Everything will be swell if we see them first."[21]

At 0112 on 19 April, the SD radar picked up Kita Iwo Jima Island of the Bonins (not to be confused with the Volcano Islands' more famous Iwo Jima) at 23 miles. This island's central mountain rose 2,680 feet high, making a nice radar target. A short time later, the lookouts could see the island.

During the early morning hours, *Seawolf* dodged sampans that were fishing with their running lights on. Bigger game was spotted

"Hope This Old Man Knows His Stuff" 233

by the lookouts at 0924. On the horizon were the masts of a small tanker on a northerly course. Roy Gross decided he was en route from Yap to Yokohama, speed 10 knots with no zigging pattern. *Seawolf* raced on four engines up the tanker's starboard side.

Two hours after sighting the little tanker, *Seawolf* dived in good position. Gross' recognition team flipped through OpNav Serial 890316, an intelligence book dated 6 March 1942, which was now already obsolete in Lt. Cdr. Gross' opinion. The best that Bill Deragon and yeoman Robert Tranquilly could find was a similar picture under the "trawlers" section, but without any data. Gross felt that his current target was larger than that in the book and had well decks both forward and aft. She was about 250 feet in length, estimated at 2,000 tons, and had the numerals 360 painted on her bridge. In this case, the estimates were exaggerated, for this ship was later found to be the 389-ton naval auxiliary *Banshu Maru No. 5*.

A sonar ping range at 1,950 yards from Paul Maley checked the ship's masthead height as 67 feet. Gross let the little *Banshu Maru* come in to 1,100 yards before firing two of John Gibson's bow tubes. These were straight shots with influence TNT heads, set to 12 feet to rip her bottom out. Both torpedoes ran hot, straight, and normal. The first, aimed forward of the bridge, hit exactly as planned 45 seconds after firing—ripping open *Banshu Maru*. Since the range was known to be accurate, the second torpedo should have hit astern. Gross wrote off this miss as a control error due to not knowing the target's true length.

Banshu Maru circled wildly after the torpedo explosion and then began to sink, bow down to starboard as her crew abandoned ship. She was gone ten minutes after the *Wolf* opened fire. As *Banshu* went under, four or five rapid explosions rumbled one minute later. They were quite heavy and were believed to be her depth charges going off as they reached their depth settings. "Made a hell of a noise down here and shook the whole boat," wrote Bud McCoy.[22]

Five minutes later, Gross ordered his boat to the surface. The battle lookouts took the shears and she began to close the wreckage. About 30 survivors were noted in boats and clinging to wreckage. *Seawolf*'s bridge guns were manned and ready in case of trouble. Swede Enslin's deck party fished out floating objects from the water with the grapnel hook. They managed to pick up three life rings, one briefcase, charts, a tide table, and other objects deemed worthy of

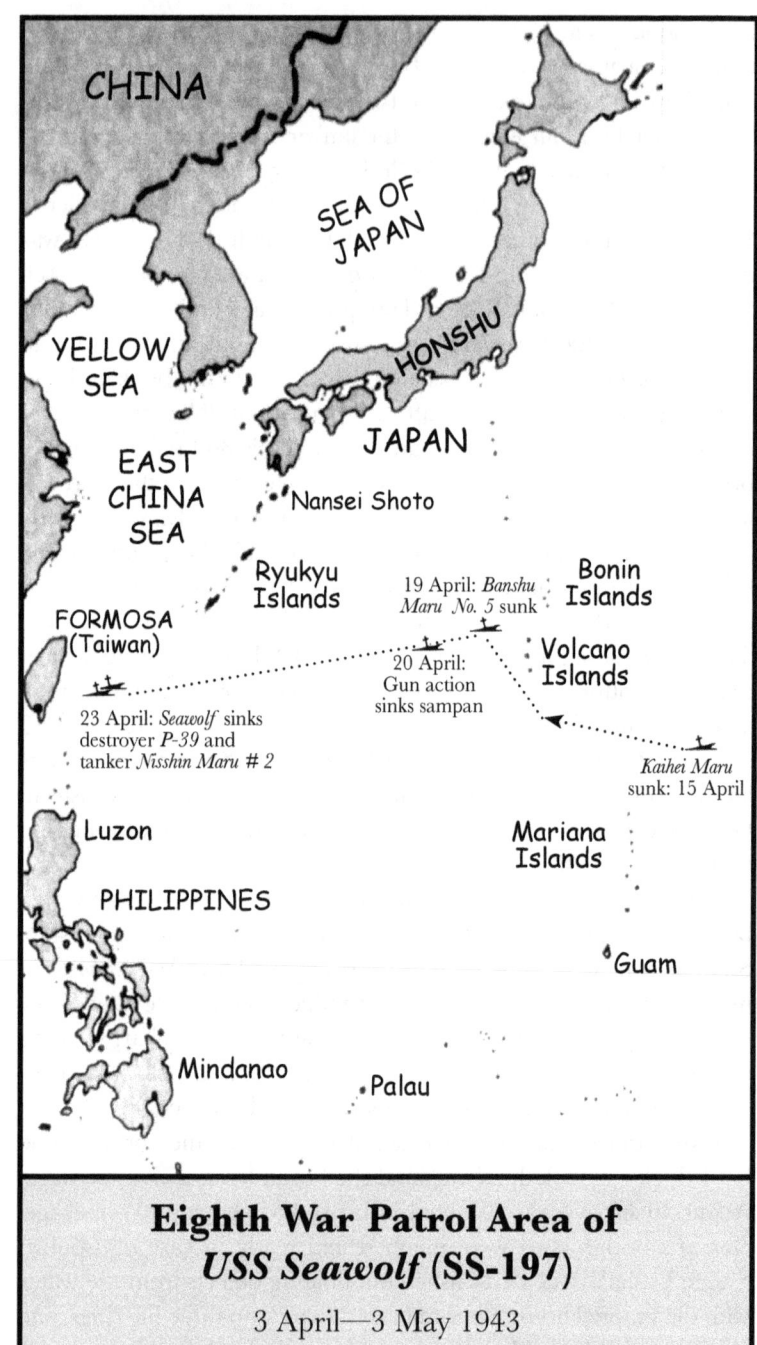

"Hope This Old Man Knows His Stuff"   235

intelligence review. Gross noted a "heavy gasoline smell, but small slick" on the surface. "Must have been empty."

Amazingly, there was no record of this ship's loss in postwar records and JANAC did not give *Seawolf* credit for a sinking. Many submarine skippers were frustrated postwar to find that ships they had witnessed to sink were not officially credited to them by JANAC due to poor Japanese records. *Seawolf* was stripped of this kill by JANAC, but more recent research by John Alden shows that the 389-ton auxiliary ship *Banshu Maru No. 5* should be credited to the *Wolf*. One of his three sources, *Warships of the Imperial Japanese Navy*, shows that *Banshu Maru* was listed as missing as of 15 April 1943.[23]

Had *Seawolf* picked up one of the Japanese sailors from the water, the postwar controversy might have been averted. "Survivors were the local problem of the skipper," Gross later wrote. "My policy was to not pick up anybody on the way out, but if it could safely be done, do so on return, for intelligence purposes." He was never given "any written, nor oral, instructions with regard to survivors."[24]

One of the life rings was marked *Bansiu Maru No. 5*, although Gross did not feel that his target resembled the same-named ship in his March 1942 OpNav recognition book. "All seemed stoical and crouched under nearest protection, no doubt expecting to be gunned," he wrote of the Japanese survivors. "Had machine guns ready but did not fire. Took pictures both submerged and surfaced. Shortly after surfacing, KGEI was heard to be broadcasting a resume of *Seawolf* past exploits, based on recent pictures in *Life* magazine."

The skipper allowed men to come up to the conning tower to take a look at the scene. For some, it was a very rare opportunity to witness the aftermath of a *Seawolf* kill. Motormac Bud McCoy was quick to come up from the forward engine room to take his view.

> The sea around for a half of [a] mile was covered with wreckage and stuff floating. I got a chance to look through the periscope and seen about twenty-five Japs in the water. Some were in a partly-filled lifeboat and some were hanging on to boards or boxes or anything that was floating. Some of the fellows wanted to shoot them, just like they do our men, but the Old Man wouldn't let them. They wouldn't miss the chance if we were in the water. If I got a chance to shoot one, stand clear for some fancy shooting.

After we picked up all the papers they wanted, we got underway for other parts. Some of the Japs drowned while we were picking up what we wanted. All maps were mostly in Japanese. They even had a map of Fremantle, Australia, in the bunch, our old home.[25]

Seawolf abandoned the area and moved through the night without further contact. She was now roughly 300 miles from Formosa. At 0614 on 20 April, one of the lookouts, torpedoman striker Dallas Malone, sighted a sampan at 5 miles. She was a three-mast version, larger than the two spotted the previous day. Gross ordered his gunners to man their battle surface stations and he charged toward this vessel at four engine speed. John Bennett, a plankowner, had never fired his deck gun against an enemy ship. Freddie Warder had been reluctant to put his men at undue risk, but Bennett found that his new skipper was more than willing to use the guns.

At 0642, Gunner Bennett's crew fired a 3-inch shot across the sampan's bow at 2,500 yards. The sampan immediately released his fishing nets and aggressively turned right towards the *Wolf* at his best speed. The Japanese opened up on the menacing submarine with their small arms—believed to be .25 or .30 caliber fire—although their shots fell well short.

"Sink her!" Gross yelled down to gunnery officer Bill Whitman.

Bennett's deck gun fired point-detonating shells that exploded at the set range. *Seawolf* made two passes, coming to within 2,400 yards on the first run and on in to 1,500 yards on the second. The sea was smooth, making fire control easy. Some misses were caused by the gun sights being full of water. After about 15 rounds, the gun crew had the little ship holed and burning. Gross estimated the sampan's size at about 75 tons. He then closed the distance and let his 20mm gunners expend six pans (360 rounds) on the vessel, while the deck gun continued to pump 3-inch shells into the target.

"Cease firing!" Gross ordered at 0700.

Shellman Red Mills felt the pains of haste. Excited to make his first gun battle, he had raced out on deck to pass shells without bothering to stop for cotton that was being passed out by the pharmacist. After standing near the deck gun for 15 rounds, his ears were ringing and his hearing would never be the same. Mills opted never to say anything to the Navy about the hearing problems he would suffer.[26]

Seawolf's gunners fired 34 rounds of the heavy deck gun shells and obtained a dozen good hits. The little ship was afire furiously amidships. "Point detonating with tracer seems to be fine ammunition against wooden craft," Gross noted. "All but one appeared to burst, mostly inside, it seemed."

Seawolf circled nearby for 30 minutes, looking for any intelligence data worth recovering. Finding nothing important, Gross opted to move on at 0730, leaving four men still alive on her stern, with a large fire and oil smoke amidships and forward. "After we sunk the ship with our deck gun, the captain handed out little bottles of brandy to the survivors," recalled auxiliaryman Paul Zimmerman.

"Someone on topside fished a big Jap flag out of the water and a glass marker buoy," Bud McCoy wrote in his diary. "There were a few Japs in the water but too far from land to swim in. So a few more will make shark bait." He noted that John Bennett's gunners worked on the water-logged gun sights after the action, insuring that "the next one will get a hell of a pounding."[27]

"Clear the bridge!"

After two days of uneventful patrolling, *Seawolf*'s SD radar made contact with a bogey at 0544 on 22 April. The plane was at 4.5 miles and closing. In the half light of mid-dawn the plane could not be seen by the lookouts, but OOD Jim Mercer did the obvious.

"Dive! Dive!"

He took the *Wolf* down quickly and the ship braced for depth bombs. Fortunately, none came. Perhaps the pilot also had trouble spotting the American submarine in the first rays of dawn. Gross returned to the surface in a half hour and searched into the night hours. The following day, 23 April, marked *Seawolf*'s fifth day on her patrol station. It would be a busy day in her history.

Red Syverson sighted a darkened ship at 0335 about 12,000 yards away. *Seawolf* began an approach on this ship and found a great target in the bright moonlight. She appeared to be a badly damaged freighter, adrift with her stern badly blown up. Her main stack had been blown off from a hit under her bridge. She had a 30° list to port and her upper works were "so mutilated that original lines not clear, but with one set of goal posts still standing aft."

This ship was being circled by one civilian tugboat, one 1,000-ton trawler and one *Wakatake*-class destroyer. Clearly, the Japanese were intent on saving this ship. Gross passed the word to make ready all seven of his remaining torpedoes. After determining that this target group was stopped, he dived the *Wolf* at 0417 to begin an approach by daylight to further study the situation.

What *Seawolf* had found was the badly damaged Japanese tanker *Nisshin Maru No. 2*. Built in the late 1930s, this 17,579-ton ship was capable of carrying 140,000 barrels of oil. Due to heavy damage to *Nisshin*, Lt. Cdr. Gross initially took her to be a freighter. Lt. Cdr. Lawson P. "Red" Ramage had been directed by Ultra to take his *Trout* to the port of Miri on the west coast of Borneo in January 1943. He blasted the 17,000-ton *Kyokuyo Maru*, badly damaging her. Three weeks later, *Trout* returned to Miri and Ramage attacked another of the giant oilers on 7 February. He hit her with two torpedoes in broad daylight and fled from the Borneo harbor. Ramage was convinced *Nisshin Maru* had been fatally damaged.[28]

Nisshin Maru, however, was repaired enough over the next two months to be made seaworthy. From Borneo, she moved toward Japan via the Philippines. When *Seawolf* found the big tanker, she was in company with a tugboat and two escorts just east of Formosa. Roy Gross knew that this big maru—537 feet in length—was important to the Japanese, regardless of whether she was an oiler or freighter.

As he began to more clearly make out and understand the size of the screening vessels, Gross weighed his options. With limited torpedoes left, he "decided freighter could wait and started jockeying for attack position on destroyer." He was circling the wrecked ship slowly and pinging on long scale. Once the destroyer was gone, Gross could fire on the derelict with his deck gun.

Seawolf carefully approached the Japanese man-of-war to point blank range. He swung to bring Bill Reiland's stern tubes to bear and let the range come down to 1,050 yards as the DD circled. The tin can was making only 5 knots. Bill Whitman on the TDC cranked in the solution for a 106 starboard track shot. At 0700, he opened fire on the 275-foot Japanese destroyer. The TNT-filled torpedoes were fired at four-second intervals with ten-foot depth settings.

Seawolf remained at 63 feet to watch through the periscope. One of the first three torpedoes made a solid hit amidships. "This hit was an influence hit, as spray came up from his far side," wrote Gross.

"Hope This Old Man Knows His Stuff" 239

The first torpedo hit at MOT, exactly as aimed, while the second and third missed. Both were heard to explode at the end of their runs.

Five seconds after the first hit, there was a boiler explosion heard aboard the *Wolf*. The destroyer then began belching white smoke from his two stacks. As the DD was not immediately sinking from the first hit, and instead making heavy white smoke, Gross ordered his fourth stern tube fired right behind the other three. He used 8 knots target speed—believing that he had sped up—and fired this one 90 seconds after the third torpedo. To his disgust, however, Roy Gross watched this fish pass harmlessly ahead of the escort.

The first hit was a good one, though. Continuing to watch the destroyer, Gross noted him to be dead in the water by 0709. Exasperated that the DD wasn't sinking, Gross decided to hit him with two more shots to expedite his sinking. Ten minutes after his last attack, the destroyer "has a slight list to starboard and very slightly down by the head."

Twelve minutes after her fourth torpedo had been fired at the DD, the *Wolf* launched one Mark 14-3A TNT-loaded forward fish, aimed as a straight shot right at the destroyer's middle. This torpedo went erratic and Gross watched the smoke and bubble track from the steam torpedo in the glassy sea take a right zig and pass ahead of the destroyer by about half a length. Upon leaving the tube, the Mark 14 had taken an initial angle 4° to the right of the angle set by John Gibson's torpedomen.

The range was a mere 1,300 yards and the ocean was calm. Keeping the depth setting at nine feet, Gross doggedly fired again after watching the previous bow fish go astray. His torpedo performance report sums up his feelings.

> After this disappointment, fired another torpedo just as carefully aimed at the middle again, nearly zero bearing and gyro angle. Target still apparently not sinking, but people dropping off astern. Second torpedo took a left zig of 15 degrees, then right 15 degrees, then left again to settle down on a course that missed astern this time, 1/2 target length. Smoke and bubble track clearly visible.

Despite two maddening misses against a sitting duck target due to faulty torpedoes, this destroyer had been fatally holed. Right after this

Three frames from a famous series of persicope photos snapped by Lieutenant Jim Mercer during *Seawolf*'s eighth patrol. The old Japanese destroyer *Tade*, redesignated as *Patrol Boat No. 39*, has her bow blown off by a *Seawolf* torpedo on 23 April 1943 and goes down with her tail in the air. In the inset, the DD takes her final dive. *Courtesy of Bob Hanson.*

last miss, he was seen to be in serious trouble as the internal flooding apparently got the best of him. Gross called for Lieutenant Jim Mercer to take more photos as *Patrol Boat No. 39*—an older destroyer originally commissioned as *Tade*—broke apart and sank during the next nine minutes. The sinking was slow and she upended, giving Mercer the chance to snap some of the war's most dramatic sinking photos. His series of shots of *Patrol Boat 39*'s demise would be widely printed and continue to surface in many a submarine book.

The old destroyer sunk by the head at 0718. Many Japanese sailors were left swimming in the water. About 30 seconds later, there was an extremely loud explosion, but no water column. Gross surmised that this was the DD's depth charges exploding at a very deep depth. "Shook the ship considerably," he wrote.

In his diary, motormac Bud McCoy wrote:

> She sunk in about twenty with a hell of a explosion—either her boilers or depth charges on its racks. There were a lot of Japs in the water and more jumping over the side. The destroyer looked like our old four-stacker, with only two stacks. From the sound of the can blowing up, I bet there weren't many Japs left. It sounded like a depth charge and shook the whole boat.[29]

Seawolf hung around the sinking site long enough for Mercer to take more pictures. At 0721, *Seawolf* then turned her attention back to the damaged tanker, which was now about 4,000 yards away from the site of the sunken destroyer. "Other escorts apparently unaware of destroyer's fate," Gross wrote incredulously. "Both [on] the other side of scene about 6,000 yards away."

At 0756, *Seawolf* fired her 20th and final torpedo at the giant *Nisshin Maru* from 1,350 yards with a 16 foot setting for her 20-foot estimated draft. Gross felt that this ship had previously been hit by two torpedoes, which is exactly what Red Ramage had claimed. Gross watched the smoke and bubbles from *Seawolf*'s final torpedo pass right under the center of the big tanker. There was no explosion, however, on impact nor was it heard to have an end of run explosion later. The upset skipper could only surmise that this faulty fish had run deep and also had exploder failure. If *Nisshin Maru* was hit by a dud from *Seawolf,* Roy Gross was unaware.

Frustrated, he addressed his crew at 0801 over the 1MC, "Prepare for battle surface!"

Gunner Bennett and his men opened the lockers and prepared their shells. *Seawolf* worked around to place herself between the freighter and the trawler escort. Gross planned to make a quick surface and then pump 24 rounds of 3" into the tanker's waterline and submerge before the trawler could bring his guns to bear, using the target ship itself as a screen.

At 0809, Gross spotted a small salvage tug heading toward the scene "making violent semaphore signals to trawler." It was apparent that the newcomer had spotted the American submarine's scope. Gross continued to jockey for a safe position to pull off his daring battle surface. At 0819, he was still approaching the good position when he saw that the trawler had turned and was heading for the *Wolf* at full speed, now 3,000 yards away.

Cursing his luck, Gross abandoned the idea of a battle surface.

"Take her deep!" he ordered as he slapped the handles against the sides of the periscope.

Red Syverson took her down to 150 feet and *Seawolf* maneuvered to temporarily withdraw from the scene to let the situation cool down. "Surprised at no planes here yet," wrote Gross. "Kirun on northern Formosa only about 100 miles to the NW."

After staying down just over ten minutes with no depth charges, Gross eased back up to the surface and took a look at 0832. The situation was still the same, but both the trawler and the tug were close aboard the freighter. "Decided to wait to see if they would open enough to allow us to battle surface."

At 0848, the situation changed again when Gross spotted a *Wakatake*-class destroyer approaching the scene from the south at 6,000 yards. Clearly, *Seawolf*'s sinking of the older destroyer had stirred up this area. Reluctantly, Gross had to decide against the battle surface. His 3-inch deck gun would not stand a chance against the newer destroyer and the armed trawler together so he retired to the east. Having sunk a destroyer, some of the crew felt the area was pretty well heated up already.

Seawolf clung to the area, remaining 5 to 10 miles away all day, taking a look every four hours, waiting to see if the escorts would leave at dark. All the while, her crew was praying that they had just a few more torpedoes to bag another unlucky tin can.

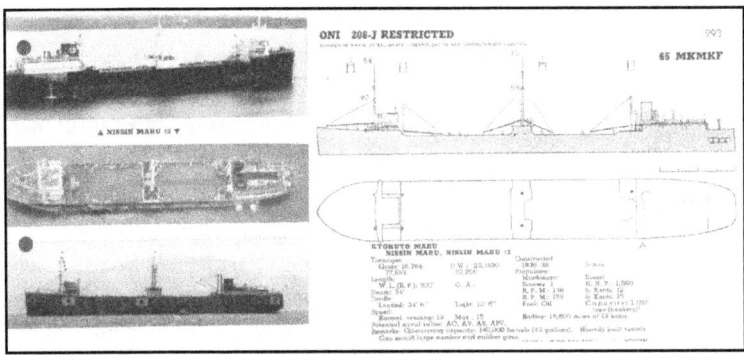

Nissin Maru # 2, badly damaged in February by the submarine *Trout*, was being towed back to Japan when *Seawolf* found her on 22 April 1943. The *Wolf* first sunk her escorting destroyer before firing on the massive 17,000-ton tanker. *ONI-208J (Rev.).*

Four distant depth charges were heard as the spooked destroyermen blasted away at a phantom sub. At 1500, with nothing in sight, Gross began closing the scene once again. The *Wolf* surfaced 20 minutes after sunset and closed at high speed, charging her batteries. By 1850, the lookouts had smoke dead ahead and she continued closing. Radar picked up *Nisshin Maru* at 15,300 yards but the bridge had the big tanker in sight even before that.

Seawolf closed to 5,000 yards during the next hour. The salvage tug appeared to be alongside the crippled tanker. Another escort was on the far side of the target ship. Gross deemed it too dark to risk a surprise encounter with an unseen escort. Moonrise in another 20 minutes would reveal *Seawolf*'s presence and require her to dive. He believed the destroyer was still lingering somewhere in the darkness, because "some ship [was] flashing signals" to *Nisshin Maru*.

In the end, Lt. Cdr. Gross decided to retire and not risk a gun engagement in this situation. Better to go home, reload with torpedoes and live to fight again another day. An SD radar contact at 2150 helped convince him that he had made the right decision. The plane was 19 miles away but was likely covering the area for just such a possibility.

At 2355, *Seawolf* radioman Paul Maley transmitted a full report, saying that the ship had expended all of her torpedoes already and was returning to Midway. "We radioed the tanker's position," wrote Bud McCoy. "One of our other boats will be on her tomorrow. That will fix her."[30]

Although unknown to the *Seawolf* crew, the 17,579-ton tanker *Nisshin Maru No. 2* never made it to Japan. In his research on submarine attacks of World War II, veteran submariner John Alden found two sources that showed *Nisshin Maru* was sunk by an unknown submarine off Formosa. Alden found that one source, *Warships of the Imperial Japanese Navy*, credited *Seawolf* with the kill—likely because she had sunk *Patrol Boat 39* in the same area. Although *Seawolf* did not score a direct hit on *Nisshin*, she certainly disrupted efforts to save her for quite some time. It is possible that her 20th torpedo, a dud, punched a hole in her patched-up side and caused additional internal flooding that could not be controlled. It is also possible that the explosions on the destroyer and the ensuing depth charges dropped in the area did not help the weakened oiler's condition. In any event, the giant tanker originally torpedoed by Red Ramage's *Trout* slipped beneath the waves—thanks at least in part to the harassment given to her and her escorts by *USS Seawolf*.[31]

Out of torpedoes, *Seawolf* moved toward her departure point at two engine speed while charging her batteries with the other two diesels. At 0628 on 24 April, Jim Mercer dived ship upon sighting two unidentified planes on the starboard bow at 8 miles away and closing. "No pip on radar," Gross noted. The aircraft passed on by and *Seawolf* resurfaced at 0647. One of the planes was smoking, as if it had been damaged in combat.

"We are headed for our base. Should get there in eight or ten days," engineman Bud McCoy wrote in his diary. "Some sunshine, beer, and rest will sure hit the spot." With the long hours of making battle approaches behind them, the crew eased back into more normal daily routines. "Have been listening to station in Frisco and also Tokyo at night," wrote McCoy. "Stories sure don't match. Don't know which is the biggest damn liar."[32]

On 26 April, Gross gave his gunners another chance to shoot up enemy shipping. At 1406, torpedoman striker Tom Warren was on watch duty in the conning tower. His job was to make periscope sweeps to pick up more distant objects that might be just beyond the sight of the bridge lookouts. While making a sweep of the horizon, Warren suddenly stopped and then called, "Sir, there's masts on the horizon."

Seawolf closed this vessel and Lt. Cdr. Gross decided she was a three-masted 75-ton motor sampan. "We tried to only go after the ones we heard or thought were using radios on them," said Delbert Mar. "It cost a lot of money to sink these sampans." *Seawolf* closed and Gross ordered battle stations gun. At a distance of 2,000 yards, she fired one shot ahead of the sampan to give him a chance to stop. The sampan kept running so at 1426, Gunner Bennett's crew was ordered to fire. They used 3-inch point detonating rounds set to detonate at 2,500 to 1,000 yards.

As *Seawolf* closed to a mere 1,000 yards, the Japanese opened fire with a .30-caliber or .50-caliber machine gun that had been hidden on his deck house behind bamboo screening. Gross noted "splashes uncomfortably close." The deck crews still could not see the sampan's gun but continued to lob 3-inchers at her while the Japanese chattered back. *Seawolf* opened the distance to get safely out of the smaller gun's range. The deck crew then landed one shot that blew away the bamboo screening and silenced the Japanese gun, visible for the first time. A few 3-inch AP shells passed through the sampan without exploding. *Seawolf* moved in closer again and let her 20mm gunners chew up the vessel with incendiaries to start him burning.

The stubborn sampan would not catch fire or sink. Gross moved around, allowing Bennett's gunner to continue pounding the ship with 3-inch shells. After 77 rounds, about 30 direct hits had been made, many along the waterline. The 20mms had cooked off 480 rounds. "The sides of it looked like a screen door with the light shining through," wrote Bud McCoy.[33]

The sampan finally sunk to his gunwales by 1433. Gross now moved close aboard and stopped firing. He noted large pieces of ice floating and many glass balls. Four men were still alive aft in the water. McCoy recorded:

> The boat was loaded with ice and maybe fish. We could see cakes of ice floating around among the dead. When we left the Japs, we could see one alive in the water about 200 miles from land and his boat almost under. Bet he got tired swimming 200 miles.

Seawolf resumed her course and speed at 1455 to clear the area. Shortly after, the bow broke off the sampan. His bowsprit showed vertically and the stubborn ship disappeared. Only her mainmast remaining showing out of the water.

The *Wolf* passed through the Bonin islands during the night. Taking an inventory, Gross found that his ship had only 30 rounds of 3-inch point detonating shells left and 40 rounds of 3-inch armor piercing. "Will expend 20 rounds more of point detonating if opportunity arises against minor target, but decided to keep the A.P. and 10 of the point detonating for any emergency. Empty magazines and empty torpedo tubes are not believed desirable."

Al Hershey and Lucien Rajotte's auxiliarymen had their share of repairs during this patrol. The SD radar mast cables became fouled. The auxiliarymen had to cut one and renew it from a spare due to the mast not coming down as fast as its cables were unwinding. The packing was loosened and greased until all was in good order again. The TDC gang of Bill Whitman, Lefty Leffingwell, and Delbert Mar found that the *Wolf*'s own course input to the TDC became more erratic as the patrol progressed. After all torpedoes had been expended, Leffingwell found that the course tracking ceased altogether. This was deemed a mechanical failure that would require lifting the top half of the unit during the planned refit at Midway.

Seawolf made an undisturbed run in toward Gooneyville during the next week. Her crew was bolstered by Roy Gross' courage and tenacity to attack the enemy. The entries in Bud McCoy's illegal diary give some insight into the thoughts of the sailors as their war patrol wound down while approaching Midway.

> April 29th: Nothing going on so far this morning and I don't think there will be. We are getting close to base and should be in about 3 days. We will be there in a rest camp for a week among the gooney birds and hairy-assed sailors. That's all there is, not a female on the island except birds and a dog; but, can get a little cold beer. Washed clothes. They were so damned dirty could hardly bend them. But that's old stuff on these pig iron sewer pipes.
>
> April 30th: Still headed in and not much doing except we are holding field day cleaning up the boat. Sure is a lot of dirt around the engine room. Have a lot of oil and salt all over them. The chow has been good this run and has been pretty cool except once or twice.[34]

"Hope This Old Man Knows His Stuff" 247

The quartermaster gang busied themselves with updating the *Seawolf* battle flag. Radar operator Ben Rogers, something of a cartoonist in his spare time, sketched out a cartoon map of *Seawolf*'s eighth war patrol. En route to station, a sketch of the *Seawolf* was labeled with: "It's a contest all the way between the engines and [Art] Lamberson to see which can stow away the most fuel!" Another area shows a little Japanese ship sinking, labeled with "tiny tanker, sunk with rancor (and torpedoes)." The little sampan, nicknamed "Campbell's Sloop" by Rogers, was sunk by the deck gun with "only 2,468 rounds."

At 1400 on 3 May, a PBY was sighted on a northeast course, position 130 miles bearing 263 degrees from Midway. *Seawolf*'s leading signalman, Chief John Beatley, successfully exchanged recognition signals with the PBY and averted the possibility of an attack.

Seawolf crossed the international date line on 4 May and rolled the calendar back to 3 May once again. At 0500, she made a successful rendezvous with Midway air escort and entered the lagoon. Liberty was liberty, even if Midway didn't offer too much entertainment. "Old place looks the same," wrote Bud McCoy. "Not a damn thing but sailors, buildings, and sand. We go up to the hotel tomorrow for seven days. I sure can use some beer and some sleep."[35]

Lt. Cdr. Roy Gross, making his first patrol, had brought his *Seawolf* back after a mere 30 days with all torpedoes exhausted and much of his shells expended. Gross' patrol report included four pages of what he felt was "thorough and impartial analysis" on torpedo performance, including prematures, exploder failures, and control errors caused by target maneuvers or poor estimates by his team. Firing all 20 of *Seawolf*'s torpedoes, he claimed a 30% hit ratio: six hits, six control errors, and eight torpedo failures. Freddie Warder was thus not alone in damning the torpedo performance of the troubled Mark 14s.

Despite "improvements" to the Mark 14, other skippers continued to report serious performance issues in early 1943. During March, Lt. Cdr. Walter Gale Ebert's *Scamp* found plenty of targets on her first war patrol. Ebert, however, was outraged at the sinkings he missed when five of his torpedoes exploded prematurely, and two

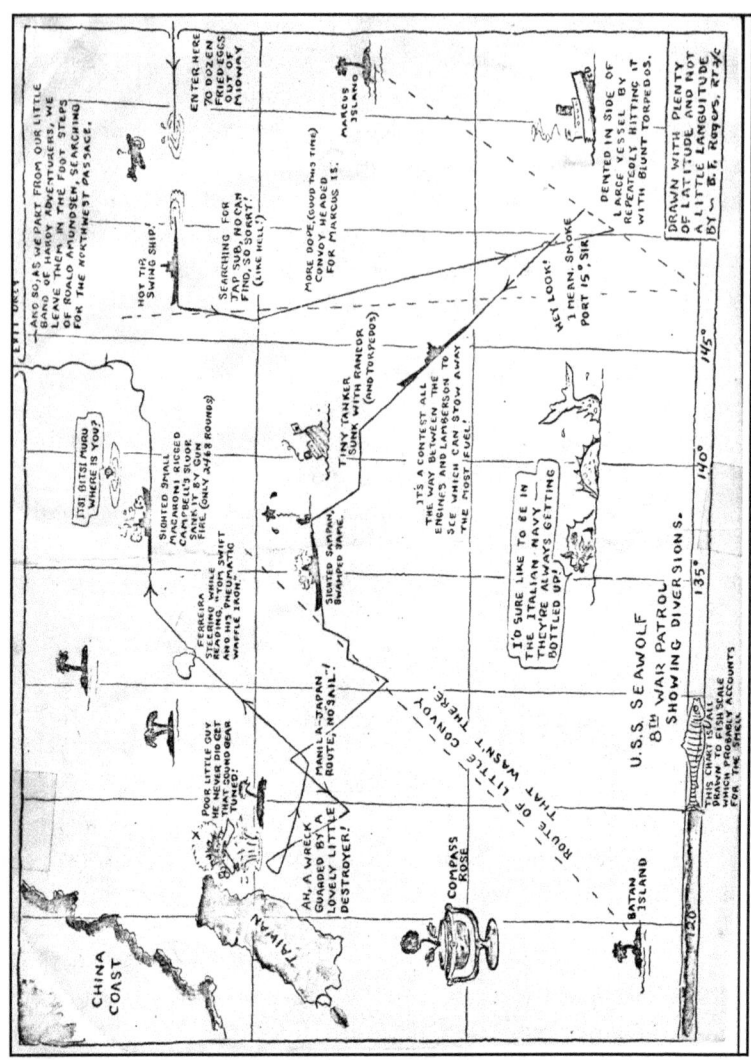

This cartoon map of the eighth war patrol of *Seawolf* was drawn by her radar operator Ben Rogers, who was later lost with the boat. Rogers made light of *Seawolf*'s gun and torpedo attacks, her faulty Mark 14 torpedoes, the eating habits of torpedoman Art Lamberson and the helm work of seaman John Ferreira. *Courtesy of Hank Thomson.*

ran erratically. "The great geyser of water is like waving a flag at a destroyer," Ebert boldly wrote in *Scamp*'s patrol report. "Most important, they break down confidence" within the crew. During the time of Roy Gross' torpedo problems on *Seawolf*'s eighth patrol, *Tunny* and *Pompano* both missed chances to sink Japanese aircraft carriers in April 1943 due to Mark 14s which exploded prematurely.³⁶

To offset possibilities that the poor torpedo performance could be attributed to his crew, Lt. Cdr. Gross noted that "exhaustive tests have been run to check out TDC, gyro setters, TBTs, to eliminate any possible source of error." He added that the "torpedo routine has been religious," thoroughly supervised by his capable torpedo officer, Bill Whitman. "It is felt that most of the torpedoes were fired under almost proof conditions," wrote Gross. "Rarely have so many torpedoes been fired against zero target speed conditions, which in itself resolves all natural and usual errors in estimates to zero."

Due to the unreliability of the torpedoes, Gross thought that other skippers should not rely solely on the spread method. He would instead opt to fire two fish at MOT and spread two others just to cover erratics or faulty exploders. "Having seen our first big ship receive four hits, 3 of which were torpex, and the stern still half afloat 12 hours later; and seen what is believed to be *Runner*'s [acutally *Trout*'s] target with two tremendous holes in her and her top hamper a shambles, still afloat several days later, it appears that an excessive number of hits actually is required to sink this type of target (large, well-built freighter)."

Roy Gross' report was reviewed by the senior commander, Rear Admiral Charlie Lockwood. He felt that Roy Gross and his *Seawolf* "continued her long and early outstanding war record with a successful and most aggressive patrol." Instead of being offended by the criticism on the torpedoes, he found Gross' detailed report of torpedo performance "most complete and helpful."

Lockwood congratulated the *Seawolf* crew "for an aggressive and tenaciously performed war patrol." He allowed:

1 freighter sunk	10,000 tons
1 Tanker (*Bansiu Maru # 5*)	2,000 tons
1 Destroyer (*Wakatake* class)	900 tons
2 sampans	150 tons
Total by Lockwood:	13,050 tons

JANAC's postwar credit for *Seawolf*—or any submarine, for that matter—did not count non-military vessel sinkings under 500 tons, so *Banshu Maru* and the two sampans sunk by gunfire on this patrol were not officially credited—although their demise was certain. With the additional 17,579 tons from the *Nisshin Maru*—aided to Davy Jones' locker by the *Wolf*'s persistence—the true tally for the eighth patrol was six ships destroyed for 23,678 tons.

Certainly Admiral Lockwood felt that the *Wolf*'s first patrol under Googy Gross—although hampered by the poor Mark 14 torpedo performance—had been carried out with the same fearless style of Freddie Warder.

USS Seawolf Eighth Patrol Summary

Departure From:	Pearl Harbor/Midway
Patrol Area:	Bonin Islands, Formosa
Time Period:	3 April–3 May 1943
Number of Men Aboard:	78: 70 enlisted and 8 officers
Total Days on Patrol:	30
Fuel Burned:	82,000 gallons (from Midway)
Total Miles Steamed:	8,954
Number of Torpedoes Fired:	20
Ships Credited as Sunk:	3/13,050 tons
JANAC Postwar Credit:	2/5,395 tons plus two sampans
Credit by Alden research:	6/23,678 including two sampans
Return To:	Midway

USS Seawolf viewed from port side on 7 March 1943 following her refit at Mare Island Navy Yard. This view was taken off California just before her departure for Pearl Harbor and her eighth war patrol. *Official U.S. Navy photo.*

11

"The Next One's Gonna Get Us"

Ninth Patrol *17 May–12 July 1943*

Submariners who had made R&R periods in Hawaii or Australia early in the war did not hold refits at the Midway Islands in high regard. Returning to Pearl Harbor was usually reserved for boats that had made several straight war patrols. For *Seawolf*—having recently spent time on the West Coast for overhaul—the Midway refit was in order following her eighth run. The Submarine Base, Midway Repair Force started to work on 3 May and a ComSubDiv 44 relief crew came on board as the *Wolf*'s crew went ashore.

The crew stayed at the former Pan American Hotel, which had served as a stopover for airplane passengers in the China Clipper days. The two islands, Midway and Eastern, were literally infested with Laysan albatross, more commonly known as gooney birds. The old Pan American Hotel—a wooden, one-story structure which became known to sub crews as the Gooneyville—did not offer the same luxuries as Pearl Harbor's Royal Hawaiian. "We are just laying around, eating and sleeping; drinking beer in afternoon from 2 o'clock to 5 and from 7 in the evening to 8:30," wrote Bud McCoy. "Sure taste good, even if it is the worst beer I ever drank."[1]

Bob Hanson also found little excitement on Midway.

> There was not really much to do but watch the gooneybirds. Pretty soon, you had people acting like them. The Marines had to go down the runway with trucks to get them off the runway before planes could take off. Some of our guys would feed the gooneys beer. They would get a little

Seawolf arrives back from patrol in early 1943. On the bridge are Lt. Bill Whitman (left) and Lt. Cdr. Roy Gross (right). *Courtesy of Submarine Force Library, Groton, Connecticut.*

tipsy. They would fall and flop on one wing—even without the beer. They were used to landing at sea so sometimes they would forget to put down their landing gear on the beach.

The clumsy, beer-drinking birds were entertaining to a point. "The fellows have a hell of a time getting the birds drunk on beer and then watching them," McCoy wrote. Delbert Mar recalled, "You've never seen such a funny thing as when the gooneybirds land. They just crash and flop around." After the men tired of watching the clumsy seabirds, they played horseshoes, drank beer, or tried to catch flights with some of the aircraft going out on scouting missions.[2]

Chief John Bilkey found Gooneyville inferior to R&R at Pearl. "Midway, as far as I'm concerned, is an abandoned island. The only advantage was that it was nice and warm and you could swim if you wanted to or play baseball with other ships," Bilkey said. "As far as living on Midway—I wouldn't!"

The refit was three days longer than had been planned due to a string of enemy air alerts during May 8–10. A Japanese task force was reported hundreds of miles from Midway, which meant that bombers could reach the island by dawn. The crew raced back to the ship and backed her clear of the dock. The whole event turned out to be a false alarm, but the island was kept on high alert for days. This meant that some of *Seawolf*'s crew had to remain aboard on duty while those off duty could stay at the Gooneyville at night. Each room could draw one case of beer per night for their enjoyment.[3]

During these days of high alert, yeoman John Burruss used the time to tackle the necessary business of transferring men to new assignments and drawing replacements. Prior to *Seawolf*'s departure for her ninth patrol, Burruss would transfer 17 and receive 17 new men. *Seawolf* lost some of her most experienced veterans, including Hank Brengelman, Otis Dishman, Rudy Gervais, John Gibson, Al Hershey, Art Lamberson, Doc Loaiza, Sandy Randazzo, and Gus Wright. Among those eager to leave was radioman Joseph Strong, who had missed his planned 30-day-leave Stateside when he was tagged as a late replacement for *Seawolf*'s eighth patrol. "I got off her at Midway," Strong stated, "which was fine with me because the beer was just as good on Midway as anywhere else."

The loss of three senior torpedoman necessitated changes in both rooms. Bill Reiland took over the forward room and the other first class torpedoman, Clarence Kibbons, took charge of the after room. TM2c Alfred Ostrander, a veteran from California, came aboard to help fill the void created in the after room.

Among the other new hands joining for the ninth patrol was 21-year-old MoMM2c Vernon Palmer Wall, a native of Council Bluffs, Iowa. He had served aboard the sub tender *Fulton* in early 1942 and then started tending submarines at Pearl Harbor. Wall, an avid boxer, had won the Silver Gloves of the Pacific and was nicknamed "Iron Duke" by his shipmates. He joined *Greenling* as a fireman and was aboard her for her third and fourth war patrols, during which time *Greenling* earned the Presidential Unit Citation. He then trans-

MoMM2c Vernon Palmer "Duke" Wall, a boxer from Iowa, joined *Seawolf* prior to her ninth patrol. Wall served on the *Wolf*'s deck gun crew and was with his ship when she was lost on her final war patrol. *Courtesy of Charles Hinman, USS Bowfin Museum.*

ferred to Submarine Division 44, from which he joined *Seawolf* on 17 May 1943, the day she departed on her ninth patrol. Duke Wall was an able engineman, serving as throttleman in his engine room, the number one loader on the deck gun crew, and assistant to the new pharmacist's mate, Bill Hadley.[4]

With the transfer of chiefs Dishman and Hershey, and with Swede Enslin as chief of the boat, Chief Lucien Rajotte was now the leading enlisted man in the engineering department. "I was chief motor machinist's mate in charge of all the machinery," he said. "During battle stations I usually stood watches as chief of the watch." In addition to being over the senior men in both engine rooms, Rajotte would direct the efforts and schedules of the *Wolf*'s auxiliarymen.

As the crew reported back aboard, they found that the relief crew at Gooneyville had things in order. Since the prior patrol had been short, the repairs had been minimal. The refit crew worked on the TDC, which had begun acting up during the patrol, and replaced the No. 2 periscope with a base spare because of spots in the focal plane. *Seawolf*'s refit was completed on 14 May and then Lt. Cdr. Gross put his boat out to sea that day to break in his new hands.

He returned to Midway that evening and then put to sea again on 15 May for a second day of training before beginning his next patrol. Roy Gross' crew had the "unpleasant experience" of witnessing a Marine SBD dive bomber fly into the drink this day. The Dauntless went into a spin while pulling up after zooming *Seawolf*. It then slammed into the ocean about 200 yards away from the ship.

His 500 pound depth bomb exploded three seconds after impact causing no damage to *Seawolf*. There was nothing left of the plane or its occupants except a spot of burning gasoline, a small tank, and a few burning carbide marker bombs. "The jar below-decks was considerable," wrote Lt. Cdr. Gross. Hank Thomson was the duty quartermaster at the time. "The zoomies were always making practice runs," he said. "This guy got a little careless."

In his secret diary, Bud McCoy noted the tragedy.

> May 15th: We were out test diving again today and came close to making our last dive. A dive bomber of ours was making practice run on us and something went wrong with the plane. He was headed at us very close to the water and crashed into the sea about a hundred yards from us. The plane blew up and sank as soon as it hit. Nothing came up but a piece of wing and fire where the gas spread on the water. The bombs he had on his wing blew it to bits. I guess the pilot seen he was going to hit us, so he crashed it in the water. It landed upside down. It sure made a hell of a noise. I thought at first we had been hit by a torpedo, as it shook the whole damn ship.[5]

Seawolf returned to Midway and began final preparations for patrol. She was loaded with Mark 14 torpedoes during the night hours, as Midway was expecting an air raid during the morning. *Runner* and *Pollack* were also loaded the same evening. During the final changes prior to patrol, Ens. Hubert Gluski was transferred to *Runner*, which had come into Midway for refit on 6 May. *Runner* departed on her third patrol on 28 May, but neither she nor Gluski were to return—presumably the victims of a minefield off Honshu.

In place of Gluski, *Seawolf* received Lt. Clary Leonard John, an Arkansas native and Academy graduate who would be making his first submarine war patrol. Gluski and Lt. Cdr. Bill Deragon were both transferred at 1200 on 17 May, the day the *Wolf* was heading out for her ninth run. A veteran of all eight previous patrols, Deragon had orders to new construction to command *Pipefish*.

In his place, Lieutenant Commander Robert Dunlap Risser reported aboard as the new Exec. Thirty-one-years old, Bob Risser had been born in Des Moines, Iowa, and entered the Naval

Lieutenant Commander Robert Dunlap Risser reported aboard *Seawolf* in May 1943. As the *Wolf*'s second wartime Executive Officer, Bob Risser would make two patrols. *Official U.S. Navy photo.*

Academy in 1930. He was commissioned an ensign on 31 May 1934. Risser's first assignment after the Academy was three years aboard the cruiser *Chicago*. In June 1937, he elected to enter Submarine School at New London, Connecticut. "My choice of submarine duty was wholly personal," he later reflected. "Mainly, I think because of what appeared to be different (from surface), exciting and challenging duty and secondarily a chance for relatively early command."[6]

Risser excelled in academics and graduated No. 1 from his submarine school class. His first year of sub duty in 1938 was on the old *S-21*. In January 1939, Lt. Risser joined Lt. Cdr. G. W. Patterson Jr. in commissioning the new submarine *Saury*. His time bringing *Saury* into commission roughly paralleled *Seawolf*'s early days of training. "We were in New London when *Squalus* went down and post shakedown overhauling in Portsmouth when she was raised," Risser recalled. "*Saury*'s diving procedures did not permit the disaster that befell *Squalus*. In *Saury* we closed all hull flappers prior to final commitment on the dive. They were hand operated beasts and I know the crew groused over the procedure but in case of the hydraulic operated main induction failing, the worst we could do was flood the induction piping outside the pressure hull."[7]

Risser attended postgraduate school in Annapolis and by 1942 had earned a masters in chemical engineering from the University of Michigan. Risser's schooling in Ann Arbor kept him "essentially isolated from submarines and the war until the spring of 1943." In May, he was assigned to *Seawolf* as her executive officer.

Risser felt that coming into the war late in the game offered him some benefits as a prospective future commander. "My introduction was under the aggressive Roy Gross and by that time we knew far more about what we could 'get away with'—such as day surface cruising, night surface attacks, [and] rather poor Japanese ASW."[8]

In relieving Bill Deragon, Risser did not at first assume the navigational duties that normally befell the executive officer. "I was Exec on that patrol but Jughead Casler navigated," Risser recalled. Casler had just been promoted to Lt.(jg). Risser found that "at least 50% of the crew were old timers. Of the officers, Doug Syverson, Jim Mercer, Bill Whitman, and Jughead Casler had made most if not all of the previous patrols."[9]

Seawolf already had a proud reputation under Freddie Warder. Gross and Risser did not look at their command as having any pressure to live up to Warder's reputation. "We never discussed it," wrote Risser, "although I opine that this must have been pertinent to Gross' achievement."[10]

The crew had confidence in their new skipper from his first patrol. "Captain Gross was a newcomer to us, but he proved to be an exceptionally fine skipper," recalled motormac Paul Zimmerman. "He knew the submarine and his attacks were just perfect."

Seawolf was underway from Midway's sub base at 1400 on 17 May 1943, with her patrol area designated as the East China Sea. Gross held training dives and battle problems on his way to station. En route, the black gang endured their usual struggles with their engines. "One engine back down already," Bud McCoy wrote in his diary on 19 May. "This damn junk heap won't run a week without something falling off or apart." The next day, McCoy added, "Not much doing except aft engine room is still trying to find out why #3 looses lube oil pressure."[11]

At 0857 on 20 May an SD radar contact was picked up at 11 miles. The bogey closed to 8 miles and *Seawolf* dived, although the lookouts never did see the plane. There was considerable doubt as to the authenticity of this aircraft contact, as the *Wolf* was 585 miles from Midway and 560 miles from Wake Island.

Two days later, an operational accident was nearly very costly. "Today came close to being the last day for this boat," Bud McCoy wrote on 22 May. In the process of making a dive, the *Seawolf*'s engine induction valve was accidentally left open, flooding the ven-

Recently modernized in 1943 and under a new skipper, *Seawolf* continued her warfare in the Pacific. The *Wolf* is viewed here bow on while off Mare Island on 7 March 1943. Her 3"/50-caliber deck gun was moved from aft to forward of the conning tower and new radar antennas were mounted atop her periscope shears. *National Archives.*

tilation system. Water ran from the induction drains into the engine room bilges for about an hour after the Chief Rajotte's engineers managed to close it off. Water was all over the deck in the after battery and the No. 2 sanitary tank was completely filled, flooding the shower room. It was "a hell of a mess, in general," McCoy noted. "About time for me to move when they start that type of diving."[12]

Seawolf ran on the surface as she moved toward Formosa. On 25 May, she began running submerged all day as she passed through the Bonins. She surfaced that night and by 2300 had sighted four intermittent lights. Officer of the deck Bill Whitman held course and passed between what were believed to be fishing sampans, out from Kita Io Shima of the Bonin group, about 50 miles away.

One of the lookouts picked up Iwo Shima at 0157 on 26 May by binoculars and the *Wolf* patrolled near this island the next day. Y2c John Burruss, standing lookout duty, sighted a 75-ton motor sampan at 1539 on 27 May with its sail up. "He was new and shiny," wrote Gross. He ordered a battle surface to polish up his crews. The *Wolf* opened fire at 3,000 yards and closed gradually, circling the vessel.

Lt. Whitman directed the 3-inch crew, with Chief John Bennett as gun captain. On the gun's left side sat Lucien Rajotte, serving as the trainer. In the right seat, pointer Big Swede Hanson fired the gun upon command. Duke Wall, first loader, and second loader Red Mills stood ready with shells in hand. On the bridge, gunner's mates Swede Larson and Rex Mickey manned the 20mms.

Gross felt that attacking such sampans was good target practice and also efficient in disabling the Japanese picket line. "We gunned the sampans because they radioed our position and they also fed the Empire with their fishing," explained quartermaster Hank Thomson. Chief Bennett's gunners fired 50-caliber, and 360 rounds of 20mm, and 32 hits from 51 rounds of 3-inch to sink the sampan. The 20mms were unable to set her afire, so the big gun holed her waterline from close range with point-detonating shells. No return fire was observed from this sampan.

Two survivors were seen in the water as *Seawolf* departed. At least one man had been killed aboard the steel sampan. "One of the Japs started up on their bridge to man their gun, but one of our shells took off everything from the main deck up," wrote McCoy. "I guess he can shoot at Davy Jones instead if he [is] still able to."[13]

The following day was uneventful as *Seawolf* approached Formosa. The SD radar picked up the 13,00-foot peaks of Formosa's hills at 26 miles at 0422 on 30 May. *Seawolf* ran submerged during the day, coasting up the east side of Formosa about 5 miles off shore. She did slow surface running at night while searching for targets. Sampans were avoided both day and night. Any gunfire with one here would give away Lt. Cdr. Gross' chance of surprise.

Seawolf spent 31 May–3 June patrolling 10-20 miles north and west of Hoka Sho light. Admiral Lockwood's staff sent an Ultra to *Seawolf* on the night of 3 June to move toward an expected Japanese convoy. "Got a message tonite; a convoy headed our way of 12 ships," wrote Bud McCoy. "If we can get into it, there will be a big time for someone. Hope we have it." Roy Gross increased speed to 18 knots and raced through the morning hours of 4 June to a point about 60 miles from the China coast. All he found that afternoon, however, was a line of sampans in pairs.[14]

Seawolf's second shipping contact of her ninth patrol came at 0419 on 5 June as she patrolled near Formosa in 200 feet of water. QM1c Hank Thomson on radar duty picked up a pip 14 miles on the port bow. This developed into an eleven ship convoy. Gross put on all engines at full power, making 19 knots. The *Wolf* pulled around their starboard flank, keeping about 20,000 yards off. "Weather was bad for surface running," wrote Gross. The lack of wind, unlimited visibility and calm seas almost guaranteed this convoy would have aerial coverage at dawn.

Gross and his lookouts could count eleven ships after daylight, hull down on the horizon, all single funnel, with size estimated from 3,000 to 7,000 tons. One destroyer appeared to lead the convoy, with only her masts visible. Other escorts were not visible due to the range. "Our 19 knots should see us ahead in about 3 hours or less," Gross wrote. At 0600, the lookouts spotted an object on the starboard bow, either a sampan or a submarine. OOD Jim Mercer dived to investigate. Gross decided to stay down until dusk, then run ahead to be in a position for a night attack by about 2000.

At 0924, three distant explosions were heard. The convoy was still in sight 10 miles on the port bow, as *Seawolf* continued to parallel their course. "Either *Tinosa* getting in or escorts are on false contact," wrote Gross. In his diary, Bud McCoy scribbled, "Close to land this morning, and a tin can is up on top. Every one quiet so they won't hear us. Not much water here. Bad place to get caught."[15]

At 0931, eleven more distant explosion were heard. Three hours later, the jumpy escorts dropped three more distant depth charges. By 1300, Gross was forced to surface in the daylight to maintain contact. He put three engines on the line and used the fourth to charge up his batteries. As *Seawolf* raced ahead, a lookout reported a destroyer on her port beam, 6 miles off at 1312. "It was no doubt he who had dropped the charges today on a contact he had dropped aft to investigate."

Seawolf turned her stern to the DD and tried to keep in pursuit of the convoy, but she was quickly spotted. At 1314, Doug Syverson's lookouts reported that the destroyer had stopped circling and the angle on his bow became zero as he headed straight for the *Wolf*. Gross increased speed to full and made gradual turns of 10 degrees to the left until she was back on the convoy's base course with the destroyer still dead astern, not closing very much.

Those who had made the previous run under Googy Gross knew that he was not afraid to challenge the enemy. For the newer hands, it was quite an experience. Electrician's mate Olin Fogle, having joined the Navy in 1942 and a veteran of previous patrols on *Haddock*, had just joined *Seawolf* for her ninth. "The Old Man was a poker player," said Fogle. "He'd go for a 50–50 chance. If he thought we had a 50–50 chance or better, he'd take us in. He wouldn't go for less."

Gross almost lost his prey in a rain curtain. He also had to pass one sampan 3,000 yards to starboard in maintaining an evading course. One hour later, *Seawolf* appeared to be holding her own against the *Wakatake*-class destroyer, but this suddenly ended. The angry escort began making smoke and built up his speed, narrowing the distance to the submarine. At 1433, with the destroyer at 10,000 yards and closing, Gross was forced to submerge, having picked up a good bit of ground on the convoy.

At 1510, the first depth charge dropped about 4,000 yards astern. Gross watched the Japanese skipper circling his point of contact. The DD unloaded three more charges, circled the spot for a few more minutes and then dropped another depth charge. Gross found that these charges "sounded very loud considering they were seen to be about 4,000 yards" away.

Radiomen Paul Maley and Ed Hinson announced that the DD was using long range pinging, but they could not hear his screws. His circles grew wider and the sea was glassy calm. Under these conditions, Gross did not consider an attack practical. So, at 1530, he ordered 200 feet and speeded up to clear the area. One hour later, the pinging was getting louder, so silent running continued. When *Seawolf* eased up to periscope depth at 1832, Jim Mercer spotted smoke dead astern where the destroyer should be. The sun was just setting, so Gross waited another 15 minutes before surfacing.

As his lookouts took the shears, he found the destroyer was now 6 miles astern, circling a suspected target. The tin can spotted *Seawolf* again, for his angle on the bow changed to zero. "He didn't have a chance this time," wrote Gross. "Opened at 4 engines and quickly lost him in the dark."

In the dark, moonless night, *Seawolf* was soon able to ease back to three engine speed. Gross set his course to intercept the convoy again at midnight. "It's this night or never as tomorrow we'll be nearly in the straits of Tsushima." Gross had Paul Maley send a contact report to *Tinosa*, in case she was waiting just a few miles too far off the line. By 2007, seven large pips were visible on Ben Rogers' SJ radar screen at 12,000 to 16,000 yards, but on the starboard bow now. This appeared to be a different convoy. By 2030 the range had decreased to 7,900 yards. Gross slowed down. "The many pips and

The control room, just below the conning tower, was the command center of *Seawolf* during depth charge attacks. The hull opening indicator panel behind the sailors—which displayed red and green status lights—was known as the Christmas Tree. *Courtesy of Bill Harlow.*

the difficulty of keeping on the same one prevented us from realizing immediately that something was wrong. Two set-ups gave us a reverse course for the convoy, so assumed radarman had gotten different pips."

At 2035, two lighted sampans forced a change of course. At 2055, with the range now opening up to 14,000 yards, Roy Gross stopped, thinking that he had overshot the convoy and that they had slowed or stopped for some reason. By 2137, Gross and his plotting team, Jug Casler and Jim Mercer, finally realized that radarman Rogers was not wrong. "We had let a big convoy, a stranger, go by on reverse course of one we were chasing," Gross wrote. "We therefore had about 20 ships within a range of about 40 miles of us at the moment, and the question was which way to jump."

Only three engines were available at the moment. Lucien Rajotte's black gang was working on the fourth, which had developed a cracked liner. The southbound convoy was plotted at 14 knots. A quick estimate showed that *Seawolf* had lost valuable time while stopped. She could not catch the northbound convoy, dangerously close to Tsushima, before dawn. She could, however, catch the southbound one about midnight, in spite of their high speed.

Gross started the long stern chase on the pip 16,000 yards away. This was the only pip on the screen, the others probably being ahead. *Seawolf* set up a "splendid radar track of the enemy now." The pip was a big one, speed 13.5 knots and not zigging. By 2240, *Seawolf* had closed the range to 9,000 yards. One escort was spotted 2,000 yards on either beam of the very large target. Gross decided to come up the target ship's port side, behind the small escort, turn in, and then fire. By 2329, Ben Rogers had picked up a third escort on radar, considered to be the size of a PT boat. Visibility was a mere 500 yards, with light wind and heavy rain setting in on a very black night. "Felt the lack of binocular assistance very much," Gross noted. "Had to rely entirely on radar."

Seawolf stopped briefly. The bearing on the PT boat then drew left instead of right. "Not understood," noted Gross.

"All ahead full," he ordered helmsman Jim Grimes.

The range to the target was now 5,500 yards and opening due to *Seawolf*'s stopping. The *Wolf* sheared over to the starboard side of the target's track and spent the next 90 minutes closing the range to 1,350 yards in the poor visibility. By 0022 on 6 June, Gross ordered his outer torpedo doors opened. "Just then sighted target loom on port bow, angle on bow 165 starboard," he noted. This checked with Bill Whitman's TDC exactly. Gross came left slowly to decrease the gyro angle as he made ready to fire. Two minutes later, the target suddenly loomed up out of the rain and darkness, very large on the starboard bow, showing a large port angle on the bow. "We had both turned left apparently," wrote Gross. The ship was drawing left across *Seawolf*'s bow on a collision course.

"All back full!" shouted Gross to his battle helmsman. Grimes threw the annunciator controls that signaled those below to reverse the screws at full speed. *Seawolf*'s screws dug in and she avoided a near collision with her would-be target. The Japanese ship then turned away right and was immediately lost to sight in the darkness. In his log, Gross wrote:

> It developed his range when across our bow was probably 1,000 yards, but it seemed closer due to his size. Did not get a clear picture of him, except high, long flat top, stack in middle, and no masts seen. Thought we'd been seen, but no offensive action taken, therefore assumed we were not

actually sighted. There were no signals on the radar detector. Used it every 15 minutes.

At 0025, Gross ordered all ahead full, taking up a closing course. The rain was still bad and the binoculars were of no use. The target was maneuvering radically and the range opened back out to 6,000 yards as *Seawolf* swung back and forth following him. Gross allowed the spooked ship to settle down. By 0115, the range was down to 3,700 yards and slowly closing. One of the lookouts had the ship in sight on the port bow at 0123, with a large starboard angle. The TBT bearing was put into Bill Whitman's TDC, with the range about 1,900 yards.

Seawolf had been in constant pursuit of enemy shipping for more than 20 hours now. Roy Gross—monitoring the action from the bridge while Bob Risser supervised the conning tower attack party—was ready to attack and did so at 0123 on 6 June. The plan had been to fire using a white light spread from the forward bridge TBT, manned by Lt.(jg) Jack Kennelly. Although the range dropped to 1,700 yards, low visibility and rain prevented Kennelly from seeing the limits of the target. Only the base of the smoke column could be seen through the binoculars in the TBT, and it took too long for him to wipe off the lenses between each mark.

Gross felt that Whitman's generated data on the TDC seemed to be very good, however. Without a speed spread ready, he opted to just fire all four bows tube at the MOT. Lt. Whitman put in a check bearing after the second torpedo had been fired. It checked well with the generated bearing. Just after the second torpedo was fired, there was a tremendous explosion ahead of *Seawolf*. The first torpedo had prematured at 15 seconds while at very shallow depth. It was becoming an irritating pattern for Roy Gross. The first war fish he had fired on his previous *Seawolf* patrol had also blown up in his face.

This torpedo exploded closer than the one on the eighth patrol. "Red-hot shrapnel from it fell all about us, landing in the water with sharp cracks which we at first thought to be machine gun fire," wrote Gross. "One officer thought some landed on deck. Fortunately, this was a small Mark 15 head."

The fourth torpedo also prematured, but it exploded at a safer depth without the potentially deadly shrapnel. One other explosion was heard from either the second or third torpedo. Analysis showed this to be a probable hit on one of the far escorts on the port side of the target at about 3,000 yards. The torpedo had likely run under the freighter. "Believe we sank one escort, type unknown. Size of pip showed it to be 1,000 tons or larger," wrote Gross. His claim of a hit was supported by radar reporting loss of this escort's pip four minutes after the explosion.

In his diary, Bud McCoy wrote, "Three of the fish were premature and shook hell out of us, almost blinding the fellow on topside. Too bad we can't get some good torpedoes for a change."[16]

The remaining torpedo did not explode. *Seawolf* had more than 40 fathoms of water, so Gross elected to stay surfaced and retire at three engine speed on a reverse course of that of the target ship. One escort and the target were bunched together on radar. The rain was continuing and the visibility was still poor. "Conditions indicated that a second attack would have more chance of success in daylight, so retired at his speed, 14 knots, for a daylight attack."

Seawolf lost her target at 13,000 yards. "Should have maintained radar contact, but having once lost it, could only retire along his probable course due to his probably radical maneuvers," Gross logged. Jim Mercer dived *Seawolf* at dawn at 0410 and commenced a submerged patrol along the target's estimated track. *Seawolf* avoided several sampans, but sighted nothing all day. "Valuable lesson learned regarding radar approaches on multiple targets, difficulty of staying on same pip, and close coordination between plot, radar and bridge," wrote Gross. For the crew, the luckless patrol was becoming frustrating. McCoy's diary for 6 June states, "On surface today, looking for something, but no luck. Damn fishing boats everywhere."[17]

On 7 June, *Seawolf* sighted several sampans during the night. She conducted a surface patrol during the day retiring slowly southwestward towards her patrol area, having run the length of the China Sea from Formosa nearly to Nagasaki. After much deliberation, Gross sent in a report on what meager information he had on this large convoy. "No sights in four days made our own position problematic, although it developed later that it was correct as used," the skipper wrote.

At 0356 on 8 June, a red light from a coal-burning trawler was sighted and avoided. At 0448, the smoke and masts of another ship were sighted. After dawn, this ship also proved to be a trawler-type vessel unworthy of attack.

They were both heavy smokers, moving at slow, variable speeds and not pinging. By 1200, both patrol vessels were out of sight. After a day of no contacts, *Seawolf* surfaced at 1906, 20 minutes after sunset. It was still raining with a solid overcast, moderate wind and occasional lightning. *Seawolf* used the poor visibility and light rain to stay on the surface during the afternoon of 9 June. Due to Jug Casler and chief quartermaster John Beatley's inability to take good navigational fixes during the past 48 hours, *Seawolf* stayed safely outside the 40-fathom curve by taking single-ping fathometer soundings at one or two-hour intervals.

Seawolf moved toward Turnabout Light during the next two days, avoiding sampans en route. Bud McCoy's diary gives an indication of how frustrating the poor weather was on the ship's navigators:

> Down all day. Still out of area and have been for over a week. Been lost almost all the time in the last two weeks. Can't see sun in daytime or stars at nite. Don't have any way to tell where we are and damn little water under us. Two more tanks of fuel left. Probably will have to make slow speed going in.[18]

Casler and Beatley were still unable to get a good navigational fix due to the continued rainy weather. With new intelligence of an approaching Japanese convoy, *Seawolf* sped eastward along the China coast during the morning. Sunrise at 0502 on 12 June showed much improved conditions: the sea was glassy, the sky was cloud-free, and visibility was perfect.

At 1421, OOD Doug Syverson picked up smoke astern, believed to be the convoy, in two columns. Roy Gross decided to continue opening at slow speed for a dusk attack due to the glassy sea conditions. By 1835, there were stacks and masts coming over the horizon. One destroyer was ahead of the convoy. There were six other ships, making 6 knots, all small except for one in the middle that was fairly large. The identification team soon determined that they had three small freighters and one larger freighter of about 7,000 tons.

By sunset, the range on the nearest destroyer was 7,000 yards. Gross reversed his course to close the convoy now. The ships were apparently zigging with constant helm. Gross decided to attack the destroyer first but had to abandon this attempt at 1935, when the closest range obtained was 4,000 yards with a 90 port angle. There was a bright half moon with no clouds, making observations quite easy. The destroyer was making 6 knots, utilizing long range pinging while keeping a patrol station ahead about 2 miles.

At 1952, Gross had angles on the bow of three ships on the left flank of the convoy of about zero degrees. With his battery running very low, he wanted to attack quickly and make his escape. At 2004, angle on the bow was still zero on two of the ships, range about 3,000 yards. "Prepare stern tubes," talker Hank Thomson called to Clarence Kibbons. His after gang—Alfred Ostrander, John Neil, Tom Warren, and Robert Miller—prepared their torpedoes and kept the doors closed, awaiting the final word.

At 2006, the convoy zigged, making the angle on the bow on one ship 70 port, with the other one still at zero, range 1,200 yards. One minute later the second ship finally zigged right for a 70 port angle on the bow at close range.

"Open after doors," Thomson called back aft.

By the time Kibbons' gang had the doors open, the range had closed to 500 yards or less and opened again. The target ship was about 3,500 tons. Lt. Cdr. Gross checked fire, considering his set-up to be "too snap. It was problematic whether we would get a shot in or get forced deep to avoid collision." Cursing his luck, he figured that he had probably kept the outer doors closed too long to avoid flooding. Second-guessing himself, he had missed his best firing potential. A disgusted Roy Gross wrote, "This is third convoy we've had contact with with no results. Beginning to prefer single ships now."

Not one to quit, Gross reversed his course at 2028 and opened the range prior to surfacing to avoid a trawler-type escort on her port bow. The range on this ship dropped to 4,000 yards and Gross suddenly realized that this "escort" was actually the eighth ship of the convoy, a mid-sized freighter, which was running about 4 miles astern of the other ships. "Should certainly have investigated to be sure what he was," Gross berated himself, "but was so anxious to surface and get ahead again that we opened him instead of closing."

At 2112, the range on this freighter was about 7,000 yards, still too close to safely surface. "We had not been suspected yet." At 2133, Gross ordered, "Bring her up! Battle lookouts to the bridge."

Seawolf surfaced and started a battery charge as Gunner Bennett and the other top lookouts climbed the shears with binoculars in hand. *Seawolf* ran around and ahead of the trailing ship to cut him off from the main convoy first. At 2152, the *Wolf* was making 100% speed on two main engines while the other two jammed juice into the batteries. Not making ground quickly enough, Gross ordered a third engine on the line at 2300 to pick up the pace.

Six minutes later, the convoy was 9,000 yards ahead, clearly visible. The *Wolf* dived, her battery about half full, and closed the target at 4 knots submerged. A quick ping range by sonar showed the ship at 1,900 yards. Gross did not dare use his radar in the bright moon and glassy sea. The ping range checked exactly with Bill Whitman's TDC's data. *Seawolf* had the enemy's course and speed accurate before diving.

At 2351, Paul Maley took another single ping range and found the distance to be 2,100 yards. Ed Hinson swept around the area on the other sonar set, keeping tabs on the presence of the Japanese escort vessels.

"Open forward doors!" Gross ordered.

Torpedomen Bill Reiland and his crew—Jim Cashero, Larry Roberts, Dallas Malone, and John Sadler—flooded their tubes and stood ready for the awaited orders. At 2352, *Seawolf* commenced firing Mark 14-3A torpedoes from all four bow tubes at this 2,500-ton freighter. Gross aimed for one for her bow and one for her stern. Using his own advice from the previous patrol to help ward off poor torpedo performance, Lt. Cdr. Gross opted to put his other two stingers midships, bypassing the preferred full-spread method.

Once again, there were no explosions. Ping ranges showed the target zigged away at three minutes after firing, indicating that he had seen the torpedo wakes. Gross still believed that at least the last torpedo should have connected. Even more baffling, there were not even end-of-run explosions from these tin fish.

As the night passed into the early minutes of 13 June, *Seawolf* continued to chase this convoy. She surfaced and headed for the distant ships' smoke. A half hour later, two ships were seen to be closing, forcing the *Wolf* to take an evasive course. At 0110, the moon set in a cloudless sky. "Could almost be heard to sizzle as it dipped into the water," wrote Gross. Although Ben Rogers had the convoy on his radar at 16,000 yards, all sight of them was lost in the darkness.

Red Syverson called for radical evasive maneuvers at 0135, when a radar pip appeared 1,000 yards off the port bow. The object was unseen, but could have been a submarine. *Seawolf* listened briefly for enemy screws, heard none, and then set course to the northeast to overhaul the convoy again while charging her batteries.

Radar picked up the convoy again at 0300, distance 13,000 yards ahead. Mason Poole and John Bilkey's electricians secured their battery charge and put the fourth engine on screw. Dawn began breaking at 0351 and *Seawolf* dived with the convoy at 7,000 yards.

At her slower submerged speed, however, *Seawolf* could not close the range sufficiently to fire. Another half hour of darkness would have insured a kill, but these ships began opening up the range. The enemy failed to offer a timely zig in their direction, so Roy Gross' crew faced yet another heartbreaking missed opportunity. He gathered his officers in the wardroom for a pow-wow. "Started analysis of what we were doing wrong to develop such consistently poor situations. Decided to place blame on the gremlins until something better turns up. Hope *Gunnel* does better."

Seawolf surfaced in a flat sea that evening. Ben Rogers found this night that his SJ was sensitive enough even to pick up living creatures. His set showed a pip at 2040, distance 1,800 yards. Nothing was in sight. OOD Bill Whitman first thought was that this might be a motor torpedo boat or a periscope. The *Wolf* maneuvered radically under battery power while listening. The pip closed to 700 yards. Another pip appeared on the other side at 500 yards. Whitman then cleared the bridge and made ready to dive.

Still, nothing was in sight. Other pips soon appeared until it became obvious these could not be a submarine. Rogers and his skipper ultimately decided that the SJ radar was picking up the oily, wet wings of sea birds. "Finally saw a couple ailing 6 inches off the water at about 25 knots, between the low, oily swells," wrote Gross. "We hadn't heard of this before, and it gave us quite a start. Although we were sure we could see 6,000 yards, there is always a little doubt when the water is so glassy that it merges with the sky to leave no horizon."

Two patrol vessels were avoided during the night. Radar contact was made at 0250 on 14 June on a ship that was tracked some 9,800 yards away. *Seawolf* went to battle stations and opened up on four engines to get ahead of this ship. Roy Gross' team plotted this ship into the pre-dawn hours but was unable to reach firing position before dawn.

By 0426, Gross identified the ship as an *Otori*-class torpedo boat. He was not zigging, but was pinging alternately between long range automatic and hand pinging. Gross watched him disappear ahead without changing course or speed apparently. "Would have been a beautiful shot," he wrote. "Again, another half hour of darkness would have put us in position."

At 0438, *Seawolf* resumed course to the southwest again at dead slow speed, planning to head for Amoy on the China coast. This position on a northeast-southwest line about halfway between northern Formosa and the China coast seemed to be where all the convoys had passed through. The long, luckless patrol continued to grate on the crew. Bud McCoy noted the lack of fresh water in his diary.

> About 9:30 at night here on watch in engine room. Fairly cool. Try[ing] to make some fresh water from sea water; low on water. Showers are turned off. Everyone smells like a goat, but you get used to it.[19]

Seawolf closed on Amoy during the next 48 hours, running about 30 miles off the China coast, due north of Formosa. She avoided sampans and large sailing vessels along the way, including 23 sampans sighted within one minute before dawn on 19 June. At 0615, OOD Jim Mercer could count 37 sailing sampans in sight.

At 0922, Lt. Bill Whitman picked up two ships through the periscope. They were in column on a southwesterly course, approaching through the haze on *Seawolf*'s port bow at 4,000 yards. By the time the captain could arrive, Whitman had called for battle stations, ordered full rudder, and twisted the ship to set up for an immediate stern shot.

These two ships were later identified as *Reiyo Maru* and *Shojin Maru*. They were southwest bound, just north of Amoy, 20 miles off shore in 20 fathoms of water. The range was coming on quickly, so Clarence Kibbons' after room made ready to fire. Using an estimated draft of 14 feet, the torpedoes were set to 12 feet. The leading freighter was believed to be similar to *Hokuyo Maru*, a 4,000-ton mast-funnel-mast transport shown in the ONI 208-J. After working eight minutes on the TDC—during all the time the ship's head was swinging fast to starboard—Whitman and assistant Lefty Leffingwell had steadied on their firing track and gyro angles.

At 0935, Gross fired all four stern tubes from 1,000 yards. Through his periscope, Gross estimated her to be 375 feet long. Figuring her speed at 6.5 knots, the firing team used a speed spread of plus and minus 2 knots, firing at her middle, bow and stern. One hit the 4,739-ton *Shojin Maru* in the stern.

"Believe contact hit as black smoke and debris was thrown into the air," noted Gross. The other three torpedoes did not explode, but this TNT-loaded fish was enough. The target ship sank by the stern in nine minutes. Her bow hung up in the air long enough to suspect that her stern must be on the bottom for a while before it finally dived. The water was a mere 18 fathoms, with only 8 fathoms under *Seawolf*'s keel. *Shojin Maru* was gone by 0944. She left many life boats in the water, loaded to the gunwales. Gross estimated 500 people to be afloat, "indicating small transport."

Trailing ship *Reiyo Maru* turned away to 8,000 yards, then appeared to stop. This ship had a gun forward and aft. *Seawolf* closed on him submerged. A few sampans also began closing on the wreckage also to pick up survivors. Many other sampans sailed on without stopping. Before Gross could get within firing range, *Reiyo Maru* turned away, made heavy smoke and disappeared in the general direction of Amoy at 1122.

"Well, our luck changed today," Bud McCoy wrote of this sinking in his secret diary in the forward engine room. "There are about 600 soldiers up there in lifeboats. If we get a chance, we will probably surface and warm up the 20mm, but so far the escort is staying around."[20]

Seawolf swept over the area where the sinking had occurred. Rex Mickey, standing by on one of the bridge 20mms, recalled:

> The first ship we sunk when I was aboard was carrying Japanese troops. They were floating all around when we surfaced. One of our officers saw some papers and stuff floating around so he swam out there to get them. We didn't let any of the soldiers come aboard, though. We took pictures of the men in the water and these were turned in at Honolulu. We made another patrol and when we came back to Hawaii,

Another torpedo victim of *Seawolf* and Lt. Cdr. Roy Gross. This periscope photo shows the last second of a maru sinking, with rafts and debris in the foreground and another Japanese ship heading over the horizon. This appears to be taken on 20 June 1943, during the *Wolf*'s ninth patrol. *Shojin Maru*'s bow hung up in the air for several minutes as her stern appeared to be in the ocean bottom 18 fathoms below. The fleeing maru is *Reiyo Maru*. *Courtesy of Rex Mickey.*

we saw a newspaper headline of "U.S. SUBMARINE RESCUES JAP PRISONERS." The photos were the ones we had taken. This is the honest truth. It that wasn't propaganda, I don't know what was!

Admiral Lockwood flashed word to *Seawolf* of yet another important convoy coming up through the Pescadores channel. Roy Gross set course across the Formosa straits to intercept. The *Wolf* moved toward the area the next day and was in position shortly after midnight on 21 June. At 0105, her seventh shipping contact of the patrol was made by OOD Lt.(jg) Jack Kennelly at 16,000 yards through the periscope. The six ships Kennelly found were moving southbound along the west coast of Formosa, approaching Pescadores channel.

Seawolf took a parallel course 10,000 yards off their track for radar tracking and worked ahead of them.

By 0211, the *Wolf* was in good position ahead and slowed down. The convoy now zigged right to show a port angle at 15,000 yards. Ben Rogers' radar picked up this zig very well. One of the escorts apparently picked up a false contact several minutes later and proceeded to drop nine depth charges as the convoy evaded.

By 0300, the *Wolf* was in good position, guarding the north entrance to Pescadores channel. The fathometer showed 15 fathoms at 0410, so Gross changed his course to the west. A sounding 15 minutes later showed 18 fathoms (108 feet). Maley and Hinson rigged in their sound heads and pit log for protection. *Seawolf* dived prior to dawn in the shallow water. The pit log and one sound head were then rigged out again after diving officer Red Syverson got a 60 foot trim. Easing west, the depth increased to 20 fathoms by 0512.

Seven minutes later, six ships were sighted on various bearings to seaward, on many courses. Apparently it was the same convoy seen during the night, reforming at about the spot where the depth charging had caused them to scatter. By 0705, the hills of Formosa were just visible. With a seventh ship now in sight, Lt. Cdr. Gross called for battle stations.

Seawolf singled out a good-sized tanker, which was followed by a 5,000-ton freighter and a smaller oiler. The larger tanker was being hugged tightly by a corvette-type escort. With the range down to 2,350 yards, Gross opened fire on the larger tanker. Bill Whitman's TDC generated a speed of 6 knots, which checked very well with the skipper and Bob Risser's periscope observations.

"Fire one!" Gross called at 0745.

Seawolf proceeded to fire her full forward spread using what the skipper believed to be an excellent set-up. Once again, he sent a fish ahead, a fish behind by one-quarter length, and two for the MOT.

A torpedo exploded at 98 seconds after firing the fourth torpedo, which checked exactly for the torpedo run of the fourth fish. This torpedo exploded one-quarter length behind the ship and threw up a tall white plume of water. "This torpedo was possibly exploded by a towed object," wrote Gross. "A line was clearly seen over the stern." No other explosions occurred. The set-up had been good, so the attack team could only speculate that the MOT fish had gone under her without exploding. There were no end-of-run explosions.

"Damn poor fish, or eye sight, or both," wrote Bud McCoy.[21]

Two minutes after the *Wolf* had fired, the escort was alerted and broke away. He turned toward *Seawolf* with a zero angle on the bow and a white bow wave rising. "Left full rudder," called Gross. "Take her to 90 feet!" This escort was a heavily laden freighter who turned to ram *Seawolf*.

Red Syverson took her down, knowing that the ocean bottom at last sounding was 120 feet from the surface. *Seawolf* hit bottom, however, at 90 feet while making full submerged speed. The force of the blow threw men against the bulkheads as the submarine bounced upward from the ocean's bottom. Dishes and pans crashed to the floor in the galley.

Gunner Rex Mickey, standing by on the torpedo reload gang in the forward room, was knocked off his feet. "We fired at this ship and the order was to take her to 90 feet," Mickey said. "We hit bottom at 87 feet!"

In the forward engine room, Bud McCoy wrote in his diary, "We hit bottom at 88. Sure bounced hard. Could hear the bottom scraping in the sand." Seconds later, he added, "Hit again at 80 ft. The escort dropped 8 ashcans. Could feel the ship kind of float up and down from the concussion while we were skimming the bottom."[22]

The starboard JK sound head was smashed in the impact. Gross had ordered both the pit log and two sound heads rigged in upon his order for 90 feet, but then ordered the starboard JK head back down again before reaching 90 feet in order to listen for the escort.

"The *Wolf* was grounded with an up angle as we hit the rudder and she spun over to hard right, where the positive stops were," said quartermaster Hank Thomson. Picking himself up before Jim Grimes had his feet, Thomson went over and "started turning the wheel slowly off the 35 degree position. The equipment responded and I checked the position in the opposite direction."

Thomson righted the ship, and thereafter "had no problem with the steering." *Seawolf* scraped along the bottom at 90-95 feet at one-third speed. Gross wrote that Lt. Syverson overcame the crash and was doing "a fine job of depth control. Steadied at 78 feet at slow speed."

The crash against the bottom helped the Japanese ship lock onto *Seawolf*, for when his first depth charges came at 0756, they were close. BOOM! BOOM! BOOM! BOOM! Four tooth-shakers slammed the *Wolf*, close aboard. Water swished through the superstructure as the force of mighty ashcans shook the submarine. At silent running, the temperatures in the boat had quickly soared to upwards of 120° as the men waited out the next explosions. Sweat poured off bodies and down the men's legs, pooling on the deck as they clutched something in preparation.

Rex Mickey, making his second war patrol, figured that his luck had run out.

> We were sitting there on the bottom and here came this attacker. Every depth charge was getting closer. It got close enough that it was shaking us pretty good and we couldn't stand up. Valves were cracking open and the lights went off. They got pretty close to us, and I thought, "Well, the next one's gonna get us."
>
> And then it quit.
>
> When it quit, we could hear the screws of the ship going from the back torpedo room kinda sideways right across the forward torpedo room.

GM2c Rex Mickey in photo from a 1943 *Seawolf* patrol. Mickey was one of the last men to leave *Seawolf* following her 14th patrol. *Courtesy of Rex Mickey.*

The escort ship's screws churned directly overhead of *Seawolf* in her most vulnerable position. In the control room, terrified eyes noted the mere 75 feet on the depth gauge. Anything rolling off the ash can racks above would spell destruction.

Nothing dropped. "They went right over us," said Paul Zimmerman, who was in the control room at the time. "We were listening for any second to hear one of their depth charges to land right on us. That was scary."

Throughout it, Mickey found Lt. Cdr. Gross to be "cool, calm, and slick" as he conned his boat to safety. "The first spit kit went right over us and he was sure close," wrote Bud McCoy. "Could hear his screws as plain as hell only 30ft above us. Bet he would be mad as hell if he knew how close he came. If he had dropped ashcans, well, here is another boat that would have been on the overdue list."[23]

The crew sweated it out for another 20 minutes before their skipper dared ease up to 55 feet for a quick periscope peek. The two escorts were not far away. One of them appeared to be a smaller, 1000-ton or larger freighter. For the next three hours, *Seawolf* ran silent to find deeper water. Gross found these foes "persistent but inaccurate. The shallow water made it very uncomfortable."

Seawolf slipped away from the pingers and by 0947 the smaller escort had moved away. A large land-based bomber appeared and circled the area to help the escorts locate the sub. The larger escort ship disappeared from sight by 1210 but had been replaced by a *Chidori* torpedo boat. He moved as close as 8,000 yards in *Seawolf*'s direction, but did not circle close enough to warrant an attack.

Gross worked his boat away from the scene of his near grounding. The escort vessel continued pinging and dropping a few depth charges into the early afternoon. The torpedo boat continued making wider circles, sometimes passing within 4,000 yards of *Seawolf* in its searches. At 1450 another escort began overtaking *Seawolf* from astern as she withdrew. The *Wolf* rigged for silent running again, and altered her course. This escort was clear by 1550 and Gross secured from silent running as the spit kit ran off, still pinging occasionally.

"Somebody was riding with us today," thought Bud McCoy. "Lady luck or something."[24]

Seawolf surfaced at 1939 and cracked the hatches to allow in fresh air. She headed northwest to clear the area at two-engine speed. "Agreed with almanac that this was the longest day in the year," noted Gross. "About all we accomplished was to lead the Japs to believe there was a submarine in the area and to expend a number of depth charges."

Seawolf had four torpedoes remaining aft and six days to remain on station to use them. Roy Gross elected to run at slow speed up the middle of the Formosa Straits to reach the edge of his area by the last day, while conserving fuel.

During 23 June, Lt.(jg) Casler was able to take another solid fix on Sento Shosho. The boat's estimated position was found to be only 3 miles off against the fix on the lighthouse some 35 hours ago. Gross noted in his log that he had 450 miles to go to the departure point. Engineering officer Red Syverson, conferring with Chief Rajotte, reported that there was 32,000 gallons of fuel on hand, which called for economy. Gross decided to continue eastward at minimum fuel consumption, patrolling close to Okinawa Gunto en route to the departure point.

Passing his time on duty in the forward engine room, Bud McCoy made brief entries in his diary as *Seawolf* completed her time on patrol station.

> June 22nd: Well, we're headed for the barn now, given the Old Man got his belly full of this place and it's time to go in anyway. Got to get out of this area or one of our own boats will bump us off.
>
> June 23rd: Same as yesterday. Not much going on, except we are running submerged. Should be on top in a day or two.
>
> June 24th: We are about 200 miles from Formosa now. 1,000 to go. Sure be glad to get in and, boy, will soak up the liquor if I can find any? Every damn thing is rationed out here, even the women. About 1,000 to one when we reach port.[25]

During her final days on station, *Seawolf* cruised slowly, patrolling along Okinawa Gunto's southeast coast. The only potential target sighted was a Japanese sampan which was fishing off the coast. The little ship, however, had a machine gun mounted forward and Gross opted not to battle surface. On 27 June, *Seawolf* departed her patrol station on schedule at 1115. She had 29,130 gallons of fuel remaining. With 2,800 miles to reach Midway, her engineers maintained a slow one-engine speed. She was on full surface running since dusk of 26 June, diving only for trim dives. Her next dive was on 29 June at 0627 when a plane contact was picked up at 10 miles. The plane was small and did not even make a radar pip.

Seawolf made only one more dive during the day to avoid two Japanese planes believed to be Millies. As the days passed on the slow return to Midway, Roy Gross began to ease up on topside restrictions. Other members of the crew were allowed to come up for short exposures to the sun. Motormac Bud McCoy took the chance to come out of the inner depths of the engine rooms on 30 June. "I was up on topside today for a few minutes," he wrote. "First time in 42 days I have seen daylight. Kinda like the old days in the first part of the war."[26]

Jim Mercer made a crash dive at 1809 on 2 July when an incoming plane—once again not detected by radar—was spotted 10 miles out. Marcus Island was 330 miles away at this point. *Seawolf* resurfaced and continued undetected except for one plane picked up on radar during the early hours of 5 July which caused her to dive. One of these quick dives almost proved fatal when the bow planes froze up on hard dive. As men were thrown from their feet, the duty team in the control room blew emergency tanks and reversed the engines to stop the *Wolf*'s descent. "Sure made a good dive today," McCoy wrote sarcastically on July 6. "Took 35-degree angle. Couldn't get bow planes to work. Plenty of scared boys aboard. Didn't feel so brave myself."[27]

Lt. Cdr. Gross continued to be surprised by the aircraft contacts, which were as far as 600 miles from Wake Island. By the morning of 7 July, the *Wolf* was within Midway's 500-mile circle. An unidentified monoplane sighting at 8 miles out compelled Bill Whitman to dive ship that morning and a patrolling PBY came within 8 miles that afternoon. *Seawolf* again used prudence, pulling one green recognition flare before diving, but going deep for discretion.

At 2400 on 8 July, the calendar shifted back to 7 July as *Seawolf* crossed the 180th meridian. The crew continued to enjoy some of the sunshine breaks topside. "Went topside today; got a little more sunshine," wrote Bud McCoy. "Crew don't look so much like dead men now." He was eager to get ashore. "I can almost taste that cold beer now. I am going to get on a hell of a drunk when we get in, for a change."[28]

Seawolf approached Midway on her second straight 8 July. At 0500, she made a rendezvous with two Midway planes and entered the lagoon. At 0721, she moored alongside the submarine *Sailfish* in a nest with *Sperry*. *Sailfish*'s executive officer was Lt. Cdr. Al Bontier, who would later become well known aboard *Seawolf*.

The *Wolf* would remain in port only long enough to top off her nearly empty fuel tanks before continuing on to Pearl Harbor. The crew was thrilled that the Pink Palace lay ahead in store for them versus more weeks of chasing gooneybirds at Midway. The engineers took on 20,000 gallons of fuel and ship's last four torpedoes were left with *Sperry*. At 1302, *Seawolf* was underway for Pearl Harbor. The return trip was uneventful and she made her rendezvous with an escort ship at 0600 on 12 July. At 1100, *Seawolf* moored alongside the submarine base, Pearl Harbor and ended her ninth war patrol.

In reviewing *Seawolf*'s patrol report, Admiral Lockwood found it "regrettable that the *Seawolf* was unable to inflict more damage to the numerous convoys contacted." Making seven contacts on five good-sized convoys, the *Wolf* was credited by Lockwood with sinking one 4,217-ton passenger freighter, with sinking one 75-ton sampan, and with damaging a 1,000-ton class unknown escort.

Gross' team reviewed their firing solutions numerous times for errors. He was frustrated "to get only five explosions, of which one was a sure hit, one a probable hit, and two sure prematures, out of sixteen torpedoes fired, [which] has caused considerable reflection in *Seawolf*." The ranges were often longer than desired, but the magnetic exploders were viewed by the *Seawolf* wardroom to be questionable whether they would fire or not in waters of 20 to 50 fathoms. "The answer cannot be found on board in any publication."

Roy Gross allowed his men to go ashore for a beer bash in the afternoon. True to his self-promise, newly-promoted Chief Bud McCoy "had about 20 cold ones."[29]

USS *Seawolf* Ninth Patrol Summary

Departure From:	Midway
Patrol Area:	Formosa
Time Period:	17 May - 12 July 1943
Number of Men Aboard:	78: 70 enlisted and 8 officers
Total Days on Patrol:	56
Fuel Burned:	107,000 gallons
Total Miles Steamed:	10,900
Number of Torpedoes Fired:	16
Ships Credited as Sunk:	1/4,217 tons, plus one 75-ton sampan via gunfire
JANAC Postwar Credit:	1/4,700 tons
Shipping Damage Claimed:	1/1,000-ton escort
Return To:	Pearl Harbor via Midway

12

"Looks Like a Long Chase"

Refit and Tenth Patrol   *12 July - 15 September 1943*

The crew was released to the Royal Hawaiian beginning 13 July for two weeks of R&R. For many, it was two weeks of wild partying ashore. The officers did their best to keep the men from getting into serious trouble. Lt.(jg) Jack Kennelly, the former lawyer, defended one crewman who was being held on charges of improper conduct with a female ashore. "Kennelly took up the case and this sailor made the next patrol," said Hank Thomson.

Due to return back to his boat in the morning, seaman Red Mills tied on a good drunk with his shipmates. He came back in the late afternoon and headed for the dock where *Seawolf* had been tied up for days. Smashed out of his mind, Mills just walked right off the dock where *Seawolf* had been. Into the drink he went, floundering around. Another famous boat, Mush Morton's *Wahoo*, happened to have docked at Pearl on 19 July, returning from a West Coast overhaul. One of the *Wahoo* sailors used the grapnel hook to help haul the drunken Mills aboard ship. As Mills later related, Admiral Chester Nimitz himself had come aboard *Wahoo* to meet with her famous skipper. CincPac glanced over at the dripping wet sailor that was being hauled aboard *Wahoo*. Mills gave him a salute and hurried on his way without saying a word.[1]

Some *Seawolf* sailors worked the bars and sought female companionship in the local brothels. Others managed to make new friends with the locals ashore. "I got acquainted with a couple who ran an ice cream shop in town," said gunner's mate Rex Mickey. "They would invite me up and I could stay at their house instead of always being at the Royal Hawaiian Hotel. They would take me and a buddy around to show us the legends and stories of Hawaii."

Back aboard ship on 27 July, yeoman John Burruss took up the task of transferring crewmen. He was among those receiving orders to new duty. CTM Clarence Kibbons and CMoMM Casey Mallough, both freshly promoted, were also among the transferees. *Seawolf* also lost eight others who had been aboard for all nine runs: chief of the boat Swede Enslin, motormac Charles "Johnny" Johnson, radioman Paul Maley, cook Bill Mallory, officer's steward Brigido Tamayo, auxiliaryman Paul Zimmerman, and the Hanson brothers. Johnny Johnson had orders to help put *Pipefish* into commission, where his new skipper was none other than Bill Deragon.

Big Swede and Little Swede Hanson were transferred due to new Navy policy. "After the Sullivan brothers went down—there were five of them lost—they separated us," said Bob Hanson, referring to the cruiser *Juneau* lost at Guadalcanal. Following all five brothers being killed in November 1942 aboard one ship, brothers like the Hansons began being split up by the Navy. "We had made nine runs together, but they sent us to different boats. Henry went on the *Muskallunge* for one run and then on to the *Tench*. I was transferred to the *Guitarro* and made four runs on her," said Little Swede.

Paul Zimmerman would end up back in a Philadelphia shipyard. He was first sent to New London for new assignment, where he ran into Lt.(jg) John Sullivan, *Seawolf*'s old yeoman. Sully was now the New London sub base's personnel officer. "Sully gave me a set of the periscope photos that Lieutenant Mercer had taken of the destroyer we sunk," said Zimmerman. "Then, of course, I saw them later on in *Life* magazine." These photos would become quite famous.

With Chief Enslin's transfer, CGM John "Gunner" Bennett took on the role of chief of the boat. Many of his former CPOs would be gone for the next run. Newly-promoted CMoMM Roy Bateman went into the hospital and would miss the next patrol. Aside from those transferred, Roy Gross exercised a special leave program in which he would send a couple of his key men stateside on 30-day leave while the ship was out on patrol. Chiefs John Bilkey and Lefty Leffingwell received transfers to enable them to return home, but both would rejoin the *Wolf* after her tenth war patrol.

Hank Thomson made chief quartermaster and transferred to the flagship *Plunger* (CSD-43) for duty on 27 July. He had been aboard the *Wolf* since the war's start but was now hoping for a change. "I had asked for pilot school," he recalled. "Captain Gross said, 'It won't go through,' but he okayed it." The division commander also okayed it but warned Thomson that the request would not go through. Aboard *Plunger*, Thomson found that the squadron commander did not approve his request and he remained stuck aboard the CSD—sans pilot school and without his beloved *Seawolf*.

As far as the new hands received aboard as replacements, not everyone who came aboard did so voluntarily. Bkr3c Fred Cucchi came on board as the replacement baker. "I was shanghaied," he admitted. "I was right out of boot camp and this bosun's mate grabbed me and another guy and told us, 'You just volunteered for sub duty.'" Whereas almost all aboard submarines had been through sub school, it became tougher to staff all of the new boats quickly. "*Seawolf* was in drydock undergoing a refit at Pearl when I joined her," said Cucchi. "They had me painting and every damn thing."

Cucchi preferred surface ships and Exec Bob Risser saw to it that he made his way to one before the *Wolf* sailed again. He and Mallory were replaced by SC3c Paul Schultz and SC3c John Cani from Detroit. Most of the others were eager to join the well-known *Seawolf*. MoMM2c Bob Curtin, a former steel mill worker pre-war, came aboard from *Plunger*, but had also made runs on *Tuna*. He had joined the Navy because he believed "it was better than going into the Army" via draft. "My draft notice showed up, so I jumped on a bus and signed right up for the Navy," Curtin said. "With submarines, we got more money and more liberty."

Another of the new enginemen was MoMM2c Lewis Sigmund Donche from New Jersey. After high school, he had worked in his home state at Eclipse Aviation, machining parts for the Norden bomb sight for aviators. At age 20, he also enlisted in the Navy to avoid being drafted into the Army. "Back then, sub service just sounded like better duty," Donche later said. With the number of crewmen aboard, he had to hot-bunk it to sleep. Lew Donche was assigned to Bud McCoy's forward engine room, where he considered the mechanical wonders of his diesel boat to be "fascinating."[2]

Twenty-three-year-old MoMM1c Mike Wiegenstein, a veteran of five patrols on the submarine *Plunger*, joined the *Wolf*'s after engine room as a senior man. Prior to *Plunger*, he had served for three years aboard the cruiser *Portland* (CA-33) before volunteering for subma-

Lt. Dougald "Duke" Robinson (left), battle stations plotting officer, seen on *Seawolf*'s bridge on patrol. *Courtesy of Claude Robinson.* Like Robinson, CMoMM Michael Paul Wiegenstein (right), joined the *Wolf* prior to her tenth patrol. *Courtesy of Amy Wiegenstein.* Lt. Robinson was transferred when he became ill after the *Wolf*'s twelfth patrol, while Wiegenstein remains on eternal patrol with *Seawolf.*

rine school. Just a young boy from a Ozarks farm family, Wiegenstein had been out on maneuvers aboard *Portland* when news of Pearl Harbor caught up with him. The sobering reality of life or death at war soon took over. Mike loved sports and once wrote to his younger brother Gene, "Boy, I sure would like to to hunting or get behind a plow again. Maybe you don't know it, but you're a hell of a lot better off there than I am."[3]

Seawolf's refit was delayed by the discovery that the shutter was missing on the No. 7 torpedo tube upon docking during the later stages of refit. This loss was unknown and unsuspected aboard *Seawolf.* "Navy Yard Pearl Harbor had considerable difficulty forming a new one," wrote Lt. Cdr. Gross. Due to this problem and other improvements, the ship would remain at Pearl Harbor for another two weeks. The men were excited because they could take liberties from 1600 to 0800 during these additional two weeks.

Following this, there was a training period of four days, with the firing of three exercise torpedoes. There was also a two-day exercise with an east-bound five-ship convoy with two escorts. *Seal* also worked with the *Wolf* on this convoy. Following these exercises, *Seawolf* was flashed and depermed to decrease her natural magnetic field. The yard crews also installed an auxiliary gyro-compass and bathythermograph.

Before the refit and training period was complete, *Seawolf*'s wardroom also had some personnel changes. Lieutenant Commander Philip Weaver Garnett, a 1933 Academy graduate, came aboard to make his PCO run. Roy Gross wrote that Lt.(jg) Jug Casler was "detached with regret after nine patrols." Lt.(jg) Kennelly was also transferred due to physical reasons. As far as navigation was concerned, Exec Bob Risser would take on this duty for the tenth run.

Kennelly's replacement was Lt.(jg) Dougald "G" Robinson, a sharp young man fresh from submarine school who had been named for his Scottish grandfather. Lean, with blue eyes and reddish hair, Robinson would be quick to earn his dolphins and qualify as an OOD on *Seawolf*. He was fine with "Dougald," although his buddies generally called him by "Duke" or "Doog." Born in Colorado, and raised in Sand Point, Idaho, Robinson had a degree in Chemistry from Whitworth College in Spokane. He was working for Boeing in Seattle when he and a buddy joined the Navy after the war broke out. Doug married his sweetheart Mary Katherine on April 5, 1942, prior to attending sub school at New London.

Happily assigned to *Seawolf*, his first sub duty, reservist Duke Robinson soon found July 1943 memorable for another reason. His wife gave birth to his first son Claude on 20 July. He became the "George" of the wardroom, or low man on the totem pole, assigned as commissary officer and assistant communications officer.

Duke Robinson had a flare for science, and he was quick to figure out when his instructors sometimes mixed up what they were teaching at sub school. He later related his time at school and coming on as the *Wolf*'s commissary officer.

> If you would open your mouth and say, "Hey, you've got that wrong," they would say, "You want to leave subs?" When I got out to Pearl Harbor, I had a captain who asked me, "Well, what do you think of sub school?"
>
> I said, "A lot of it is good, but a lot of it is just plain horse manure."
>
> He said, "Do you want to leave submarines?"
>
> He had been in charge of it!
>
> As commissary, you had to have enough food to keep those guys going for sixty days or better. Our one problem was that the refrigeration kept dropping off. It's pretty hard

to fix out there with nothing to fix it with. We had no refrigerant. We had a few bottles of fire retardant—CO2—but that wouldn't do much for the refrigeration system.

Seawolf departed Pearl Harbor at 1310 on 14 August, bound for the East China Sea. With 71 enlisted men, eight officers, and 20 Mark 14-3A and Mark 16 torpedoes, she was fully loaded and ready for action. After a trim dive off Barbers Point, *Seawolf* headed on and released her escort after dark. In between training dives and drills, Roy Gross and his officers enjoyed card games in the wardroom. In Kelly's Pool Hall, the crew enjoyed some light betting as well, payable at the end of the patrol. "I wasn't much of a poker player, but then there was Dallas Malone," said gunner Rex Mickey. "That man could shuffle the deck and deal himself three aces, slicker than a whistle! He wouldn't win much money, but he always won. I wasn't that good. I mainly stuck to acey-deucy to pass the time."

At 0851 on 16 August a Navy PBY made a close call of *Seawolf*. Lookouts sighted the patrol plane at 10 miles on the port quarter. "At 7 miles he headed right at us," wrote Gross. At 6 miles, the *Wolf* ceased zigging and steadied on her base course. Bill Whitman pulled the after flare, but the PBY continued to approach aggressively. Radar finally picked up the inbound PBY at 5 miles at an altitude of 3,000 feet. The bridge was then cleared by Lt. Whitman for safety's sake but the boat did not dive. The Navy plane turned away slightly and then circled to cross over *Seawolf* twice to examine her while dropping to as low as 500 feet. "Feel sure we would have been bombed if we dove," recorded Gross. The entire time the plane was approaching, one of the quartermasters was challenging him with the 12-inch searchlight with a one-letter challenge. No reply was seen to his challenge, however.

Chief Bud McCoy, keeping his secret diary in the forward engine room, found the outbound patrol leg to be routine. "Still making test dives on way to Midway. Everything working so far, except aft engine room. Broke down, as usual."[4]

Seawolf moored at Midway at 0850 on 18 August, pulling up alongside *Pompano* who was moored to the dock. Radar did successfully pick up Midway's harbor entrance. Due to a continual heavy rain, the OOD had to rely upon radar and the fathometer to negotiate this tricky entrance. The island, escort planes, and patrol vessels could not be seen in the rain until the *Wolf* was 3 miles south of the entrance to Gooneyville.

The time in port was brief enough only to take on fuel. "Duke Wall was a big guy and when we came in for fueling, he was our fuel king," recalled new hand Bob Curtin. Midway's repair crews also repaired *Seawolf*'s faulty air conditioning circulating water line and worked on a new low reading on the SJ radar. The *Wolf* was underway at 1704, moving out of the harbor entrance and standing out to sea for Roy Gross' third command patrol.

En route he conducted the usual training drills. He had the drill "intensified this time due to excessive number of new people." Crossing the international dateline, *Seawolf* deleted 19 August from her calendar. Still a little spooked by the PBY incident coming into Midway, *Seawolf*'s OODs took the cautious approach thereafter. On 20 August—with the ship about 260 miles west of Midway—Red Syverson made a crash dive to avoid another patrol plane. Prior to departing Midway, the *Wolf*'s officers had been warned of a Japanese plane that had been spotted recently in this area. Radar did not pick up the plane and Lt. Syverson didn't wait around long enough to identify. It was safer to dive first and sort it out later.

Chief David Goudy—a native of the little town of Eldon, Iowa, where Chief Bud McCoy had also been raised—was the new senior radioman aboard *Seawolf*. Goudy and his radio gang—RM2c Sam Ottaway, RM2c James Call, and striker Jim Johnson—decided to liven up the patrol a little bit with a ship's newspaper. They copied down interesting stories from their various intercepts, added cartoons from radarman Ben Rogers, and inserted jokes to keep the crew entertained. The outside news of the world picked up by the radio operators was most enlightening. "As far as news aboard the ship itself, it was all old," said motor machinist's mate Bob Curtin. "News travels pretty fast on a submarine."

Seawolf entered her patrol area during the early morning hours of 29 August and dived for a submerged patrol at 0500 around the north end of Kikai Jima Island. She moved off the coastline in search of shipping. There was not even a sampan sighted, however. Through the periscope they could see what appeared to be a lookout station on the north end of this island on a ridge.

At 1955, *Seawolf* surfaced 10 miles off Naze, the largest town on the island. The SJ radar could follow the land out to 33,000 yards, a good distance for *Seawolf*'s cranky set. "Today is the first all day dive," wrote Bud McCoy on 29 August. "This trip getting close to Tokyo's back door. About 150 miles off of Nagasaki."[5]

On 30 August, the SD picked up a contact at 7 miles and closing. Bill Whitman dived ship and used the SD radar at one-minute intervals for five seconds each time. He rigged an alarm clock to give periodic warnings on a buzzer. The day was quiet save for an SD radar contact at 1350. Red Syverson made a crash dive when this plane closed to 6 miles, but again no bombs were dropped.

Due to operating so close to Japanese mainland, *Seawolf* was required by Admiral Lockwood's staff to make special accommodations. "Water pretty deep for a change; that [is] one thing nice about this place," Chief McCoy noted in his diary. "Still pretty dangerous place; enough so that we left some important equipment in port so it wouldn't fall in the wrong hands." This included communications officer Jim Mercer's decoding machine, which made his job even more challenging as new receipts came in via radio.[6]

The sun rose at 0530 on 31 August, and Lt. Mercer sighted smoke on the horizon at 0655. This developed into a nice six-ship convoy plus one *Chidori* torpedo boat. One of the six was an engine-aft small trawler believed to be an escort vessel. Four of the freighters were estimated to be in the 5,000 to 7,000 ton range. The fifth, larger freighter was estimated to be 8,500 tons.

Roy Gross made out their smoke in three columns on *Seawolf*'s port bow, range 30 miles. The tracking team quickly plotted the convoy on a northeast course making 9 knots. The *Wolf* raced ahead on the surface at 17 knots, closing in on their track while also running ahead. At 0745, he dived his ship once the angle on the bow to the target ships had reached zero and the range closed to 18,000 yards.

In the conning tower, Lt. Cdr. Gross handled the scope and offered occasional looks to Lt. Cdr. Weaver Garnett, the PCO aboard to learn the ropes of command. Assistant approach officer Bob Risser and SM2c Frank Romito stood by to hit the firing keys and call out the periscope bearings from its dials. Forward, QM3c Jim Grimes was the battle stations helmsman. In the absence of Hank Thomson, Y2c Roger O'Brien stepped in to be the battle stations talker, relaying the orders from the skipper to other departments. Jim

Mercer, the most tenured *Wolf* veteran on the plot team, was seated at the little plotting table, assisted by Lt.(jg) Duke Robinson in tracking the convoy's course.

Below, in the control room, Red Syverson was the diving officer monitoring the men at the planes stations. Chief of the boat John Bennett stood sentry at the Christmas tree hull security panel. Bill Whitman sat at the old TDC, assisted by FC3c Edward Morris and Delbert Mar, the fire controlman striker. Mar later described his duties with the TDC.

> I was responsible for helping to lay down all the information for the torpedo data computer, almost like a secretary. There were just two people running it, the officer in charge and our leading fire controlman. That old machine was the size of a dining room table. Now, they shoot a torpedo with something as small as an adding machine. I don't know how we ever hit anything with it!

Gross settled on the largest ship, the well-decker of about 8,500 tons, which appeared by its waterlines to be moderately loaded. Searching through ONI-208-J, the ID team could not find any ship that closely resembled this one. These six ships would later prove to be *Durban Maru, Kokko Maru, Fusei Maru, Banshu Maru, Harushima Maru, Shoto Maru*, and the torpedo boat *Sagi*.[7]

The *Chidori* was pinging but he passed well clear to port of *Seawolf* while patrolling ahead. The convoy's formation was very loose with no regular pattern apparent. It was using irregular zigs—some large, some small—to foil any lurking enemy. Lt. Cdr. Gross' log picks up the action here.

> 0906. Found ourselves in good position for bow shot, port track, at selected target but left flank ship headed right at us, angle on bow slight port, range about 1800 yards.
> 0910. They zigged left, putting us right between two ships. Fired four bow tubes at main target, range 1300 yards, 25 port track, 350 gyro angle, torpedoes set to run at 6 feet, speed spread to cover only 400 feet target length, two at middle, one 200 feet forward of middle, one 200 feet aft of middle.

Swinging his stern tubes quickly to bear, Gross fired all four Mark 16 aft fish into the convoy from 1,300 yards. Picking the second largest freighter of the convoy, he estimated her to be about 450 feet long. She was moderately loaded, her weight guessed at 7,500 tons. She had a large numeral "3" painted on her side.

Forty-eight seconds after firing, Frank Romito timed the sound of the first explosion, which rang clearly throughout *Seawolf*. Two hits were heard from the bow tubes even as *Seawolf*'s stern tubes were being trained on the target. One more hit was made aft on *Shoto Maru* just aft of her stacks. After this hit was observed, Gross swung his periscope back to observe his first target. He could see that her stern was already under the water up to her stacks.

Gross then shifted his view back to the stern tubes target, and saw her down by the stern but not sinking. The bow tube target—the 5,253-ton army freighter *Shoto Maru*—was then seen to sink exactly two minutes after the first torpedo hits.

By 0914, the Japanese knew of *Seawolf*'s presence. Gun fire from the stern tubes target and other ships found the *Wolf*'s range. Shots were landing about her periscope. By 0918, the second torpedo victim——5,486-ton freighter *Kokko Maru*—was gone. *Kokko Maru*, when last seen, was seen to be 5 degrees down by the stern with with a 10° starboard list. The 840-ton torpedo boat *Sagi* also suffered medium damage. *Kokko Maru* was carrying 8,500 tons of iron ore from Yulin, Hainan Island, for delivery to Yawata in northern Kyushu. Her lookouts spotted three torpedo tracks just 200 meters in front of her. Although *Kokko*'s skipper tried to avoid, one of *Seawolf*'s torpedoes hit her amidships on the port side, causing violent flooding and an immediate list. The 5,486-ton freighter went down about one minute and 20 seconds after being hit, with 19 of her crewmen being lost in the sinking. The torpedo boat *Sagi* then made a depth charge attack on the *Seawolf*, unleashing 16 ashcans.[8]

Two minutes after firing the last fish, Gross ordered the ship to rig for depth charges. Doug Syverson took her as deep as possible, which was only 175 feet. Soundings on the fathometer chart showed 31 fathoms. As *Seawolf* attained depth, she was rocked at 0921 with the first of six depth charges. This time, the explosions were close enough to break light bulbs.

For Lt.(jg) Duke Robinson, this was his baptism by fire under depth charge attack. "I just stood there and it would go 'BLAM' and the lights would pop," he stated. "They are not a nice thing to sit and listen to. I kept telling myself things could be worse. On a surface ship, you get airplanes shooting at you and dropping bombs."[9]

Seawolf rigged in her pit log, the QC head, and ran silent. The sound operators continued to hear a heavy gun firing on the surface above for about five minutes. Fortunately, there was a slight temperature gradient working in the *Wolf*'s favor this day, and she was able to use it to help deflect the sonar from the enemy ship.

Seawolf gradually evaded the pinging *Sagi* by keeping her stern pointed at him. At 1048, Gross started up to take a look-see. As he did, six depth charges exploded in one pattern, fortunately none too close. Three minutes later, another four depth charges exploded. Gross took her back down to 175 feet. Chief Goudy's sound operators reported many strange noises and screws on sound.

"The escort dropped 15 depth charges," wrote Bud McCoy. "5 of them were pretty close. Broke a few light bulbs and shook us up plenty. The rest were a little further away, but could feel them. Had to stay down all day. The escort boat was trying to find us."[10]

Seawolf remained at silent running for another hour as the temperature in the boat grew hotter. At 1158, Goudy reported that the patrol boat seemed to make contact on the *Wolf* with long scale pinging. After a tense quarter-hour, *Sagi* apparently lost contact and resumed a sweeping of the area.

Gross eased back to periscope depth at 1255 and found his assailant about 4,000 yards astern, making heavy smoke and moving in large circles. He opened up the distance and secured his crew from silent running and depth charge quarters at 1321. A low wing small monoplane could be seen circling the torpedo boat, looking to be a Zeke fighter. With air cover and an angry escort, Gross smartly cleared this hot spot during the afternoon.

Seawolf surfaced at 1900—45 minutes after sunset—and ran at 15 knots back to the scene of the morning's attack to see if her cripple was still there. She made no contact and Gross assumed that his crippled ship had sunk. He then changed course to the northeast to overtake the remaining ships, which should be about 100 miles ahead of him. He did not believe *Pargo* and *Snook* to be in the area yet, so Gross opted not to send out a contact report yet.

At 2050, the SJ radar made contact at 6,700 yards. *Seawolf* closed to 5,000 yards to identify a *Chidori* torpedo boat on a southerly course. "Believe this was the one hunting us and now en route Kirun or some other port for more depth charges," wrote Gross. "He had probably been standing by damaged ship. His departure indicated that damaged ship had sunk. Continued chase at 17 knots to the northeast."

At 2305, Gross opened up on all four engine to 18.5 knots. His instincts were right on. At 0425 on 1 September, Jim Mercer announced that his lookouts had smoke on the horizon in the early dawn. *Seawolf* maneuvered to put the smoke on her port bow to run around it while keeping the rising sun behind her. By 0500, the smoke could be identified as three columns, but still at least 20 miles ahead of *Seawolf*. "Looks like a long chase," wrote Gross. The sun rose at 0518 into a cloudless sky and glassy sea. Roy Gross decided that his lookouts should be able to spot enemy aircraft at a good distance, so he stayed on the surface to continue his end-around.

By 0635, he could make out the convoy to be four ships on *Seawolf*'s port beam showing a 90° starboard angle on the bow for a base course of about 010(T). This differed from the convoy's course of the previous morning, so Gross presumed that the ships were either in evasive course or long zigging legs on their regular course. The *Wolf* continued drawing ahead slowly. The convoy was tracked as making 9 knots. Gross kept the four ships in his sight using the No. 2 periscope, with only their stack and masts above the horizon.

At 0942, the convoy finally zigged right for a zero angle on the bow, putting them on a course of about 050—just as Roy Gross had expected. He dived *Seawolf* eight minutes later some 15 miles ahead of the Japanese ships. Red Syverson found a temperature break at 75 feet—where the temperature abruptly dropped from 78° to 56°—making a great barrier for a submarine to hide under.

"Battle stations, torpedo," called out yeoman O'Brien as the Bells of St. Mary's bonged throughout the *Wolf* at 1036.

As *Seawolf* awaited the oncoming ships, Gross was careful with his observations. The sea was absolutely glassy and his scope would be easily seen. At 1046, the enemy's masts and stacks were just beginning to show over the horizon.

Gross felt certain that he was attacking the same convoy from the previous morning. Four ships were present, three marus and the same little trawler with engines aft. The *Chidori* torpedo boat was now absent. "It appears the one damaged yesterday, not being present, nor being at scene of attack, must have sunk as it is unlikely that she could get underway with a torpedo hit aft," thought Gross. "The torpedo boat leaving the scene at dark bears out this supposition."

The plotting team of Mercer and Robinson had the convoy all with zero angles on the bow. Sound operators David Goudy and Sam Ottaway could hear nothing, as their sound head was in colder water than the enemy's screws were in. At 1140, the range was down to 1,500 yards and angle on the bow had shifted to 30 port. The skipper had Chief Goudy take a ping range, as sound still could not pick up the targets. All ships were moving across *Seawolf*'s stern.

"Open after torpedo tubes," called O'Brien to Alfred Ostrander and John Neil's crew.

Starting at 1146, *Seawolf* fired three stern tubes in a spread at the nearest target. They were set to 6 feet and had only a 900-yard run distance. Roy Gross used a speed spread, his first two fish set to hit MOT and the third to hit 200 feet ahead of the middle. His target was a standard well deck freighter of about 6,500 tons, about 420 feet long. This ship was later identified as *Banshu Maru*.

As he fired, Gross noted that *Banshu Maru* had a large No. 1 painted in white on her stack. He held fire on the fourth torpedo and swung in a new bearing for the middle of the next ship astern, a well-decked freighter of about 5,000 tons. He fired this fourth torpedo 19 seconds after the third after fish had departed. Due to reduced target speed data, he quickly realized that this fourth fish would miss astern. In fact, all four torpedoes missed.

In analyzing his mistakes, the skipper decided that the target speed for the fourth torpedo was set at 6.5 knots instead of the proper 9 knots that the ship was traveling. As for the first three misses, he reasoned that the enemy ship "possibly maneuvered to avoid." *Banshu Maru* had been seen to be turning while firing, but the set-up looked very good. Periscope exposures during the latter stages of approaches were never more than seven seconds and sometimes only five seconds, as timed by Lt. Cdr. Garnett in the conning tower.

"It is believed that a close range shot in such a sea condition is doomed to failure," wrote Gross. "In addition, our periscopes operate too slowly. This was a very disappointing attack, after having carefully chased convoy for fifteen hours at best speed."

After firing her fourth stern tube, *Seawolf* went deep to avoid counterattack. The trawler was last seen to be 1,500 yards away and headed in *Seawolf*'s direction. Only one light explosion followed, and this was assumed to be from shell fire on the surface. *Seawolf* had 30 fathoms of water in which to operate. Red Syverson had a time with the temperature layer. At about 120 feet, he had to flood the negative tank to completely get under this heavy layer of cold water.

Shortly after noon, shell fire could still be heard on the surface, but there were no more depth charges. Gross waited until 1306 to ease back to periscope depth. There was nothing in sight except for smoke on the horizon. "Amazed at lack of counterattack," he thought. "Trawler may be out of depth charges. Expected plane search never materialized either."

He moved toward the convoy's bearing during the afternoon, but waited until 1615 before surfacing to continue the chase again. He put one engine on battery charge and opened up on the other three, running down the last true bearing of the convoy's smoke—although that bearing did not take the convoy toward any known port. Five minutes later, lookouts spotted the smoke. The radio gang also reported "chattering" on the radio.

At 1725, the SD radar began acting up, but was readjusted by Ben Rogers and company in short order. The nearest land was Danjo Gunto, some 130 miles away. Chief Goudy's men continued to listen to radio chatter from the Japanese, who were apparently sending out distress calls. Gross secured his battery charge at 1751 and put the fourth engine on line for emergency speed of 20.1 knots.

The sun set at 1802 with the convoy's smoke still in sight but the smoke was soon lost in the darkness. At 1945, faint smoke traces were again picked up as the moon set in a clear, starlit night. The SJ soon had three pips at 11,800 yards. Their evasive easterly course after the attack was followed an hour after sunset by a return to their base course of northeast.

At 2000 on 1 September, Roy Gross called for battle stations, torpedo for the third time in two days on this convoy. *Seawolf* was now 8,000 yards from their track and 45° on their port bow. She began closing their track now while presenting only a minimum silhouette for enemy lookouts to potentially spot. There were now only three ships in sight and only three ships showing on Rogers' radar. At 2017, the fourth pip was picked up, certain to be the engines-aft trawler. "Had hoped that he had been by some chance the big explosion this afternoon," Gross wrote jokingly.

At 2053, *Seawolf* fired four bow torpedoes at a range of 3,700 yards, with a depth setting 6 feet, as the leading ship was overlapping the next ship. Both were estimated to be 6,500 ton freighters and were believed to be the same two that were fired on earlier in the afternoon. From the bridge, the officers and lookouts could see the torpedo wakes streaking toward the convoy. Gross saw one certain hit under the foremast of the leading target. He believed this was the fourth torpedo fired and that the first two had either passed between the two ships or had been duds. The third torpedo had likely passed astern of the trailing target. *Seawolf* turned away at full speed after firing to open up the range to 7,000 yards.

Gross' estimation on his performance was right on. Postwar records would later show that one of his torpedoes had hit the 2,256-ton freighter *Fusei Maru*, formerly the British *Fausang*. *Fusei Maru* was struck in her port side in the No. 1 hold. Another ship, the freighter *Durban Maru*—loaded with 9,000 tons of white rice for planned delivery at Yokohama, Japan—was hit in the bow by one of *Seawolf*'s dud torpedoes.[11]

Gunfire from all ships could be seen and heard. It was inaccurate, although some splashes were seen in *Seawolf*'s direction. The ships were clearly in sight through binoculars on the *Wolf*'s bridge but Gross believed that his ship was not seen. He noted that the other three Japanese ships were running away, leaving their cripple behind. "Decided to finish off this ship rather than chase the others as she might get underway while we were gone," he logged. "Four torpedoes remaining forward, none aft."

Seawolf came in for a one-torpedo attack after plotting *Fusei Maru* as stopped. "Our plan was to take good aim with one fish and finish off the dead one," wrote Duke Robinson. Lt. Clary John had his binoculars locked in to the target bearing transmitter, which would feed info directly into Bill Whitman's TDC below decks. Gross and PCO Weaver Garnett were also on the bridge for the attack, while Bob Risser supervised the firing from the conning tower.

At 2129, *Seawolf* fired her No. 1 tube at 3,600 yards range. The gyro was zero and the bearing was zero. Depth setting was 6 feet, everything checking on Mercer's plot and Whitman's TDC. The torpedo ran straight for *Fusei Maru*'s middle and made a white splash under her stack. "It could be heard to run up to the ship and stop running, but no explosion," wrote Lt.(jg) Robinson.[12]

Another Mark 14 dud!

At 2132, the maru opened up with heavy gunfire, the nearest of her point-detonating shells landing about 500 yards away. Roy Gross felt certain that the Japanese ship was ranging his guns down the torpedo wake, which *Seawolf* cleared at full speed right after firing. With three torpedoes remaining, Gross pulled back out of gunfire range. He decided to move in closer by diving at 4,000 yards and using the periscope and radar from the safety of 40 feet depth.

At 2217, *Seawolf* dived at 4,600 yards. The target was now very difficult to see. At 2,800 yards, various pips were picked up on radar and screw noises were heard on sound. Through the scope, however, there was nothing else in sight but the stopped *Fusei Maru*. Gross thought of the different pips being life boats but he could not understand the Japanese abandoning a ship apparently in no danger of sinking. "Don't like running at 40 feet without being able to see what's going on." Duke Robinson, assistant communications officer, also noted the radar pips. "There were a bunch of radar pips that went rocking away from it," he said. "We figured, 'there goes the crew to get some dirt under their feet.'"

Insecure with the visibility and setups, Lt. Cdr. Gross ultimately opted to return to the surface at 2322 to make another attack. He ran around her to fire on a starboard track "to change our luck." Gross ordered Bill Reiland's forward crew to open the door to the No. 4 tube. This time he opted to fire from further out—from 3,800 yards—to make sure the fish had time to set. Again, Lt. John was on the bridge TBT, feeding the data to the TDC below.

This torpedo took a small right angle, and missed one-quarter length ahead of *Fusei Maru*. Gross swung away and moved at high speed to avoid shell fire, cursing his bad luck again. No enemy fire developed, so the Japanese apparently were oblivious that they had been fired upon again.

With only two torpedoes remaining, *Seawolf* turned for a third try at this cripple with the gyro settings on zero. This time, Gross would not use his TDC or regulators. It was the early minutes of 2 September, the third day of chasing this convoy, as *Seawolf* approached again. This time, Gross came in to 2,800 yards. Using the bridge TBT, he had Clary John line up on the ship's stack and make sure the TBT read zero. He fired at 0008, another single fish set at 6 feet. This torpedo also took a slight left angle of about 5 degrees and missed astern by about 1/4 length! *Seawolf* again turned away from expected gunfire, but none developed.

No pips were on radar. "Couldn't understand this unless they again didn't know they had been fired at. One torpedo remaining. Opened out to consider problem. Decided to try last torpedo submerged at close range. Settled down on firing course using TDC for a 90 starboard track again, closed slowly, checking bearing while approaching, target still showed stop on TDC and plot."

Seawolf dived and moved in slowly. Her radar again began picking up smaller pips around the target. At 2,800 yards, she checked the range with a ping range by radar. Gross went to 60 feet to avoid whatever the small pips were that were seen in the vicinity of the target. Although the target was very hard to pick up in the darkness, he eventually made her out, estimating her to be 420 feet in length.

Gross moved his submarine in to 1,100 yards distance before firing his 20th and final torpedo. He had the target in his scope filling his periscope window at high power. The relative bearing was zero, depth setting 6 feet and an 86 starboard track. Sound reported the torpedo running straight on zero bearing and at 50 seconds sound reported the fish had stopped running. Nothing was seen through the periscope. There was no thud either by sound or heard in the boat, but Ottaway and Goudy heard a high-pitched squeal at that time. "Must have been a dud," a dejected Gross wrote.

Submariners being a superstitious bunch, they now had more scuttlebutt for Kelly's Pool Hall. The old timers who had been aboard for the fifth run could gripe of the torpedo performance that they had another PCO jinx on their hands. In the forward engine room, Chief McCoy scribbled notes of the action of 2 September.

> Sept 2nd: We are still after these ships of some kind. It's so dark that we can't see what they are. We fired all the torpedoes we had left and hit a large 8,000-ton freighter and stopped her dead in the water. Two other fish hit the target but didn't go off. The other two ships are running like hell. We haven't any more fish, and the ship isn't sinking, so I think we will battle surface and use our deck gun.[13]

Seawolf secured her tubes, turned away and opened out after firing her last tin fish.

"Prepare for battle surface!" Gross ordered.

The *Wolf* surfaced at 0145, 30 minutes after firing her last faulty torpedo. The target ship was in sight and was the only pip on Ben Rogers' radar at about 4,500 yards. At 0154, Gross called, "Bring up the gun crew."

Gunner John Bennett, the chief of the boat and captain of the 3-inch .50-caliber gun, had his men ready. First loader Duke Wall inserted the first shell and stood ready with the next as the *Wolf* slowly closed the stricken *Fusei Maru*. Gunnery officer Bill Whitman was down on deck to help direct the fire from range estimates that were called to him from the bridge. Within moments, the freighter could be seen through the gun sight telescopes at 4,500 yards.

"Commence firing!" came at 0200.

The *Wolf*'s 3-incher immediately barked out its first shell. Bennett's team planned on firing five ranging shots to see if the enemy's returning fire was accurate. The men had difficulty staying on the target through their telescopes, but they appeared to score some hits after six shots. There was no return fire, so the *Wolf* closed, continuing to fire deliberately using radar ranges and radar spots.

By 0230, *Fusei Maru* could be seen to be clearly abandoned. Her bridge was now on fire from hits which set off pyrotechnics, but she was not burning very well. Lt.(jg) Robinson, on the bridge helping to spot for Whitman and Bennett's gun crew, later recalled:

> We were shooting it with our 3-inch gun right at the water line. A fire started right on the bridge. We could sight the gun on that fire and we could adjust it down to the water line. That ship was listing toward us from the torpedo hit. There were little pools of fire that made fun targets. There was enough fire that you could see the deck. After a while, things were tumbling off the deck into the water. A three inch hole will take a lot of water, but it took a while to sink the ship.[14]

Lt. Cdr. Gross had his gunners continue shooting, closing the range to 1,700 yards. On the SJ screen, the short shots could be easily seen to throw up a splash but the shells that passed over were much more difficult to pick up. The overs tended to silhouette the target. Rogers and Robinson coached the gunners up any time they were landing short of the ship.

By 0304, just over an hour into the gun attack, *Fusei Maru* was burning fiercely and had taken on a large starboard list. Some of the 71 point-detonating shells fired were observed to actually penetrate the ship instead of bursting immediately on contact. Others seemed to burst outside of her hull. "Those bursting on deck had good incendiary effects," noted Roy Gross.

Seawolf checked fire at 0315 and circled to inspect her target. She was listing heavily and burning well. Gross felt certain this was the same ship he had fired on the previous day. He could not, however, see the tell-tale painting on her stack, which was now blistered and damaged. To the delight of those topside, *Fusei Maru* rolled over on her beam and sank without exploding at 0332. "A very impressive sight," wrote Gross in his log. The sky was just lightening in the east. Bennett's gun team had fired 125 flashless rounds, of which about 70% were hits. Eight hangfires from the deck stowage were tossed overboard. These were all suspected to have received excessive dampness. One stuck projectile was cleared with a short charge. "The shooting and spotting were excellent," wrote Gross.

Chief McCoy's diary entry summed up the battle surface that polished off this maru.

> Before we started firing, the target was shooting at us with a four or five-inch gun, but have stopped. Japs in lifeboats are shooting rifles at the guys on the deck. When we made a submerged run at the target, one of the Japs was hanging onto our periscope, but we went down to a 100ft and got him off. We set the target on fire and it's light of day up there and the target just went under. We started out of here in a hurry, as there are two more ships headed toward us and we don't want to tangle with a destroyer with no fish.[15]

The *Wolf* stayed down on 2 September, and all was quiet. She cleared *Pargo*'s assigned patrol area that night. At 2305, Chief Goudy resent the patrol message regarding results and another advising *Pargo* to stay clear of the previous night's position for a few days. Gross set course for Midway. His crew was exhausted, having chased and attacked the enemy convoy day and night for three days.

At 0012 on 3 September, lookouts sighted a correctly lighted hospital ship—believed to be Russian. She was not zigging but had two escort vessels. The initial range picked up on radar was 10,000 yards but bridge personnel had her in sight several minutes before radar saw her. Gross avoided this neutral ship but closed her to 6,000 yards just to inspect her escort ships. Remaining unsighted, he resumed his own course to the southward.

Seawolf cleared the Nansei Shoto on 4 September, dodging an enemy patrol vessel during the early morning. The familiar mariner's mark of Sofu Gan (Lot's Wife) was sighted at 0506 on 5 September. The tall rocky pinnacle was 20 miles distant when spotted and *Seawolf* altered her course to keep the rock about 20 miles away.

At 0530, however, Ben Rogers called out, "Skipper, I'm getting radar interference on the SD from that rock."

Roy Gross and his officers discussed the possibilities. "Although Sofu Gan is a sheer pinnacle rock, thought possibly a radar outpost upon it," wrote Gross. "Interference flashed across at about 3 or 4 second intervals. Also got send noises out of the ARC-1, but not conclusively radar noises."

Gross headed for the rock to investigate. At 0723, the bridge lookouts sighted two sampans with small boats, but they were apparently fishing. Deciding that these sampans must be radar-equipped, Gross called for battle stations, gun at 0746.

At 0750 a pip was reported on radar and *Seawolf* dived. "We dived, thinking it was an airplane," related Lt.(jg) Duke Robinson. "After we had time to think about it, we decided that the rock was just causing an echo or something. So, we didn't need to dive."[16]

Sofu Gan was now just 3 miles away, and Ben Rogers was able to confirm that its surface was causing the echo return on his unit. Gross ordered stations for battle surface. Both sampans were now scurrying about collecting their little boats. Nothing could be seen on the pinnacle.

Seawolf closed to one mile from Lot's Wife before Gross called for battle surface again at 0758. Gun captain John Bennett and gunnery officer Bill Whitman directed the fire from the deck, with direction from the bridge spotters and radar. Lucien Rajotte and gunner's mate Kenneth Newman operated the gun as pointer and trainer, as Duke Wall and Red Mills loaded in the shells that were passed up from the ready lockers.

"Looks Like a Long Chase" 301

A deck gun attack by *Seawolf* left this Japanese sampan sinking beside the nautical landmark Sofu Gan ("Lot's Wife") on 5 September 1943 during her tenth patrol. *Official U.S. Navy photo, courtesy of Mike McCoy.*

Bennett's 3-incher opened fire on the first sampan at about 3,000 yards. The sampan was hit within the first five shells and stopped, an impressive record for the *Wolf*'s able gunners. They immediately shifted to the second sampan but only got in three shots before he put the giant pinnacle between himself and the submarine.

Unphased, the gunners returned to blasting the first sampan and scored 19 hits out of 23 shots from the deck gun. At 0809, the SD air-search radar picked up a pip at 32 miles. This bogey closed to 27 miles. The large and steady pip had the appearance of a flight of planes. Roy Gross called for the gunners to clear the decks and he dived. At 0850, the first sampan was observed to sink through the periscope. Gross worked around while submerged to put the second sampan between his ship and Lot's Wife. At 0910, he ordered battle stations again. The gunners gathered their shells and stood ready at the hatches.

At 0936, the *Wolf* surfaced 3,500 yards from the rock and the gunners hit the deck as the water was still washing over the sides. They found the sampan working around to the far side of the rocky pinnacle. Lt. Cdr. Gross feinted to the left for a clockwise chase around the rock. As soon as the sampan disappeared from sight, he wheeled *Seawolf* back to the right to head off the little ship.

The moment the sampan's bow came out from the other side of the rock, *Seawolf*'s gunners surprised him by opening fire with their 3-incher. "We switched directions and came right dead end at him and we shot the thing to pieces," said Duke Robinson. Bennett's men quickly landed five 3-inch direct hits before the sampan could turn away to hide again.[17]

The deck gun continued a deliberate fire, pounding the Japanese vessel. The Japanese sailors abandoned ship and swam for Lot's Wife. Lt.(jg) Robinson noted one of the Japanese survivors head for the rock at record speed. "He was almost planing—he was really pouring it on." They climbed up the steep rock and began waving a white surrender flag. "There wasn't much to hang onto, but they got up there on it somehow," said gunner Rex Mickey. "There was nothing else around but that one big rock."

Chief Bud McCoy recorded in his secret diary that *Seawolf* had "run the other one around the rock and sunk her with 15 hits out of 17 shots. Damn good shooting."[18]

Once the sampan was demolished, Roy Gross allowed the Japanese sailors to safely climb the rock. "The thing I liked about our skipper, he would just pull away," related Duke Robinson. "Not all of the skippers would do that. Some would rip into lifeboats and shoot them all to pieces."[19]

The sampan's decks were soon awash and *Seawolf* left it astern, ceasing fire as she departed the area. "This sampan had a radio antenna showing, and had had plenty of time to use it," Gross noted. *Seawolf* took evasive course to the southwest until she was out of sight of the survivors. She then turned and resumed her course for

Midway at 17 knots. The second sampan was not seen to sink but was last seen to be low in the water and badly broken by 12 point-detonating shells out of 20 shots.

Rogers reported that he was still picking up radar interference on the SD although no radar had been seen on Sofu Gan. Two hours later, the SD radar line was jumping down occasionally, leaving a clear pip at 17 to 19 miles for an instant. The bridge lookouts also reported lightning visible in the distance at the same time the receiver picked up the static and the SD pip jumped. Rogers and his team learned to disregard these jumps. The lightning phenomenon was one that they had not been trained on with their early radar.

As *Seawolf* ran toward Midway, her next shipping contact came at 1800 on 7 September. Lookouts sighted a large trawler about 6 miles away, plotted as stopped. Gross searched the area, finding fishing nets, floating lights, and a small sampan with running lights. He realized he had stumbled onto a fishing fleet. Gross then started a square search of 7 miles but by 2108, no further shipping contacts had developed. He then moved back toward the sampan and the lights of the fishing nets. He believed this sampan to be on a reconnaissance station located 385 miles from Marcus Island and that "his fishing was just to kill time."

Deciding that this vessel was out to report U.S. submarines, Roy Gross called for his fourth deck gun action of the patrol. At 2128, Bennett's gun opened fire with 3-inch point-detonating shells from 1,700 yards. The target was in the path of the moon and was caught by complete surprise. *Seawolf* fired ten shells—hoping to hold some in emergency reserve—scoring one and possibly two hits. She then fired another five, getting a total of four hits.

The common shells passed right through the sampan without bursting from past experience. The Japanese returned the fire with machine gun and rifle bursts. Spotting the shots from the bridge, Lt.(jg) Duke Robinson recalled, "This small boat had a machine gun on it. One guy ran for the machine gun, and they brought the old 3-incher on him, and he just disappeared." All point-detonating shells had been expended now and *Seawolf* had only seven common 3-inch shells remaining. Gross ordered a cease fire of his deck gun.

The sampan was still making 10 knots. With only seven shells left, Gross then began maneuvering for 20mm and .50-caliber firing. "We kept them stored down below in the ordnance room near the galley," said gunner Rex Mickey, in charge of the 20mms. "At battle stations, we carried them up to the bridge. Our 20s would shoot 2 or 3 miles. We could hit something at least as far as you could see."

From 1,000 yards, *Seawolf* opened up with her 20mm bridge guns. Neither got off more than three shots without a stoppage. Both broke the lip on the extractors and gunner's mates Newman and Mickey had to shift the guns. They finally got off about 500 rounds in spasmodic bursts. Very few hits were scored and the sampan returned fire every time the *Wolf* closed within 800 yards.

The visibility advantage in the dark night was now with the Japanese vessel due to the *Wolf*'s larger size. The sampan could not be seen until within 500 yards. *Seawolf* manned her .50-caliber and closed to 500 yards in a rain squall, firing one belt of 100 rounds, without effect. The sampan again returned fire with machine guns and rifles but neither sub nor sampan scored hits on each other.

Ten minutes later, *Seawolf* fired another belt of 100 rounds of .50-caliber at about 500 yards. Once again, the Japanese crew opened fire in return. Gross thought, "Only way to get him would be to close to 100 yards and give him everything. Decided that the risk of losing a good submariner was not worth the target's risk."

Seawolf secured from battle stations gun and resumed her course for Midway. No signals were heard on the radio during the entire attack. Since he was likely to have been carrying a radio at this remote picket station, Gross assumed that one of his hits in the deck house aft may have wrecked the radio.

"Had a little fun tonite," logged Bud McCoy. "Hit the big one about 4 times and then ran out of the right kind of shells, so we quit the big gun and sprayed him a little with the 20mm and let him go. He was smoking the last we seen of him."[20]

Crossing back over the international dateline, *Seawolf*'s 11 September became 10 September again. Through a break in a rain squall, Midway Island was sighted at 1600 at 5 miles. No escort planes were contacted so Exec Bob Risser conned his submarine in without an escort. *Seawolf* moored alongside *Skipjack* at 1725 and began taking on 20,000 gallons of fuel.

She stayed overnight and got underway at 1110 on 11 September for Pearl Harbor. *Seawolf* carried five extra passengers back to Pearl, enlisted men who had received orders to new assignments back in the States. The voyage was routine and she arrived on the morning of 15 September, docking at the Submarine Base, Pearl at 1100.

On her tenth run, *Seawolf* sank 12,996 tons of shipping by JANAC's postwar count, excluding two 75-ton sampans sunk by shellfire. In his report, Gross counted only four good hits of 20 torpedoes fired. Two others were duds and two went erratic after being fired. He blamed himself for missing with 12 of his torpedoes but added that "more torpedo failures are suspected." His endorsers considered him too hard on himself and noted that some of the ships had likely maneuvered out of harm's way. Commander Submarine Division 43 wrote: "The Commanding Officer is a bit severe in himself when he states that twelve misses were the result of control errors."

In short, the endorsements for Gross' third patrol were high. Spending just five days in her patrol area before expending all her torpedoes, "the tenth war patrol of the *Seawolf* was brief, aggressive and successful." Credit was given for sinking one 8,500-ton freighter, another 6,662-ton freighter, and for two sampans sunk by gun. Lockwood allowed 15,312 tons total.

Using contact heads instead of the old magnetic exploders, Gross had expected better results. En route home to Pearl, he did receive a dispatch offering instructions on using an oblique track angle for firing on contact, but this was too late. "It is gratifying to learn that the exploder problem is well underway to solution, and that the cause of many of *Seawolf*'s misses did not lie wholly in the fire-control equipment, tactics, and personnel," Gross wrote.

Gross commended his deck gun crews for shooting up four enemy ships and wished that his *Wolf* could be fitted with a larger caliber deck gun. In three patrols, Roy Gross had fired 56 out of 60 torpedoes, sinking six ships. He was on his way to becoming one of the war's high scorers. Bob Risser, after two runs on *Seawolf*, knew that his ship had missed on several sinkings due to the torpedoes.

> During those two patrols we fired 36 torpedoes, Mark 14-3A. Nine of these hit but two of them were duds. Of the

other 27, two were prematures which I personally witnessed. One exploded in the wake of the target and Gross called two erratic. I also witnessed these last two. They were fired from the surface at a stopped target—one just missing to the right, the other just to the left.[21]

Fortunately, Charlie Lockwood was on the case and his team would soon get to the bottom of the problems. Lt.(jg) Duke Robinson later wrote of the issue to the Naval Weapons Station in California.

When we arrived back in Pearl, Admiral Lockwood had a test going where he was dropping torpedoes against plates equivalent to the skin of a merchant ship and he found the warhead was crushing and seizing the firing pin before it could strike the primer. One other item which should be mentioned is the fact that the unmodified torpedoes were apparently better when the target was moving than when there was a stopped target and a 90° track.

The neglect of the Navy's ordnance department to test their most valued weapons had cost the Silent Service—and the *Wolf*—many opportunities to shorten the war before the problem was fixed.

USS Seawolf Tenth Patrol Summary

Departure From:	Pearl Harbor/Midway
Patrol Area:	East China Sea
Time Period:	14 August—11 Sept. 1943
Number of Men Aboard:	79: 71 enlisted and 8 officers
Total Days on Patrol:	32
Total Days Submerged:	5
Number of Torpedoes Fired:	20
Ships attacked with deck gun:	4
Ships Credited as Sunk:	4/15,312 (including 2 sampans)
JANAC Postwar Credit:	2/12,996
Shipping Damage:	840-ton torpedo boat *Sagi* and one sampan
Return To:	Midway

13

All Torpedoes Expended

Eleventh Patrol   *5 October–27 November 1943*

No one was happier than Hank Thomson to see *Seawolf* tie up to the submarine base at Pearl Harbor. He had transferred over to *Plunger* prior to the *Wolf* departing on her tenth patrol in order to attend pilot school. The chief quartermaster's request had stalled out and he thus remained aboard *Plunger*. He wasted no time in finding the yeoman and working himself back aboard ship on 26 September. Thomson later felt very fortunate to get back aboard.

> The captain of the *Runner* was making an effort to transfer me to his boat at the time *Seawolf* pulled in from patrol. I talked to Captain Gross and told him I would rather make more patrols on the *Wolf*. He took care of the deal. This was very fortunate because the *Runner* did not return from her first patrol.

Lieutenant Syverson assigned Chief Thomson back into the conning tower attack team and also to the 3-inch deck gun. Returning to *Seawolf* from 30-day leaves were two other CPOs, Lefty Leffingwell and John Bilkey. Chief Roy Bateman, who had been hospitalized prior to the tenth patrol, also returned. With their return, Roy Gross transferred two other senior petty officers—Chief Bud McCoy and MoMM1c Edward Chapman—both of whom would return to the *Wolf* before her twelfth run.

Seawolf's tenth patrol had been a quick one and thus her time at Pearl Harbor was quick. Division Forty-Three and base personnel conducted a ten-day refit on the boat while the crew enjoyed some time off at the Royal Hawaiian.

As usual, the *Seawolf* crew enjoyed hitting the bars and letting off steam. Duke Robinson served as duty officer during some of the *Wolf*'s refit period. He remembered one particular sailor who would "drink like a fish" and then pick fights with everyone. Robinson pulled this sailor aside and advised him that he was to stay out of trouble and out of the brig while at Pearl this time.

"Lieutenant, I'm done with drinking," swore the sailor, who pledged to stay on the straight and narrow this time. "Just call me the Coca-Cola Kid!" he said.

That evening, Robinson was not surprised when he was called on to bail out several drunken sailors accused of fighting, including the "Coca-Cola Kid." Robinson and others hauled them back to the Royal Hawaiian—where he tossed the "Kid" on his bed to sleep it off. No sooner had Robinson departed again, than the inebriated sailor awoke and charged toward the "door" of the room to settle a score with his adversary. The "door," however, proved to be an open window. The sailor plunged through it and only escaped serious injury by crashing down through a palm tree. The thought of the "Coca-Cola Kid's" near-tragic charge would bring Duke Robinson to laughter for many years.

The Navy's rotation policy required a number of seasoned hands to rotate on to new assignments. Roughly 20 percent of the crew turned over prior to *Seawolf*'s eleventh patrol. Leading ship's cook David Fenwick was promoted to chief commissary steward. CRM David Goudy—one of four kids raised in rural Eldon, Iowa—had joined the Navy for opportunity. Having completed one run on the *Wolf*, Goudy found the military to be a worthwhile venture, as Lt. Cdr. Gross promoted him to officer status. *Seawolf*'s latest mustang received a new leading radioman, 21-year-old RM1c Gerald Bekke to fill his shoes in the radio shack. A child who had grown up hunting and trapping on a South Dakota farm, Bekke joined the Navy after high school in 1939 and was stationed at Pearl Harbor at the time of its attack in 1941.[1]

The *Wolf* still had four of her plankowners, those who had been aboard since her 1 December 1939 commissioning: Roy Bateman, John Bilkey, Lefty Leffingwell, and Mason Poole. Leaving prior to *Seawolf*'s eleventh patrol was her longest tenured veteran. Chief

Lucien Rajotte, the senior motormac aboard, had been ordered to *Seawolf* in the summer of 1939 before she was even launched. He had orders to join *Skipjack* as a warrant officer, but Rajotte found leaving his beloved *Wolf* a tough thing to do. "I was the last person from her launching to leave her," he said. "I had spent four years, four months and four days on her."

Another plankowner being transferred was CGM John Bennett. His place as chief of the boat was filled by newly arrived CMoMM Leonard Caillier, who had most recently served under Cdr. Roy Davenport on *Haddock*, on which he had made four patrols and been awarded the Bronze Star Medal. From the wardroom, Lieutenant Jim Mercer was transferred after having made ten runs on *Seawolf*. His role as engineering and diving officer was now filled by Lt. Clary John. Mercer reported aboard *USS Aspro* on 2 November 1943.

During refit in September 1943, Exec Bob Risser was also transferred pending his upcoming orders. He would next command *Flying Fish* for five war patrols. "Roy Gross was a fine naval officer and I have nothing but admiration and respect for him," Risser later wrote. "My two patrols with him were a source of inspiration for my next five in *Flying Fish*." Lieutenant Red Syverson, a veteran of all ten prior patrols, fleeted up to Exec of *Seawolf*.[2]

Duke Robinson received notice of his promotion to full lieutenant. With the transfers, he was *Seawolf*'s fifth senior officer and was able to pass his commissary duties to newly promoted David Goudy. Robinson took over as communications officer, and as such would be the battle stations plotting officer for torpedo attacks.

Robinson and a number of men earned their dolphins while on patrol and were now fully qualified in submarines. A longtime tradition in the service called for friends of those newly initiated to toss them into the "drink." When Lieutenant Robinson's fellow officers tossed him over the side, he took a deep breath and swam underneath *Seawolf* to her opposite side. Leaving everyone momentarily wondering if he had drowned, Robinson enjoyed the last laugh.

In addition to the newly-promoted Ensign Goudy, two other new officers joined the wardroom for the eleventh patrol. Lt. William Lee Smith from Oklahoma came aboard from another submarine to serve as *Seawolf*'s first lieutenant and gunnery officer. Ensign Raymond Eugene Clevenger from Indiana was assigned as electrical officer and as Clary John's assistant engineer.

Fireman William R. Harlow from New York joined *Seawolf* prior to her eleventh war patrol. *Courtesy of Bill Harlow.*

Among the new men was TM1c Donald Joseph Naze, who would take over the after torpedo room. Born in Jamestown, North Dakota, he had enlisted in the Navy at the age of 21 and served on a submarine tender after boot camp. Naze volunteered for submarine duty, attending the New London school and then the Rhode Island torpedo training. Naze helped commission the new boat *Pogy* and made her first two war patrols, during which Naze helped her sink three confirmed Japanese ships.[3]

Another of the new hands was fireman second class William Robert Harlow, a native of Amsterdam, New York, who enlisted in the Navy in August, 1942. Fresh out of sub school, Harlow had two brothers in the Army who were serving overseas. "I joined the Navy when I was 20, about a year after high school," he said. As opposed to his brothers, Bill Harlow did not care to be shot at. "I made up my mind that if I got it, I didn't want to lay on no battlefield. I would come back in one piece, or not at all."

In final preparation for patrol, Lt. Cdr. Gross and Red Syverson conducted a three-day training period to work out their new fire control team. They fired three training torpedoes at friendly targets. *Seawolf* returned to the sub base and Ensign Goudy's commissary department commenced loading stores. On the night of 4 October, a new SJ radar range unit and transmitter were received aboard ship, but not the PPI screen for viewing contacts.

Seawolf was underway at 1300 on 5 October 1943, bound for the East China Sea. En route to Midway, Gross conducted the usual training dives and battle problems to work out his green hands. The *Wolf* moored to the port side of *Saury* at the north side of Midway's submarine pier at 0828 on 9 October.

Doc Hadley had reported ten cases of sore throat on board en route from Pearl. *Seawolf* was thus held for a 24-hour observation period at Gooneyville. The Pearl Harbor Submarine Base had 400 cases of streptococcus sore throat hospitalized when the *Wolf* had departed. *Seawolf* remained quarantined for one full day except for workers who were absolutely necessary onboard. For these people, Gross made sure that Doc Hadley "liberally sprayed" their throats before they came on board.

Seawolf was underway at 1500 on 10 October. The sore throat cases were somewhat improved and none had developed fevers which could indicate the risk of further spreading the sickness. The run to the South China Sea was uneventful, with only one enemy plane being spotted along the way.

Clary John's watch spotted a sampan at 1439 on 17 October, some 240 miles from Marcus Island. This potential picket boat was avoided. Roy Gross changed his course to the south and southeast to investigate the area to the northeast of the Marianas Islands. By 19 October, this area had been thoroughly swept and he swung the *Wolf* west towards the south end of Taiwan.

Doc Hadley monitored those with sore throats during the first two weeks of the patrol, with several persisting off and on. One man came in with a swollen jaw, followed by a different man the next day. Hadley—the *Wolf*'s only medical help—was worried about a mumps epidemic, but the two cases quickly improved with sulfathiazole treatment. Among the pharmacist mate's other patients on the eleventh patrol were a bad case of conjunctivitis, several catarrhal fever cases, and one spinal cyst that had to be cut for relief.

The *Wolf* entered her patrol area on 24 October and crossed the Bashi Channel off Taiwan on 25 October. On 26 October, the south end of Taiwan was sighted about 20 miles distant. That evening, nothing but sampans had been sighted so Gross set course across Formosa Straits for a point on the China Coast about 70 miles north of Hong Kong. RT1c Ben Rogers reported the SJ radar out of commission during the night.

A pip on the SD at 0800 on 27 October closed to 6 miles, forcing the OOD to dive ship. Bill Whitman had just passed the watch to Lt. Duke Robinson, newly qualified in standing officer of the deck watches. When his lookout called out, "Aircraft," Robinson duly ordered his men to clear the bridge. He continued peering through his own binoculars, however, trying to make out the aircraft. As *Seawolf* headed into her dive, Lt. Cdr. Gross finally called out, "Dammit, Robby, get down here!"

Seawolf remained submerged all day while Rogers' radar team repaired the SJ. Strong winds and rough seas made it tough going for the lookouts on the surface that night. The SJ remained down, but the heavy seas made repair work topside too dangerous. "This is the new transmitter and unfamiliar to our radar technicians," wrote Gross, "although our radarman spent the whole day in Midway getting what information he could on the new gear."

Seawolf remained submerged the next day, as the sea was too rough to even maintain periscope patrol, except for altering course during observations. Rogers and company continued working on their SJ. Seas and wind were force 6 from the northeast. After *Seawolf* surfaced that evening, Rogers reported his SJ back in commission at 2100, now working better than ever. Twenty-five minutes later, he picked up the China coast on the SJ at 55,000 yards.

When not repairing the radar, Ben Rogers continued to work on his cartoons. "Rogers had his own comic strip for us on the ship," recalled Rex Mickey. "He would draw a different one every week for us and paste it on the board in the crew's mess to keep us entertained."

Seawolf dodged several lighted sampans throughout the night and started a patrol about 16 miles off the Chinese coast in 16 to 20 fathoms of water. Her cold luck finally changed almost immediately. At 0132 on 29 October, the radar watch picked up a lone good-sized pip at 16,900 yards. Roy Gross called for battle stations torpedo at 0145. By 0201, two smaller pips were seen on radar, indicating escorts with the larger ship. They appeared to be trailing the target ship, as the range came down to 6,880 yards. Those topside believed this ship might be a man-of-war, due to a long, flat superstructure and one low stack seen in the amidship section. Gross finally decided this ship must be a minelayer of some sort approximately 350 feet in length, as determined by field width against radar ranges.

All Torpedoes Expended 313

Gross let the range decrease for nine more minutes and then commenced firing all four bow tubes at 0210. Gunnery officer Bill Smith was stationed on the forward TBT to feed bearings to the TDC below. In the control room, Bill Whitman and Lefty Leffingwell operated the TDC, while Duke Robinson handled the plot in the conning tower and Red Syverson served as assistant approach officer. The torpedoes were set for 8 feet, with the target speed entered as 8 knots. Using the white light from the bridge TBT, the fish were fired on a 90 starboard track from 2,000 yards.

Two hits were observed aft on this freighter, later identified as the 3,222-ton *Wuhu Maru*. The AK began settling by the stern as *Seawolf* opened the distance out to 4,000 yards to watch her sink. "Escorts apparently baffled as to direction torpedoes came from, or else just didn't care," wrote Gross. By 0220, *Wuhu Maru* was apparently standing on her stern in 16 fathoms of water with her bow sticking out vertically. *Seawolf* circled her victim at about 4,000 yards to keep it between her and the escorts.

By 0258, the freighter's bow finally disappeared as radar reported the pip fading out. The crew was secured from battle stations torpedo at 0312 as the ship opened the coast slowly to the southeast. Gross gave plenty of praise to "the officer and two men who spent 3 days getting the SJ radar working" which had led to this sinking.

The force 6 seas and 30-knot winds continued all night and the following day. When *Seawolf* surfaced at 1810 on 30 October, the weather had only improved slightly. Running 14 miles off the coast, the *Wolf*'s radar team picked up a large pip at 12,500 yards at 2115. The Bells of St. Mary's rang for the second consecutive night as the tracking party took station in the control room and conning tower.

This ship developed into a coal-burning small freighter making about 8.5 knots, apparently en route from a northern port to Hong Kong. No escorts were seen, so Gross again remained on the surface to attack this ship. He closed to 2,000 yards and fired all of TM1c Don Naze's four stern tubes at 2219, spread via white light from Bill Smith's bridge TBT. All torpedoes missed, however. Gross believed the ship was actually smaller than figured and that the 8 foot torpedo settings had been too deep.

Seawolf continued to maneuver for a potential second firing on the ship. The night was very black, with a heavy overcast and rough seas. Although Lt. Cdr. Gross strived to keep the *Wolf* outside of 3,000 yards from this ship, he at one time found the ship closing him to as little as 1,700 yards before he could pull away. Gross ultimately decided to let this little ship go instead of wasting any more torpedoes on such a small ship. In the process of tracking this vessel, he did gain valuable knowledge of using the new SJ radar in estimating a target's size. "This target would have not appeared on the screen until within 8,000 yards on the old one," he wrote.

Seawolf spent two more days in this area, fighting rough weather. New diving officer Clary John found during this patrol that the stern planes failed regularly during diving. Inspection showed that the brake was set too tight and a low voltage issue was preventing the brake from properly engaging. No sooner had Chief George Garrett's auxiliarymen fixed this issue than the bow plane transfer switch's copper finger broke off while shifting from rig to tilt. There were no spares, so Garrett's machinists fabricated a new one.

Roy Gross abandoned this area on 2 November and set a southwesterly course at one-engine speed for a point 20 miles due south of the entrance to Hong Kong. Numerous sampans were dodged en route and the SJ went out of commission again that night. *Seawolf* spent the next two days patrolling off Gap Rock light at the entrance to Hong Kong, staying just far enough away to avoid any possible minefields.

By the time *Seawolf* surfaced at 1827 on 4 November, Ben Rogers and his assistant, Norman Coon, were able to report that their SJ unit was back in operation. The wire cables had flooded out, and new leads were run using the 1MC wiring and gyro-pilot leads. "This again shows the poor engineering practice of making all leads of any sort whatever pass outside the pressure hull in going from the control room to the conning tower," wrote Gross.

With no action off Hong Kong, Gross left the area and proceeded at one-engine speed towards Hainan straits' entrance, an area he believed had not been exploited. This was the extreme western limit of his patrol area and was more than 1,000 miles from the eastern edge of his area, or some 3,700 miles from Midway.

Three hours later, the newly-repaired SJ radar had a pip at 5,000 yards at the same time Duke Robinson sighted a vessel. Although

All Torpedoes Expended 315

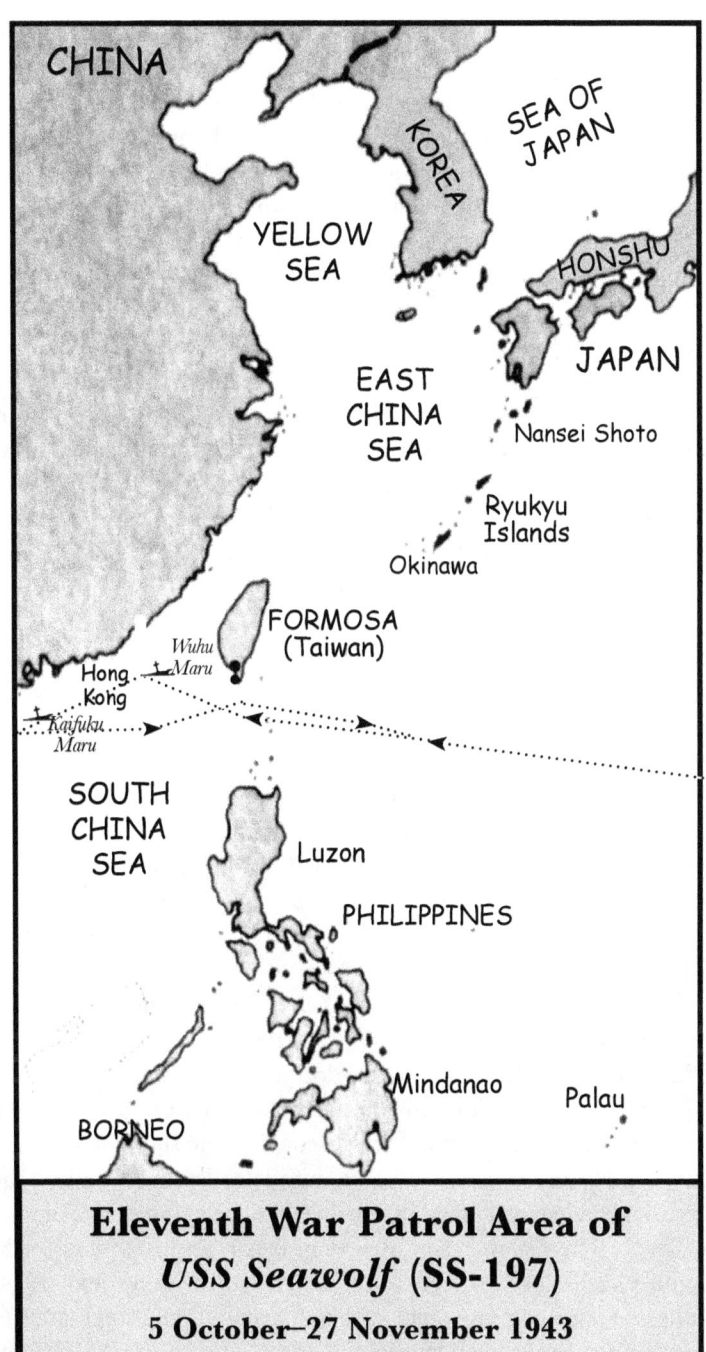

Eleventh War Patrol Area of USS *Seawolf* (SS-197)
5 October–27 November 1943

this ship was assumed to be small, *Seawolf* closed to investigate. The men were surprised to find a large heavily laden freighter making 6 knots into a moderately heavy sea with no escort vessel around. This ship was sailing straight for the Hong Kong channel, with some 35 miles to go. The radar was working, but certainly not well tuned due to the size of this ship not being picked up until at 5,000 yards.

Seawolf had a half moon to work with this night but a relatively simple fire control problem. Roy Gross confided more and more with his senior men in determining fire control and attack solutions. With no escorts present, everyone was fully confident in making another surface attack using Bill Smith's forward bridge TBT. Two fish from Chief Bill Reiland's forward room, set for 10 feet, were aimed for the MOT, another one at her bow, and one at her stern. Firing commenced at 2229 from 2,200 yards. Three solid hits were seen to rip this ship apart. Although the men on the bridge perceived her to be at least 10,000 tons, postwar records would show this ship to be the 3,177-ton freighter *Kaifuku Maru*.

Kaifuku Maru had loaded 4,080 tons of rice and tapioca at Saigon and departed on 19 October. She planned to join one of four convoys in the Cape St. Jacques area. She did join a convoy on 24 October, but became separated from it later in rough weather and was forced to take shelter in Camrahn Bay on the night of 26 October. She then headed for Hong Kong independently, hugging the coast. She was 130 kilometer southwest of Hong Kong on the night of 4 November when she was hit by three *Seawolf* torpedoes. She sank quickly, taking down with her two dozen crewmen, 13 ship's gunners, and 29 passengers of an army mountain-gun unit who were to be delivered to Takao on southern Formosa.[4]

Two of the Formosa gunners and 22 of the crewmen—including the ship's captain—survived the *Seawolf* torpedoing and subsequent sinking of *Kaifuku Maru*. These two dozen survivors reached *Kaifuku*'s No. 1 lifeboat and drifted for the next two days. On 6 November, the lifeboat fetched up on the China coast on the shore of Kwangsi Province. Having landed in an area outside of Japanese control, the survivors headed for a town they believed to be about 1,000 kilometers away. The *Kaifuku Maru* men then began an incredulous eight month journey toward civilization which ended on 30 June 1944. Along the way, one crewman and one army gunner attempted to desert and were executed by shooting. The other gunner did eventually manage to desert and was never heard from again. The others continued a torturous journey through the mountainous area and three more Japanese sailors died of disease along the way.[5]

Finally, 18 *Kaifuku Maru* sailors fell into Chinese hands and became prisoners of war. Five of them eventually died of starvation. Only 13 *Kaifuku Maru* sailors were repatriated on 6 April 1946. They arrived at Shanghai on 5 June and were returned to Kagoshima in southern Kyushu four days later, two years and eight months after *Seawolf* had sunk their ship.

The ship went down in eleven minutes and was gone from the surface by 2240. *Seawolf* moved over to investigate, but in the darkness could only make out the white water disturbance of the sinking ship and two dark objects that were thought to be lifeboats. The *Wolf* was secured from battle stations and resumed her course and speed for the Hainan channel entrance with two more kills to add to her battle flag.

Four more days would pass before *Seawolf*'s next victims would be found. The Hainan peninsula did not bear any targets so Gross followed his patrol orders to move to a point about 70 miles southeast of Pratas Reef. This was an Ultra sent out by Dick Voge in Pearl Harbor. In a letter to Admiral Ralph Christie, Charlie Lockwood said, "We were expecting this convoy and had ordered *Seawolf* to intercept...the convoy was about two hours late on its schedule."[6]

The Ultra information was often quite good. Other times, it did prove to be errant when a convoy changed its course. *Seawolf* was on station as ordered by 1912 on 8 November, awaiting this important Japanese convoy. Labeled as Convoy HI-14, it included tankers *Awa Maru No. 6*, *Kyokuei Maru*, *Hokuroku Maru*, *Amatsu Maru*, *Aki Maru*, and *Kaga Maru*, and coastal defense ship *Tsushima*. A neighboring submarine, *Bluefish*, hit this convoy first, sinking *Kyokuei Maru* on 8 November as *Seawolf* awaited their arrival.

At 0008 on 9 November, Clary John reported that his lookouts had sighted ships. They were tracked to be on a northeast course, speed 12.5 knots with moderate zigs. The initial sighting range was about 8 miles but the ships did not appear on radar until at 6 miles. After tracking these ships for an hour, Gross opted to race ahead for a submerged moonlight attack before the moon set at 0300. The appearance of rain squalls forced Gross to change his plans to a surface attack. *Seawolf* worked up ahead and to port of the convoy. Jerry Bekke and Sam Ottaway on sonar kept close tabs on the enemy ships, but could hear no pinging from the Japanese. At 5,000 yards, the convoy zigged towards the *Wolf*, putting her dead ahead. Gross worked over to the westward to get back on their port flank.

At 0350, lookouts reported what appeared to be a *Chidori*-class torpedo boat dead astern. This ship was swinging toward *Seawolf*'s course and was only 2,000 yards away. *Seawolf* kept on her course, running at the same speed as the convoy, hoping to use the old German U-boat tactics of pretending to be part of the convoy. The ship eventually zigged away and Gross was relieved to find that it was not an escort ship but apparently a smaller freighter.

The absence of pinging was a good sign that there were actually no escorts with this little convoy. Most of the ships zigged back to a northeast course at 0357 as Gross picked out the nearest large freighter—estimated at 5,000 tons—as his target. With Doug Syverson overseeing the approach with the periscope from the conning tower, Gross called the shots from the bridge. Bill Smith's binoculars were locked into the aft TBT, and his bearings were fed to Bill Whitman on the TDC below.

"Open the after tubes," Gross ordered.

Hank Thomson relayed the command to Don Naze's men in the after torpedo room. As the range dropped to 2,500 yards, *Seawolf* fired two stern tubes at 0402. The target ship was clearly in sight via binoculars on the bridge at the time of firing. One minute later, Lieutenant Smith's TBT was shifted to the left and the other two stern tubes were fired at another ship ahead of the first one. The distance to the second ship was about 3,500 yards.

Signalman Frank Romito counted down the seconds of the torpedo runs on his stopwatch, but all remained quiet. The torpedo results were disappointing this time. No explosions were seen in the vicinity of the first target. There was one large, deep explosion two minutes and 47 seconds after firing the last torpedo. This was seen at the second target, accompanied by a yellow flash.

This torpedo flash occurred shortly after the first target had opened up with gun fire, and was at first thought to be a depth charge or gunfire or both. Three minutes later, however, *Seawolf*

Chief quartermaster Hank Thomson, who had joined the ship in late 1940, became the *Wolf*'s assistant navigator during his eleventh patrol. *Courtesy of Hank Thomson.*

could hear and feel "several unmistakable depth charges which indicated clearly that the first explosion was not a depth charge, but must have been a torpedo hit."

Japanese records would later show that *Amatsu Maru, Hokuroku Maru* and *Aki Maru* all dodged *Seawolf* torpedoes in this first attack. Sailors on the 10,567-ton oiler *Amatsu Maru* watched one torpedo streak by ahead of her bow and then felt the thud of two dud warheads slamming into her port side below the bridge. Tanker *Hokuroku Maru* spotted *Seawolf* on the surface and opened fire, with other ships joining in.[7]

Gross was puzzled by these four torpedoes. For his fourth torpedo to hit, he found it "strange that the worst-aimed shot found its target." His guesses on missing the first ship were either "a large course error, insufficient spread, or poor torpedo performance." The speed and range were exact and Bill Whitman had analyzed a constant course three times. "Our confidence in torpedo performance, built up sufficiently to risk divided fire and no spread by five previous hits out of eight, fell off sharply."

By 0407, the convoy erupted into gunfire from all directions so Gross opened out his range to the west to 4,000 yards to select a new target. "We were on the surface, running along with their ships," recalled motormac Bob Curtin, one of the battle stations lookouts this night. "They were shooting over us. It was quite a spectacle."

Seawolf now had only four torpedoes remaining forward. Her tracking party settled down on a large radar pip at 4,250 yards and began closing in on her. At 3,000 yards, the ship was made out to be an engines-aft ship, estimated at 490 feet in length. Lt. Cdr. Gross guessed her to be about a 10,000-ton tanker. "This one would make up for our stern-tube misses," he wrote.

This final attack was again made using the white light method and the forward TBT. Two fish were aimed for the MOT, one at the bow, and one for her stern. *Seawolf* fired at 0443 from only 1,600 yards, with a 10-foot depth setting. Bill Reiland's torpedomen counted the seconds after their fish had whooshed from the tubes, hoping to add another stripe to the *Wolf*'s battle flag. The target ship was fully visible from the bridge. Upon firing, Gross ordered Jim Grimes to turn away right at full speed. In the process of turning, the *Wolf* passed her target ship beam to beam at 700 yards radar range.

The distance to the enemy was closer than anyone topside had ever come to their opponents while surfaced. "I could see the skipper on that ship," said Bob Curtin. "He was so close I could have hit him with a baseball!"

"Clear the bridge!" Gross called out as a precaution.

The ships passed close to one another before *Seawolf*'s diesels began opening the distance. At 1,200 yards, the Japanese spotted the submarine that was racing away and they opened fire. The target ship fired its after gun a few times. Fortunately, his aim was lousy and no whistles or shell splashes were noted.

Unfortunately, there were no torpedo explosions, either. Four white tracks were seen after firing on what looked to be the right lead angle. This attack was later plotted out and it appeared that the target ship was possibly turning towards *Seawolf* at the time of firing. Even so, the angle did not compute for all fish to have missed. "The only explanation is that surely No. 4 torpedo, and possibly Nos. 1 and 2, were duds or ran deep or erratic," wrote Gross.

Gross would be more frustrated to know that he had again fired on *Amatsu Maru* and she had again escaped due to dud torpedoes. She had been hit by two dud torpedoes in the first attack. In this attack, one dud torpedo struck her amidships on the starboard side and two thumped against her stern. Divine intervention must have been with *Amatsu Maru* for her to absorb five of *Seawolf*'s faulty torpedoes in one day without damage![8]

More gunfire and depth charges were heard as the *Wolf* retired from the attack area. Gross found this to have been "a most disappointing evening." Dawn was breaking as he secured his crew from battle stations. All of his torpedoes were now expended. "A muster of radar pips showed all present" from the convoy.

Seawolf kept tabs on the convoy into the morning hours and was able to count six ships. The radiomen sent contact reports to Pearl Harbor, which took nearly all day to clear that they had been received. The convoy continued moving at 12.5 knots into the Pescadores channel. The one ship *Seawolf* had managed to damage was able to keep up with the others. After dark, Gross had Gerald Bekke send additional contact reports to other boats to the northward. By 1000, the convoy was out of sight to the southward, so *Seawolf* set course for Bashi channel and for home on four-engine speed.

Clary John's engineers had to keep tight watch on the fuel consumption during the return voyage, as *Seawolf* had stretched her patrol boundaries and her fuel. By 11 November, the engineers had cut her to one-engine speed for the balance of the return trip.

Seawolf passed between Kita Io Shima and Io Shima in the Kazan Retto on 15 November. Head seas and strong currents reduced her advance to about 180 miles per day. The weather calmed as *Seawolf* passed near Marcus Island on 18 November and passed the usual fishing sampans. The ship made two crash dives over the next two days as Japanese planes closed in. No bombs were dropped and little time was wasted in resuming the course for Midway.

Seawolf fell in with her Midway plane escorts at 1410 on 22 November and was moored in the lagoon two hours later. She stayed overnight, taking on fuel and a few necessary provisions before shoving off for Pearl at 1000 on 23 November. After an uneventful four days, she moored at the submarine base, Pearl Harbor at 1015 on 27 November.

Her eleventh patrol had been very aggressive, as Roy Gross had made all of his attacks on the surface at night using radar and TBT to assist. He was still disappointed in torpedo performance and considered his final four torpedoes fired on 9 November against the large tanker to be a "heart-breaking attack." He was certain that at least one of the fish had run erratically. "Six hits out of twenty torpedoes is just exactly half of the 60% we had hoped for," wrote Gross. Postwar records would indeed show that A*matsu Maru* was hit by no less than *five* dud torpedoes from *Seawolf* in one day.[9]

In spite of his disappointments, Lt. Cdr. Gross and his *Seawolf* were continuing to build an enviable reputation as a hot boat.

USS *Seawolf* Eleventh Patrol Summary

Departure From:	Pearl Harbor/Midway
Patrol Area:	South China Sea
Time Period:	5 Oct.—27 Nov. 1943
Number of Men Aboard:	79: 71 enlisted and 8 officers
Total Days on Patrol:	53
Total Miles Steamed:	13,495 (Pearl to Pearl)
Total Fuel Consumed:	117,338 gallons
Total Days Submerged:	11
Number of Torpedoes Fired:	20
Ships Credited as Sunk:	2/14,000 tons
JANAC Postwar Credit:	2/6,399 tons
Shipping Damage Claimed:	1/5,000 tons
Return To:	Pearl Harbor via Midway

14

"Battle Stations, Gun!"

Twelfth Patrol   *22 December 1943—27 January 1944*

"Nobody knows how to party like a submarine sailor," motormac Bob Curtin stated. He went ashore with fellow engineman Howard "Dinky" Denemore and ship's cook John Cani to raise hell in Hawaii. The crew made the most of their two weeks. Their eleventh patrol had been 53 days, as opposed to their relatively quick tenth patrol. Submarine Division Forty-Three and the submarine base relief crews took over as all hands made their way to the Royal Hawaiian Hotel—the "Pink Palace."

During the overhaul, No. 3A and 3B main ballast tanks were converted to carry fuel, which would give *Seawolf* an additional 16,400 gallons to extend her next patrol area. The base also overhauled her main diesel engines and renewed all main engine mufflers. Technicians also struggled with installing a new, proper PPI screen to go with the radar set. Roy Gross would have to wait for a Stateside overhaul to have a proper PPI screen installed for his radar. He was also disappointed that the Pearl Harbor yard could not extend his conning tower. *Seawolf* was still operating with the old Mark I TDC in the control room. "It is expected that a major Navy yard overhaul will be directed after this patrol during which we hope to modernize *Seawolf*," he wrote.

Following the yard work, yeoman Roger O'Brien sorted through the usual paperwork of transferees and new men reporting for duty. Among the new hands reporting aboard *Seawolf* before her twelfth patrol was motormac Arnold Frank Bargenquast, a soft-spoken man born and raised in Manila, Iowa. War had come quickly

Arnold Frank Bargenquast, one of the *Wolf*'s senior motor machinist's mates, was fortunate to have survived wounds he received at Pearl Harbor in 1941. His good fortunes ran out during his fourth patrol aboard *Seawolf.* Courtesy of Charles Hinman, USS Bowfin Museum.

to Bargenquast, who had been stationed aboard the battleship *Pennsylvania* on 7 December 1941 when Pearl Harbor was attacked. Wounded during the Japanese attack, he volunteered for sub duty after his recovery. He met his wife Louise while in sub school at New London and found that Louise's brother, seaman Michael J. Quarto, had been a battleship sailor who was killed at Pearl Harbor when *Arizona* blew up. The Bargenquasts were married in April 1943. Arnold swore to his new wife that he was through with battleships and surface craft. "If I'm going to die, it's going to be on a submarine," he told her.

The wardroom had its share of changes. Old-timer Bill Whitman received new orders, as did ensigns David Goudy and Ray Clevenger. Lieutenant Robert Leon Cox, a reservist from Boston, came aboard to fill the role of first lieutenant and commissary officer. Ensign Marion Lee Asa, a reservist from Illinois, took over the duties of communications officer. Ensign Edward John Szendrey, a 1940 Naval Academy graduate who had a wife in Los Angeles, became Lt. Duke Robinson's assistant torpedo officer and assistant battle stations plotting officer. Lt. Bill Smith, now gunnery officer, would run the TDC in the control room, assisted by Ensign George "Lefty" Leffingwell, the new assistant first lieutenant. Leffingwell had joined the mustang club, another of a long line of *Seawolf* CPOs promoted into officers' country.

Exec Doug Syverson split the watch schedules from the previous four-hour rotations to three-hour shifts. His most experienced officers—Lts. Clary John, Bill Smith, Bob Cox, and Duke Robinson—would alternate as the OODs to begin the twelfth patrol.

"Battle Stations, Gun!"

CRM Gerald Edgar Bekke, a farmboy from South Dakota, was also at Pearl Harbor when the Japanese attacked in 1941. Bekke became the *Wolf*'s senior radioman and editor of the ship's newsletter. *Courtesy of Charles Hinman, USS Bowfin Museum.*

Seawolf conducted a three-day training period at the completion of her overhaul, firing three exercise torpedoes. She was deemed ready for sea on 22 December 1943, and Lt. Cdr. Gross drew his patrol orders from Admiral Lockwood. *Seawolf* was headed for the East China Sea in what was expected to be Roy Gross' final patrol. By late 1943, U.S. submarine skippers were being rotated—with few exceptions—after having completed five consecutive patrols on one boat. The stress of command was considered too great to keep even the best skippers out for more than five patrols. Sam Dealey and Mush Morton—two of the war's best sub skippers—pushed to keep making more patrols and were both lost with their boats.

Bill Smith had the conn as the *Wolf* was underway at 1305, guided out from the Hawaiian Islands by an escort vessel until dark. The crew drilled in battle problems, simulated radar problems, and dives en route. *Seawolf* celebrated Christmas 1943 by serving egg nog to all hands with a regular Christmas dinner while Christmas carols played over the loudspeakers. Rough and squally weather made leading ship's cook Leroy Edmonds' meal bumpy but enjoyable.

The new hands fell into their new duties. Some drew the dreaded duties of laundry detail or mess cooking. Fireman Bill Harlow, one of Chief Roy Bateman's young enginemen, found the duties of being low man on the totem pole trying at times. "Cleaning the tank of those Winton engines was a dirty job," said Harlow. "It was scary as hell, like climbing into a dark tunnel with very little room to move around."

Seawolf docked at Midway early on 26 December for refueling and minor work from the tender *Bushnell*. She was underway for the East China Sea by afternoon, ready to add stripes to her battle flag.

Seawolf conducted the usual battle drills and practice dives en route to her patrol station. By 7 January 1944, she was about 200 miles east of the Nansei Shoto and was experiencing rough weather. Lt. Cdr. Gross dived ship to have his torpedo gangs routine their torpedoes. Should their be any failures this patrol, he would not allow neglect of maintenance to be a factor for command to criticize.

The *Wolf* continued running on the surface as she entered enemy waters. The SD radar picked up the patrol's first aerial contact at 1103 on 8 January. The bogey was at 12 miles and closing, so Duke Robinson prudently dived ship to avoid. The weather continued to be rough as the ship entered her patrol area the following day. Roy Gross' luck as a hot skipper continued, for his first big contact came during the morning of his second day on patrol station.

Lieutenant Robinson had taken the watch at 0800 and he helped his lookouts keep a sharp eye out for enemy shipping. It was a bright morning, which allowed for great visibility for the lookouts—and also for Japanese aircraft searching for submarines. An hour and a half into his watch, Robinson made a sweep and settled on "something that looked like a cloud."[1]

A quick study of the cloud, and he knew *Seawolf* was onto something. Robinson quickly shouted down the hatch for the skipper, "Smoke on the horizon!"

Robinson called for full right rudder and conned the *Wolf* in the direction of the smoke. Within a few moments, two more plumes of smoke appeared on the horizon and Robinson knew that they were onto a convoy. Roy Gross arrived on the bridge, "almost as excited" as his young lieutenant.[2]

Seawolf had discovered a convoy of seven freighters with two destroyer escorts about 20,000 yards ahead of her making 8 knots on a southwesterly course. These ships—tracked and plotted via the high periscope and SJ radar—were in the East China Sea, about 90 miles west-southwest of Amami O Shima.

At 1030, the radar picked up an aircraft at 9 miles and *Seawolf* submerged for the remainder of the approach. By 1220, *Seawolf* found herself on the convoy's starboard flank and she singled out the nearest ship for attack. She was a medium freighter of about

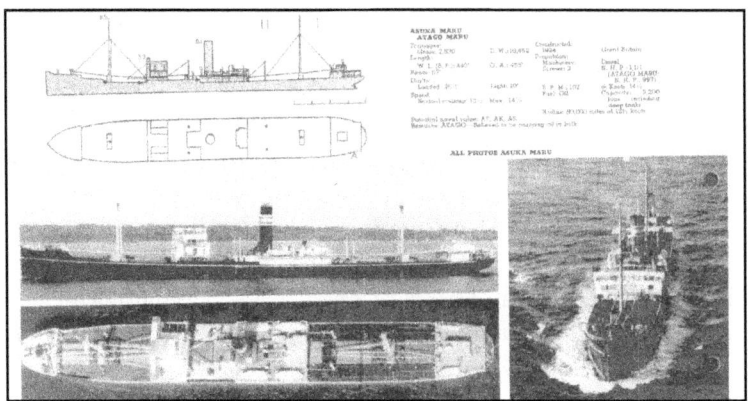

Asuka Maru was one of three ships sunk by *Seawolf* out of Japanese Convoy No. 27 in a series of daring attacks. *ONI-208J (Rev.)*.

5,000 tons. Gross made final firing observations during the next few minutes as Bill Smith fed the data into the old Mark I TDC in the control room. He was assisted by Ensign Lefty Leffingwell—the longtime leading fire controlman—and Delbert Mar. Red Syverson, who had spent many months on the TDC, was now the assistant approach officer in the conning tower. Near him at the plotting table, Duke Robinson and Ensign Ed Szendrey worked out the convoy's course and speed.

The skipper called for Bill Reiland's forward gang to use eight-foot depth settings. In the course of opening the doors, Reiland found that the outer door to tube No. 2 would not open more than 60 degrees, so it was ordered secured for the time being.

As the range reached 1,950 yards, Gross called out, "Fire one!"

Seawolf's target was later determined to be the 7,523-ton army freighter/transport ship *Asuka Maru*. Three torpedoes were sent in her direction and, for a change, the results were outstanding. Chief Hank Thomson timed the first hit in her stern at 1 minute and 49 seconds after firing. Within 90 seconds, two more explosions ripped the transport ship. With improved torpedoes, the *Wolf* had scored three for three on torpedo hits for the first time in her career.

Asuka Maru was hit between her engine room and No. 4 hold. She began listing to starboard immediately and sinking. The word to abandon ship was given even as her bow gunners fired on *Seawolf*. The seas were cold and mountainous as the sailors scrambled overboard. *Asuka Maru* sank from sight at 1239, ten minutes after being torpedoed. She took down her load of shells, calcium carbide, dynamite, one lookout, 20 crewmen, 15 passengers, and two ship's gunners. Convoy ships *Getsuyo Maru* and *Karukaya* managed to rescue 105 sailors by 1600.[3]

After blasting *Asuka Maru*, Roy Gross found that the "whole convoy zigged towards us except the destroyers, which remained on the far side. Much gun fire." He then ordered battle stations helmsman Jim Grimes to swing the boat hard to port for a stern tube shot. Torpedoman Don Naze's gang commenced firing all four of their stern tubes at 1232. Hank Thomson timed two explosions just over two minutes later. The *Seawolf* crew believed these to be hits, although they were not observed, as the skipper was sweeping the convoy and watching his first victim sinking. *Seawolf* had indeed found the freighter *Yahiko Maru* with one of her torpedoes. Japanese records show that *Yahiko* was hit in the stern and disabled but remained afloat.[4]

At 1236, Thomson timed two more explosions at three minutes 16 seconds and four minutes eleven seconds. Gross logged:

> Observed no change in attitude of stern tube target, but saw huge cloud of black and white smoke on a distant ship also in line of fire, allowing for lead angle. Type of ship could not be made out as it was obscured by the smoke. At the moment, thought it could be a ship pouring on the coal but the character of smoke was quite different. It is believed that one of the stern shots reached the other flank and hit a ship which from previous observations was a M.F.M. [mast-funnel-mast] of 4,000 tons. We hope that somebody else gets in on the rest of these ships and is able to determine whether damaged or sunk.

One minute later, Gross watched his bow tube target, *Asuka Maru*, sink stern first. The ship made a loud noise in breaking up, sounding "very much like a string of fire crackers over a period of two or three minutes." Swinging back to his stern tube target, Gross could not see any changes in her condition. He could see smoke from the far ship which had been hit by "pure luck," in his words.

"Medium freighter sunk, large freighter stopped, other freighter with way on," Clary John summarized in *Seawolf*'s deck log. The other ships were running away from the scene of the attack. Gross did not have time to count the remaining ships, as the larger destroyer was seen approaching fast from 3,000 yards out. He boldly stayed at periscope depth but rigged for depth charge attack. The DD dropped nine depth charges at 1253, but he was still 3,500 yards distant when he began dropping.

The sea was rough, making depth control a bit of a challenge for the planesmen. *Seawolf* swung her stern toward the destroyer and opened out at 4 knots, watching him. "A quick look around showed all ships going over the hill except the big goal post ship, still showing the same angle on the bow, apparently still stopped."

The *Wolf* opened out her range to 6,000 yards during the next half hour to escape *Karukaya*. The other escort, *Minesweeper No. 27*, stayed with the remainder of the convoy. Six more depth charges rumbled in the distance. By 1350, Bill Reiland's forward torpedo gang had completed their reload as Gross hoped to finish off the stopped ship, *Yahiko Maru*.

At 1414, a two-engined land bomber appeared over the stopped ship, which was about 17,000 yards distant. The destroyer was out of sight but still audibly pinging for submarines. An hour later, three float-type Japanese planes appeared over the stricken ship. Gross decided to hold visual contact with the ship until dark, when the aircraft would have to leave. The crew was secured from battle stations. *Seawolf* crawled away dead slow on her batteries all afternoon while the sonar watch listened to the sound of *Karukaya*'s pinging.

Seawolf surfaced at 1827, and Ben Rogers quickly picked up the motionless *Yahiko Maru* on his radar at 30,000 yards. With nearly a full moon rising, there was unlimited visibility. "This must be a bigger ship than thought as *Bushnell* was lost at 20,000 yards when we tracked her leaving Midway astern of us," Gross mused.

Rogers manned his SD radar at one-minute intervals while radioman Jerry Bekke sent out a contact report to Pearl Harbor. At 1850, the lookouts spotted a searchlight in the direction of the target. The *Wolf* closed slowly to make visual contact.

"Bogey, 14 miles and closing," Rogers called out 20 minutes later from his SD.

The Japanese aircraft slowly approached to 4 miles before Lt. Bob Cox pulled the plug and dived ship. He brought her back up ten minutes later when a radar sweep showed the plane had passed.

Fate of Japanese Convoy No. 127
January 1944

Ship Name	Type	Tonnage	Attacks Against/Notes
Asuka Maru	AK/AP	7,523	Sunk by *Seawolf* 1/10/44
Yahiko Maru	AK	5,747	Sunk by *Seawolf* 1/10/44
Getsuyo Maru	AK	6,440	Sunk by *Seawolf* 1/11/44
Nikki Maru	AK	5,900	Sunk 2/22/44 by *Balao*
Hokoku Maru	AK	4,000*	
Kinrei Maru	AK	5,945	
Rokko Maru	AK	2,600	
Ikutagawa Maru 8	AK	2,500	
Karukaya	DD	1,300	
Minesweeper 27	W	600*	

* Approximate tonnage.

Soon, Roy Gross could see that the larger ship—*Yahiko Maru*—was still stopped. Another freighter—later determined to be *Getsuyo Maru*—was ahead of the crippled ship, with the destroyer still hovering around, pinging low scale. Duke Robinson and Ed Szendrey plotted the group to be on a southerly course and determined that the smaller freighter had taken the big cripple under tow. *Getsuyo Maru* was hoping to tow *Yahiko* to Naha, Okinawa.[5]

Rogers continually reported and tracked planes on his radar screen, but Gross opted not to dive as long as they stayed beyond 5 miles. By 2116, the *Wolf* had crawled to a position dead ahead of the ships and prepared to attack. The Bells of St. Mary's rang out as the crew raced to battle stations for a submerged attack. *Seawolf* was on the Japanese ships' starboard side about 5,000 yards out. The towing ship, *Getsuyo Maru*, was now astern and to starboard of the crippled *Yahiko Maru*. *Yahiko* was stopped once again as the tow had been discontinued. The destroyer was not in sight but could still be heard pinging.

Studying *Yahiko Maru*, Gross found her down by the stern by about five degrees. He periodically eased the boat up to 40 feet for Rogers to take a radar range on the ships. By 2230, he had closed to 2,600 yards and was on the starboard flank. With a good view of his enemy, he now figured the towing ship *Getsuyo Maru* to be about

"Battle Stations, Gun!" 331

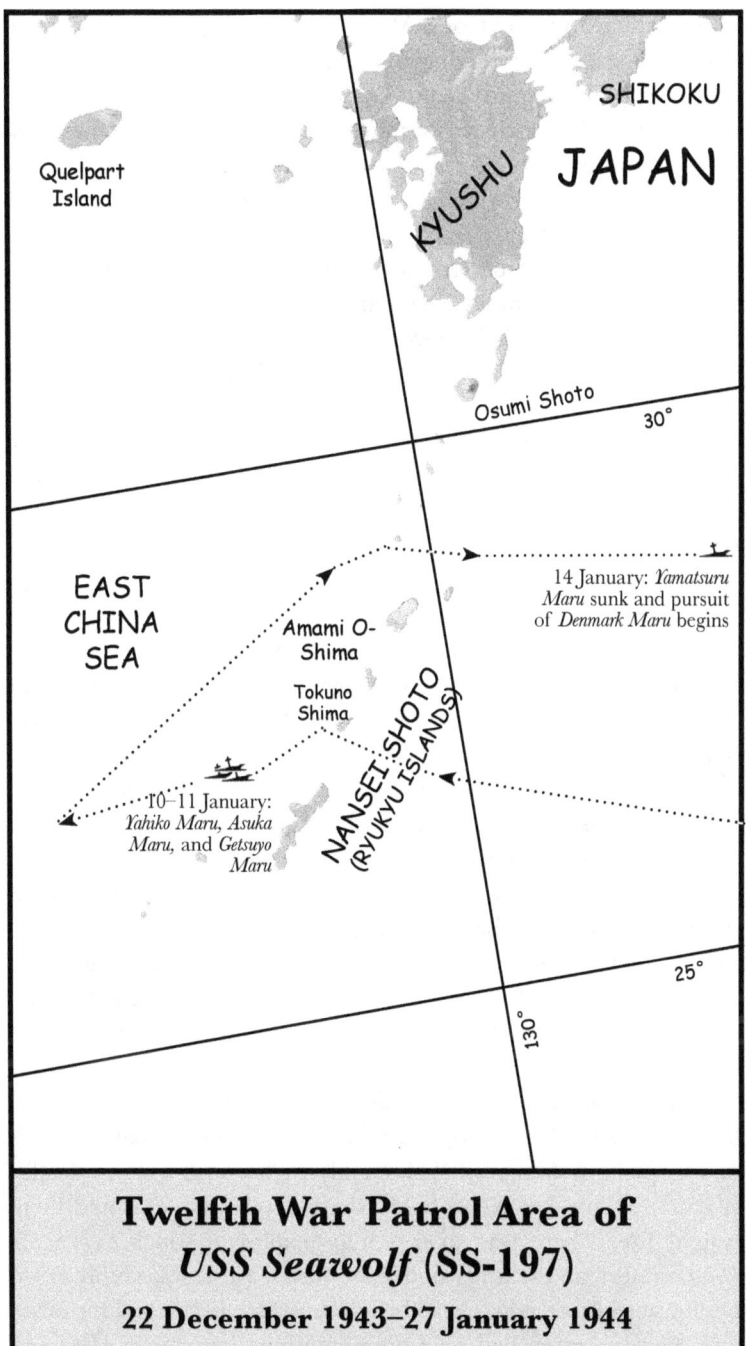

380 feet long and 4,500 tons. Although the cripple was still stopped, *Getsuyo* appeared to steadying on a course to resume the tow.

The fire control party fed the ranges and bearings into the TDC as the forward tube doors were opened. Tube No. 2 remained disabled. From a range of 1,500 yards, *Seawolf* fired her other three tubes at 2251 at the towing ship, *Getsuyo Maru*. The fish were set for six feet, as *Getsuyo* seemed only moderately loaded. One of the fish was seen to broach en route to the target, although the ocean had only moderate swells. One minute after firing, there was a solid explosion from one hit that Gross observed to strike her stern. *Getsuyo* was indeed hit in the stern and began to settle, according to Japanese records.[6]

"Bring the stern tubes to bear," the skipper ordered as soon as the first three had cleared. Helmsman Jim Grimes turned hard to port and brought the ship around as Gross watched the gunners aboard *Getsuyo Maru* open fire against the horizon. Fortunately, the destroyer was still not in sight.

Two depth charges exploded in the distance as the *Wolf* set up to fire on these ships again. This time, Gross planned to spread all four of his after torpedoes. The range was 1,900 yards to the near ship—the now-crippled *Getsuyo*—and 2,600 yards to previously-damaged *Yahiko Maru*. At 2257, *Seawolf* commenced firing from aft. The first Gross sent at the MOT of *Getsuyo*, now down by the stern. The second he fired at the MOT of *Yahiko*, the third at the bow of *Yahiko*, and the fourth back at the stern of *Getsuyo*.

It had been six minutes since firing the bow tubes, so Roy Gross frantically searched for the destroyer while the torpedoes sped toward their targets. Three solid explosion were heard, although none were witnessed due to scanning for the tin can. A fourth explosion was heard at 2302, possibly an end of run explosion from the fourth torpedo.

Three minutes later, towing ship *Getsuyo Maru* had disappeared, leaving only the larger, original cripple *Yahiko Maru* afloat. *Seawolf*'s one torpedo to *Getsuyo*'s MOT could not have been more deadly. It slammed into her No. 6 hold, where depth charges stored there caught fire. There was soon a huge explosion which broke off *Getsuyo*'s stern and sent her under the waves. *Yahiko* was visibly down by the stern more now. Gross believed two more fish had hit *Yahiko* and that one more had hit *Getsuyo*, finishing her. *Yahiko Maru* was

"Battle Stations, Gun!" 333

carrying personnel and equipment of an army air force workshop. Of her passengers and crew, 141 passengers were killed, along with 23 ship's gunners and 33 crewmen. *Getsuyo Maru* suffered four passengers and 16 crewmen lost when she sank.[7]

Five minutes after blasting these ships, *Seawolf*'s skipper spotted the destroyer *Karukaya* about 5,000 yards away on his submarine's port beam. "Everything looked fine for evasion at periscope depth," wrote Gross. Swinging the periscope around again, he suddenly froze. Filling the full field of his scope was another destroyer on his starboard beam, only 1,000 yards away.

"Take her to 280 feet! Full left rudder!" Gross shouted.

Seawolf was already at silent running, but Clary John took her down quick. The second escort, *Minesweeper No. 27*, had obviously been called back to assist with the towing of *Yahiko Maru*. Fortunately, she was making slow speed and was not pinging. Gross decided that this escort must have been alongside the crippled ship out of his periscope view the entire time. He simply could not have appeared over the horizon so suddenly. "The hits in the ship he was alongside of may have damaged his sound gear," the skipper mused.

Luckily, neither escort zeroed in on *Seawolf*, although *Karukaya* was still pinging. The *Wolf* stayed down and crept away from the area during the next hour. When she returned to periscope depth at 0044 on 11 January, no charges had been dropped and all was clear except for distant pinging.

Gross brought his boat to the surface 25 minutes later and began charging his batteries. Two targets showed together on the SJ some 10 miles away. The *Wolf* began tracking them, but only closed the distance slowly, as she needed time to charge her batteries. The electricians secured the charge at 0300 and four engines were put on the screws. With a full moon, the lookouts could soon make out the big target ship *Yahiko Maru* stopped, with two or three escorts present.

At 0312, the *Wolf*'s crew went back to battle stations. As she approached, a nervous escort dropped two more depth charges on a phantom submarine. *Seawolf* passed through the wreckage of towing ship *Getsuyo Maru* en route to attack. "Nearly rammed three large life-boats full of Japs, one large raft loaded with them, and one lone Jap passed close aboard sitting on a plank," wrote Gross.

He slowed his speed to avoid damage to the ship in case he struck a life raft. By 0355, another destroyer was picked up on *Seawolf*'s starboard beam in addition to the two other escorts on her port beam. The new destroyer joined the screen and began circling the stricken ship at about 10,000 yards distance. The two escorts nearest the target ship patrolled at slow speeds about 4,000 yards out from *Getsuyo Maru*.

Duke Robinson and Ed Szendrey's plot still had the target ships stopped as *Seawolf* closed the range to 6,000 yards. The nearest destroyer was seen to flash mast head blinker signals to other DDs, but none of the ships reacted in alarm. Minutes later, however, the destroyer began signaling another and changing course. Realizing that he had most likely been discovered, Roy Gross decided to fire right away. The range was still 5,200 yards, as he fired three forward tubes at 1402, spread to hit MOT, bow, and stern. From the bridge, the men could see *Getsuyo Maru* without binoculars and he still looked very big at that range. As the torpedoes were being fired, another destroyer on the *Wolf*'s port quarter answered the blinker signals from the closer tin can. As soon as the third tube was fired, Gross yelled down, "Full right rudder, full speed ahead!"

The *Wolf* wheeled around and all four diesels kicked in to help her race clear of the alerted destroyers. Gross detailed Lieutenant Cox to watch only for hits as he paid attention to the danger at hand. As *Seawolf* fled, she raced right through the wreckage of *Yahiko Maru* again, frantically dodging the life rafts full of Japanese sailors.

At 0407, five minutes after firing, junior OOD Bob Cox saw three torpedo hits on *Getsuyo Maru*. He saw one white plume at her middle and two large black plumes aft. The explosions were also heard by the bridge personnel and by men below decks. The range from *Getsuyo* was now opening rapidly to about 7,000 yards, with the target ship still in sight.

Robinson and Szendrey timed the torpedo hits and realized that the fish had slowed rapidly after passing the 4,000 yard mark. The longest run to target had been five minutes, 31 seconds. Ben Rogers monitored the target ship's radar pip as the *Wolf* raced out to 10,000 yards and beyond. At about 12,000 yards, the pip disappeared and did not reappear. The destroyers were still all in sight, signaling to each other, but the nearest remained some 8,000 yards off.

At 0446, Gross and his lookouts noted one of the destroyers begin firing rapidly in the direction of another DD. "Apparently he was not satisfied with the reply to his challenge," the skipper wrote. "Too far away for us to claim a damaged destroyer." A few minutes later, another destroyer was picked up visually and then by radar, forward of the starboard beam at 7,500 yards. The newcomer had a zero angle on the bow and began closing *Seawolf* slowly, although the *Wolf* was racing at 19 knots. At 0510, Gross commenced maneuvering to try and shake the new destroyer, but the Japanese skipper maintained a zero angle on the bow and closed the range to 6,000 yards. "Apparently they had someone over there that could see," Gross wryly noted. "He also had another destroyer following him several thousand yards beyond."

By 0518, the range had closed to 5,000 yards. This tin can meant business. *Seawolf* turned hard to port to present a 90 port angle to her attacker and then dived. Gross had the rudder put over full at 50 feet and he then reversed course as his diving officer crash dived to 290 feet at full speed. At 290 feet, the speed was reduced to 2/3 and the trick seemed to work. Both destroyers came charging down the resultant diving swirl on the surface and attacked along the angle of the dive, while the *Wolf* had doubled back past them submerged.

At 0529, the first two depth charges exploded, but none too close. Sonar operators Jerry Bekke and Jim Call tracked both destroyers as *Seawolf* opened out from them at 4 knots. The two hunter-killers worked together, with one pinging while the other listened. "Heard dead metallic clicks at some intervals as regular pinging," Gross noted. "Possibly the damaged sound gear on the destroyer that surprised us." The soundmen believed that three ships were pinging, as two clear pingers were heard plus the one with the damaged gear.

Fireman Bill Harlow, making his second patrol, was still getting accustomed to being attacked.

> In the engine room, we could feel the torpedoes being fired and we could hear the explosions when they hit a ship. We could definitely hear and feel the depth chargings. I was a little anxious on my first close depth charge attack. I don't know how serious they were, but to me they were all serious.

Incredibly, the trio only dropped the two depth charges. Two hours later, *Seawolf* had slipped far enough away that their screw noises were gone. At periscope depth, all was clear by 0727. The crew was secured from battle stations as the ship remained submerged and opened out from the scene of attack to the west and north.

The crew was proud to have put three ships under the waves thus far. The last victim, *Getsuyo Maru*, had taken as many as seven torpedo hits before finally going down. During the late morning and early afternoon, a number of depth charges were heard in the distance as the destroyers found something to work over. *Seawolf* surfaced that evening and ran toward the China coast, sending out a message to *Saury* on the short-range radio.

Jerry Bekke's radiomen sent out an attacks result summary to ComSubPac. In short time, Lockwood sent back a "reassuring message, which was very much appreciated," wrote Gross.

Seawolf ran a periscope patrol in the middle of the China Sea on 12 January as her radar gang worked on their SD set. That evening, fuel king Duke Wall and an auxiliary gang went topside to convert the No. 3A and 3B fuel tanks into ballast tanks. This involved men going down under the decking into the superstructure with little lights. Even slowing to 4 knots, the gentle swell of the sea washed cold water over the working party.

The conversion of the fuel tanks to ballast tanks was completed in the early minutes of 13 January. Wall reported her to still have 74,000 gallons of fuel remaining, plenty to keep on station as planned. Lieutenant John made a dive to flush the remaining fuel out of his new ballast tanks and to compensate his trim for the new weight of the salt water.

Roy Gross ran on the surface through the day, diving only for aircraft contacts on radar. His boat cleared the Nansei Shoto—just north of Amami O Shima—during the night and no radar signals were detected by her radar detector. Just through the pass, the lookouts sighted a fast torpedo boat or destroyer moving at high speed off the *Wolf*'s starboard bow.

During the early morning hours of 14 January, *Seawolf* received an Ultra from Dick Voge to intercept another convoy. Communications officer Marion Asa brought the highly classified message to the skipper for decoding. Roy Gross immediately left his regular patrol area in pursuit. Shortly into his run toward the Ultra-reported convoy,

"Battle Stations, Gun!" 337

Gross' lookouts sighted smoke on the horizon to the north, which *Seawolf* turned to trail. She had stumbled upon another nice convoy, which proved to be four freighters and two escorts, zig-zagging along at 9 knots.

"Although assigned mission was important," Gross wrote that he "decided to proceed on the bird-in-hand theory." Later, he offered his thoughts on the Ultra intelligence.

It is my recollection that very many Ultras were false alarms, whether from errors in decoding, uncooperative convoy commanders who changed their mind (and course) after the decoding, or maybe we on the boats made our own errors in copying the FOX schedules and/or in also decoding. It is always better to have some information, even if not always good, since otherwise it was just searching the broad ocean area assigned, more or less blindly.[8]

Seawolf completed sending a contact report to ComSubPac at 1123, letting Lockwood know of their contact with the convoy and the fact that they had only three forward torpedoes remaining. This convoy consisted of: *Yamatsuru Maru*, *Denmark Maru*, *Narita Maru*, *Minesweeper No. 18*, minelayer *Nuwajima*, and auxiliary minesweeper *Tama Maru*. Convoy O-105 had departed from Saeki on northeast Kyushu on 11 January 1944, bound for Palau. *Sturgeon* torpedoed and sank *Erie Maru* that day, forcing the convoy to flee back to Saeki. O-105 departed again from Saeki at 0200 on 13 January, still bound for Palau, passing south of Kyushu without incident that day. The convoy's luck would run out on 14 January, however.[9]

Seawolf's radiomen sent another contact report at 1340 for any other submarines that might be in the area as she continued to run ahead. Bekke's gang received no return receipt, however. By 1356, *Seawolf* was 27 miles ahead of the convoy's track and dived for a dusk attack or a night surface attack. "His smoke must be visible at least 30 miles," Gross wrote. The radar gang worked feverishly on the SD, which had gone out of commission. At 1603, Bill Smith brought the boat up to 35 feet and it worked. *Seawolf* resumed periscope depth tracking, running in the direction of the convoy.

By 1706, the convoy was in full view at 11,000 yards and the *Wolf* went to battle stations, torpedo. The sun set 30 minutes later and Gross dejectedly noted that the convoy was getting too far away unless they zigged back radically soon. Via periscope, the cargo ships looked to be three medium freighters, and one larger freighter.

Seawolf surfaced at 1823 and posted her battle lookouts on the shears. The ships were 17,000 yards ahead of her, and were clearly visible on radar. Ben Rogers reported no signals via the radar detector, which meant low-technology escort ships. The engineers put three engines on line and charged the batteries with the fourth. *Seawolf* charged in on the convoy's starboard flank, crossing the stern of the starboard side destroyer at 5,000 yards, giving herself a large starboard track on the nearest ship.

By 1943, *Seawolf* was 1,800 yards from the nearest Japanese ship, with all four ships on her port bow. She was ready to fire when the convoy suddenly zigged left. Gross kept the ships all on his port bow and crossed astern on his first target ship, when the same zig zag suddenly threw off his aim on his second selection. "Their zigs are certainly confusing," he wrote. "We are now 1,500 yards astern of them on their course." He suddenly had one ship on his starboard bow and the others on his port bow. *Seawolf* took the convoy's course and speed, hoping to prepare for their next radical zig. At 2021, another left zig gave the ship on the *Wolf*'s starboard bow a 120 port angle on the bow.

Electrician Olin Fogle, on duty in the maneuvering room, was busily listening to the play-by-play from above. "We attacked a six-ship convoy. The Old Man was up on the bridge and had the annunciator down so we could hear throughout the boat what was going on," recalled Fogle. "There was no loafing. It was business."

This time, the plotting team was on the opportunity immediately. "Good set-up gotten fast," Gross wrote.

He steadied on a freighter he estimated to be about 4,500 tons. The ship was clearly in view from the bridge without binoculars. With Lt. Bob Cox on the forward TBT to feed the info to Bill Smith on the old Mark I TDC, Gross fired his last three torpedoes.

This ship was *Yamatsuru Maru*, a 3,651-ton naval auxiliary. Two of *Seawolf*'s last three torpedoes found their mark with violent results, as Roy Gross logged.

The first hit was in the bow, which ignited the whole forward part of the ship, by the light of which the stack could be seen collapsing to port, the mainmast to starboard. The second hit was in the stern a few seconds later. This set off the stern and nothing but flame from the water's edge for several hundred feet up was visible. Must have been loaded with gasoline. Both hits were muffled, like hitting a pillow with the fist, and there was no noisy explosions of any sort. This illuminated us clearly, but we pulled clear before any guns could train on us. In fact, six miles later ten depth charges were dropped. We retired on the reverse course of the convoy, circled the fire at 10,000 yards which burned for two hours with an occasional spurt of flame high into the air. This may keep some of the Rabaul fliers grounded for a while.

Seawolf secured from battle stations at 2040 as her lookouts watched the aviation gasoline burn in the distance on the doomed *Yamatsuru Maru*. "Target immediately burst up, burned fiercely and sunk," OOD Bob Cox logged. Those below listened to the picture being painted by the words from those above. "One of our torpedoes hit a ship in the first line that must have been filled with aviation gasoline," recalled Olin Fogle. "That damn thing lit up everything."

Stationed in the maneuvering room on the starboard side, Fogle became excited with the rush of the moment as Lt. Cdr. Gross ordered full left rudder to escape from the convoy. "I hate to admit it," Fogle said, "but I put the power on full and overdid the engines. I squashed three out of four diesels."

His first class boss, George Needham, was on the situation in an instant. "I got all four engines back on line quickly!" said Fogle. *Seawolf* pulled away from the exploding ship as the moon rose. The escort ships raced around wildly, dropping depth charges on phantom submarines. The radiomen sent out a contact report on the remnants of this convoy. Gross felt that "it was urgent that somebody show up quickly to knock over the remaining three."

Japanese records show that *Yamatsuru Maru* had indeed loaded aviation fuel at Tokuyama in the West Inland Sea and that she was to deliver it to Rabaul. One of *Seawolf*'s torpedoes found her No. 2 hold on her port side, causing a heavy explosion. This led to great fires, which caught her aviation fuel on fire and made it begin to cascade into the sea in flames. A second torpedo struck between *Yamatsuru*'s engine room and No. 3 hold, causing her to list to port and sink a few minutes later. Four passengers, two guard force members, 30 crewmen, a load of ammunition, and even a midget submarine being towed behind the ship were lost when *Yamatsuru Maru* went down.[10]

At Pearl Harbor, operations officer Dick Voge took stock of the *Seawolf* contact report. The *Wolf* was out of torpedoes, yet remained in pursuit of the convoy, hoping to draw in other submarines. Voge alerted Lt. Cdr. Albert Collins "Acey" Burrows, who was patrolling nearby with his *Whale* (SS-239). While *Whale* hustled to catch up, *Seawolf* pursued the convoy and continued to send contact reports.

Those aboard *Whale* picked up on the *Wolf*'s contact reports to Pearl, as *Whale* radioman Vernon T. Miller later recalled.

> Shortly after midnight on January 14, 1944, I received a message from the *Seawolf* on our distress frequency. Once decoded, we learned that *Seawolf* had expended her torpedoes on a convoy but was still trailing it, and furnished coordinates on the distress frequency, in hopes of getting assistance from other subs in the vicinity. The *Whale* was en route Midway from Empire waters but only hours away from the *Seawolf* and the enemy convoy.[11]

Seawolf kept the convoy's smoke in sight, hung back at about 19,000 yards, and sent another contact report to Pearl Harbor at 0023 on 15 January. Once this had cleared, *Seawolf* closed in on their port flank. Gross looked over the nearest ship and found her to be a smaller, engine-aft tanker. Figuring that she might also be full of gasoline, Gross figured "we might touch off" this fuel with his 3-inch gun. He hoped one good hit might send her up in flames.

"Battle stations, gun!"

The call over the 1MC sent the deck gun crew into action, breaking out the heavy shells from the ready ammunition locker. Gunner's mates Dale Kropp, Frank Ballard, and their assistants spilled out onto the deck and limbered up their 3-incher. Ensign George Leffingwell took his seat as setter and Chief Hank Thomson jumped into the pointer's chair. Duke Wall and his loaders began passing the shells up from their storage locker.

Seawolf put all four engines on the line and charged in for a gun duel, completely out of torpedoes. At 0118, the *Wolf* was 6,500 yards from the target ship. There was no escort visible—only the three

Seawolf's gunners had considerable action during the twelfth patrol. In this photo from 8 May 1944 off Hunters Point, California, GM2c Frank Ballard test fires the *Wolf*'s aft 20mm bridge gun. *Official U.S. Navy photo.*

freighters. Lt. Cdr. Gross now believed that there was an escort on the far flank.

"Captain, radar contact on the SD!" Ben Rogers called out.

For a moment, Roy Gross debated on whether he should dive. The pips, however, showed on the SD at 3 to 4 miles distant—the same as that of the other convoy ships. This was the first time this phenomena had occurred on *Seawolf*, where the SD aerial search radar was showing what were actual surface contacts. "The pips didn't look exactly like plane pips, so we didn't dive," logged Gross.

By 0137, *Seawolf* had pulled to within 4,300 yards of the small tanker. Gunnery officer Bill Smith would direct the fire of the crew on deck, while communications officer Marion Asa and his radarmen would help spot and coach the results.

"Commence firing!" Gross ordered.

The 3-inch deck gun sent her first slow, ranging shot in the direction of the tanker. The target was very clear in the sights and there was considerable roll. The spotters called the distance to the splash and then the second shot was adjusted. Slowly, the *Wolf*'s 3-inch gang walked their shots in closer to the tanker. By the sixth shot, however, all three Japanese ships had opened up with small caliber and heavier caliber guns. With shells whistling by, Gross ordered, "Secure the deck gun! Get the men down below!"

Japanese records show that *Seawolf* had taken on *Denmark Maru* and that two of her shells struck the sea near *Denmark*'s stern. Escorts *Tama Maru*, *Nuwajima*, and *Minesweeper No. 18* charged after *Seawolf* and opened fire on her.[12]

Gun captain Kropp's gunners secured their gun and hauled the remaining shells below as gunfire lit up the night. No Japanese shells landed close enough for concern. "No hits were seen to burst on the target, but believe from observation of tracers and radar spots that we got one or two hits," logged Gross. "Considered this a good idea that didn't work and secured from battle stations gun."

Seawolf continued tracking this ship well clear to port. As she did, the sound of gunfire and even one depth charge were heard periodically over the next hour from the rattled Japanese sailors. At 0240, Gross contacted *Whale* on 450 kilocycles to update her on the situation and the convoy's course. At 0300, *Seawolf* sent another contact report to *Whale* which was receipted for by radioman Vernon Miller. At 0614, *Seawolf*'s radiomen sent yet another report.

As the sun rose, *Seawolf* stayed astern of the small convoy, keeping their masts in sight with her high periscope, while maintaining a range of about 15 miles. *Seawolf* was trailing along nicely at noon, astern of the convoy's base course, when a Rufe float monoplane and a Nell medium land bomber were spotted. *Seawolf* dived and followed the convoy submerged for the remainder of the afternoon.

Gross surfaced again one half hour before sunset—feeling that "this is the most we can afford to let them get ahead." The *Wolf* made 19 knots on the best guess of the enemy's course. Gross sent another contact report to ComSubPac on losing the convoy but he was also unable to raise *Whale*.

At 2003, radar contact was made on the starboard bow at 24,600 yards. The *Wolf* closed and tracked the convoy at a visual range of about 15,000 yards, noting that the enemy ships all smoked heavily. One escort was seen lagging astern. "Apparently they don't like strangers to join their formation," Gross noted. *Seawolf* was now about 50 miles south-southeast of her earlier scene of attack, having run three hours at nearly 20 knots.

At 2342, Jerry Bekke cleared a new contact report to ComSubPac, but was unable to raise *Whale*, although he continued sending blindly to her. At 0245 on 16 January, *Seawolf* sent her latest information and received a receipt. "This was encouraging. Didn't relish the idea of trailing this all the way to Rabaul," wrote Gross. By daylight, *Seawolf* was holding periscope contact on the convoy at 15 miles via her high periscope. Radiomen Bekke and Call continued sending four-letter contacts to *Whale* at three- and four-hour intervals, getting a welcome receipt each time. "Things looking up," wrote Gross.

In the mid-afternoon, several explosions were heard over the sonar, indicating the enemy was nervously dropping depth charges. By sunset at 1735, all of the ships were still in sight. It soon became apparent that *Whale* had found the convoy. At 1807, there were several explosions and then gunfire from the convoy. At 1817, there was another explosion.

Seawolf closed in and picked up the convoy at 27,000 yards. Three minutes later, there were more explosions. The convoy was not visible from the bridge but was faintly showing on radar. One group of two had stopped, one of them appearing to be an escort. Records would later show that *Whale* had damaged a tanker and fatally holed the 5,869-ton freighter *Denmark Maru*—*Seawolf*'s gun target.

Four depth charges were heard at 1900 and *Seawolf* called for *Whale* but received no answer. Gross assumed correctly that she had made a dusk attack and was submerged, avoiding escorts. Radar showed one lone pip running to the southeast at 10 knots. The tracking team kept tabs on this vessel and *Seawolf* moved in on her. Depth charges and explosions continued to rock the night world. *Seawolf* pursued the lone ship, in hopes of calling in *Whale* once she managed to return to the surface. Gross "decided to try and divert him back towards the *Whale* with the deck gun, with the further chance that a lucky hit will blow him up if he is full of gasoline."

At 2112, *Seawolf* once again went to battle stations gun. Lt. Bill Smith, gun captain Dale Kropp, and their 3-inch crew made their weapon ready. The lookouts noted blinking from the enemy ship's stern where his gun was believed to be. Gross closed the enemy ship to 2,800 yards, telling his gunners not to fire, but to wait for the rising moon to better illuminate their target ship.

Seated on the gun as pointer was chief quartermaster Hank Thomson. "The moon was up a few degrees, but behind a low cloud," he recalled. "We were at gun stations on his port beam." The ship showed as a pip on Ben Rogers' SD radar again at 3 miles, but experience now proved that this was not an airplane on the aerial radar but the ship ahead.

The target ship—the 4,865-ton armed transport *Tarushima Maru*—was silhouetted against the rising moon, and with smoke drifting back towards his stern, Gross decided it was time to open up. Radar had the range at 4,300 yards.

"Captain Gross told me that he wanted me to fire five rapid shells and have them spotted before firing again," said Thomson. No one on deck could make out the maru well enough to fire accurately, so Thomson climbed out of the pointer's seat and walked back to ask his skipper to move the ship closer. "He told me in no uncertain terms that he would move in 100 yards closer and then I had better see the damn target!" said Thomson. "I got the message."

Seawolf closed the distance a little more and then Gross hollered out, "Commence firing!"

Thomson finally felt that he had the enemy ship in his sights, "but Lefty—the sight setter—and the gunnery officer would not confirm. I decided to fire at Captain Gross' orders."

Without Lt. Smith's word, Thomson pressed the button and fired the first 3-inch shot. Smith and gunner Francis Ballard then had their deck gun crew commence a rapid firing on this ship. Radar reported a few short, but Gross felt certain that his gunners did score some hits on *Tarushima Maru*. "Saw two hit the side and burst outside," he wrote. "Shells were point-detonating, set on delay." The armed transport did not turn away, but opened fire with a big gun both forward and aft, and also with 20mm.

"The ensuing duel consisted of the enemy seeing me when I fired and me seeing him when he fired," admitted Thomson. "I can remember seeing a couple of shells passing overhead—aim correct but range long."

The first shots were not close, but at about *Seawolf*'s 30th shot, the Japanese gunners had shells whistling by overhead.

"Cease fire!" Gross yelled. "Clear the topside!"

He was aggressive, but not willing to lose good men to the enemy fire. "I was told some of the crew had dropped to the deck when the shells started passing over," said Thomson. "Frankly, I wanted to hit him and was too busy to pay much attention. I don't think anything concerned me except hearing the gun being reloaded and my trying to hit the son of a gun."

Officer of the deck Duke Robinson turned the ship and opened out the range rapidly as the gunners scrambled for cover. *Tarushima Maru*'s gunners continued firing blindly for some time. Afterwards, Hank Thomson sought out his skipper and apologized for not being able to sink the Japanese ship. "He told me it was okay and that the enemy was doing his preconceived plan."

Robinson and his lookouts had a few anxious moments under fire as *Seawolf* cleared the range. Motormac Lew Donche, one of the lookouts during the gun actions, did not bother to stuff protection in his ears before the surface actions. Like many submariners, he suffered long-term damage to his hearing from the deck gun blasts.

The radio gang below noted "much chatter" over the radio at this time as the frantic Japanese were obviously pleading for help. *Seawolf* stayed out of gunfire range but Gross secured his men from battle stations gun at 2315, "thankful that no one was hurt. Target was circling. Must have him worried."

Gross' bold strategy worked. The target ship, *Tarushima Maru*, then turned northwest and began heading back in the direction of *Whale*. The ship was zigging radically as *Seawolf* took position 5,000 yards behind her while sending out contact reports to *Whale*. "Our gun job had succeeded in changing this Jap's mind about proceeding to Rabaul alone," Gross noted. "He must have thought the seas were full of submarines."

The night passed into the early hours of 17 January, with *Seawolf* still pursuing this lone ship. She sent a radio message to *Whale* at 0126 which was immediately received. From this moment on, communications between *Whale* and the *Wolf* were excellent. The *Wolf* radiomen did away with formal call-ups, as *Whale* now immediately answered each with a "roger."

By 0313, there was a radar interference seen on the SJ, indicating that *Whale* was close aboard this target. *Seawolf* continued trailing the ship visually at 8,000 yards astern, sending out the enemy's new course every time he zigged. When Gross finally figured that the firing time was close as dawn approached, he finally quit with the course changes. During this return track, the fleeing *Tarushima Maru* passed within 11 miles of the stopped ship from the previous *Whale* attack, *Denmark Maru*.

With great satisfaction, *Seawolf*'s crew heard an explosion at 0524. *Tarushima Maru* was then observed to begin circling. During the next 12 minutes, there were several more explosions, most appearing to be end-of-run torpedo explosions. Gross believed that only one *Whale* torpedo had struck home. As the sun rose over the wounded *Tarushima Maru*, she erupted into gunfire while still making circles.

By 0600, *Tarushima*'s boiler fire died down and smoke stopped coming out of her stack. Plotting officers Robinson and Szendrey confirmed her to be dead in the water. *Seawolf* opened out to 12,000 yards to give *Whale* plenty of room to deliver the coup de grace. By 0620, Gross realized that *Tarushima* was actually shooting at *Seawolf* as the light increased. Nothing was landing close, so he held his distance to watch the show.

At 0623, Gross and his bridge crew watched "a beautiful hit in the middle by *Whale*." Smoke belched from *Tarushima Maru* and rose to twice the height of her stack. The ship began to settle forward now. During the next 12 minutes, her stern lifted into the air and she slid under the waves—a joint kill between the efforts of *Seawolf*'s persistence and her gunners and the torpedoes of *Whale*.

"It was an aggressive situation to find ourselves within reach of three more ships, but out of torpedoes," Roy Gross logged. "Contact tracking under these conditions lacks the stimulus of actual attacks, but to see the *Whale* sink our tracked target was ample reward."

Seawolf put on four engines to clear the area for *Whale* to continue her hunt. After a grueling, two-day chase and attack, the *Wolf*'s crew was dead tired. Roy Gross noted the stopped ship but moved on by to give the *Whale* plenty of operating room. Well clear of the area by 0817, he ordered OOD Bob Cox to dive ship "to give all hands a rest." He also fully expected airplane activity in retaliation, although the nearest Japanese base was some 400 miles away at Chichi Jima.

Shortly before noon, the sounds of explosions and depth charges were picked up back in the vicinity of the stricken ship. "Hope this is *Whale* accounting for the last of the four ships," wrote Gross.

Seawolf surfaced that evening and set course for home on two engine speed. Navigator Doug Syverson and Hank Thomson were able to make a good fix on their position the next evening when Kita

Iwo Shima was sighted at 53 miles. The *Wolf* went up to four engine speed to clear the Nanpo Shoto.

At 1729 on 22 January, Bill Smith's lookouts spotted smoke due south between rain squalls in the setting sun. Gross closed to investigate but found nothing but a ship's wake. The seas were condition 7 on 24 January as *Seawolf* approached Midway Island. Gross received the much welcomed orders from ComSubPac to bypass Midway this time and bring his boat straight back into Pearl Harbor. No enemy aircraft from Wake or Marcus disturbed her triumphant return from her twelfth patrol. En route home, Red Syverson rotated newly qualified deck officers Ed Szendrey and Marion Asa into the OOD schedule. This brought relief to the wardroom, for with six officers on the watch schedule pool, each only had to stand two-hour shifts.

This was altered, however, when Lt. Duke Robinson soon had to come off the watch schedule. He developed bleeding stomach ulcers that Doc Hadley could do little to remedy. "Nothing hurts as bad as an ulcer," Robinson would later relate. Fortunately ship's cook Leroy Edmonds and the two Filipino officers stewards, Tomas Rosete and Saturnino Rocaya, catered to him with scrambled eggs and other specially-prepared mild dishes. Lieutenant Syverson took Robinson off the watch schedule and recommended that he be hospitalized upon the *Wolf*'s return to Pearl.

Seawolf picked up her escort at 0650 on 27 January and proceeded into Pearl Harbor. Marion Asa stationed the maneuvering watch as *Seawolf* entered the channel at Pearl shortly after 1000. She docked at the Submarine Base at 1030 at Berth S-2, and Asa moored her portside to *USS Grayling*. On the dock, Admiral Lockwood had his band blasting out greeting tunes as the *Wolf* was secured.

Lockwood was thrilled with this patrol by *Seawolf*. In a private note to Gross he wrote: "Excellent, aggressive and long headed patrol." In his official endorsements, he commended the skipper and crew on a "most aggressive and outstanding war patrol." He found that the "use of the gun by the *Seawolf* showed excellent planning and careful consideration." Lockwood gave *Seawolf* credit for sinking four ships and damaging three others in "a series of tenacious attacks." Further, steering *Tarushima Maru* back toward the *Whale* with gunfire so that she could be torpedoed "was a masterful performance."

For his "masterful performance," Roy Gross was awarded the Navy Cross for *Seawolf*'s twelfth patrol. He in turn, recognized his other key personnel with awards for their contributions. Silver Stars went to gunnery officer Bill Smith, engineering officer Clary John, and radarman Ben Rogers. Bronze Stars were awarded to Bob Cox, fire controlman-turned-mustang George Leffingwell, communications officer Marion Asa, and quartermaster Hank Thomson. Navy Letters of Commendation were also written up for several men, including Lt. Duke Robinson and fire controlman Delbert Mar.

Roy Gross' wartime credit was dead on with the postwar credit. Lockwood gave Gross credit for sinking four ships totaling 24,000 tons. After the war, JANAC would officially give him four ships for 23,361 tons. In addition, Admiral Lockwood gave *Seawolf* half credit for the sinking of *Tarushima Maru*, which *Whale* had teamed up to help destroy. This upped her score to 4.5 ships sunk for 25,793 tons, one of the best patrols of the war. Gross' contact report had given Lt. Cdr. Burrows the chance to sink 1.5 ships for an additional 8,322 tons. Aside from the 25,793 tons sunk, *Seawolf* was credited with damaging three more ships for an additional 14,000 tons.

Although not working as a wolfpack officially, *Whale* and the *Wolf* had written a new chapter in coordinated kills. "Acey Burrows and I had many a good drink together after our successful team-up," Gross later related.[13]

USS *Seawolf* Twelfth Patrol Summary

Departure From:	Pearl Harbor
Patrol Area:	East China Sea
Time Period:	22 Dec. 1943—27 Jan. 1944
Number of Men Aboard:	81: 73 enlisted and 9 officers
Total Days on Patrol:	36
Total Miles Steamed:	10,715 miles
Total Fuel Consumed:	120,794 gallons
Total Days Submerged:	4
Number of Torpedoes Fired:	20
Ships Credited as Sunk:	4/24,000 tons
JANAC Postwar Credit:	4.5/25,793 tons
Shipping Damage Claimed:	3/14,000 tons (Lockwood)
Return To:	Pearl Harbor

15

Lifeguard League

Thirteenth Patrol *4 June—7 July 1944*

This time, the stay in Hawaii was brief. The scuttlebutt was that the old *Wolf* was due to go back Stateside for a long overhaul. Roy Gross, having made five runs as skipper, was expected to be transferred to a new assignment. The crew made the most of their liberty during the two days at Pearl Harbor. Some looked forward to spending their latest pay or poker earnings.

"Large sums of money changed hands during the poker games," recalled Hank Thomson. "Pay out was on the last night of the patrol with the winners taking their loot. If he sent it home, he was lucky. After the 12th patrol, a gunner's mate had won $5,000 and took it ashore. He was rolled that night and lost his money—he said."

Seawolf departed the Hawaiian Islands on 29 January 1944, bound for Hunters Point, California, for a major workover. The five day run gave the men plenty of time to make plans for home. Yeoman Roger O'Brien began making out leave papers for all hands who would continue to patrol on *Seawolf*. Each man could chose between the first or second leave period, with the more senior men having first choice of leave period.

Once again, the crew enjoyed lengthy leaves back home. There were more marriages and many happy reunions. Fire controlman Delbert Mar's family lived nearby in Vallejo, enabling him to visit frequently during *Seawolf*'s overhaul. Ensign Lefty Leffingwell had married his high school sweetheart from Pasadena. Now, his wife was expecting their first child in December. New orders meant that Leffingwell would remain behind when *Seawolf* left California.

Torpedoman Donald Naze spent three weeks with his wife Etta in Providence, Rhode Island, and also visited his parents for the first time in four years.[1]

Fireman Bill Harlow was in the first leave group. "When we pulled back in to California, I had the opportunity for 30-day leave and new construction," he said. "So, I jumped right at that. I made leave with another friend from New York, Wasil Politylo." Following their three weeks, Politylo returned to *Seawolf*'s electrical gang, but Bill Harlow found orders transferring him to *Tench*, a new construction boat at New London.

Chief quartermaster Hank Thomson—who had been aboard ship since before the war—also left *Seawolf* in California. "At the end of our time in San Francisco, I asked for new construction in the West Coast because I had a fillie there," he stated. Chief Reiland, one of Thomson's closest friends aboard ship, asked him to make one more patrol together. "Bill Reiland, a very good friend, begged me to make one more patrol with him, but I had orders for new construction and a lady to see."

A number of other *Seawolf* veterans left while Stateside. Electrician Olin Fogle was transferred to new construction aboard *Devilfish*. Motormac Bob Curtin caught an Air Force transport flight into Kansas, and then took a train into Chicago to see his new wife. Like many others, Curtin found new orders at Hunter's Point. Chief of the boat Leonard Caillier was also transferred, and his spot was filled by chief torpedoman Bill Reiland, who had made all but one of *Seawolf*'s previous patrols. He had sat out the twelfth patrol on leave, but came back as the senior enlisted man.

"He got the nickname 'Wild Bill' before my time, but Reiland was a good guy," said gunner's mate Rex Mickey, who was also returning after having gone on leave for Christmas and missing the twelfth run. "The crew respected Reiland as chief of the boat. He didn't have to be tough on the guys," Mickey related. "Your life depended on the guy next to you and so you treated everybody with respect because your life was in their hands."

During the time in the Hunters Point yard, the old *Wolf* changed significantly. "The overhaul was a good one," *Seawolf*'s patrol report noted. Her conning tower was lengthened and completely renewed. The interior arrangement conformed to the best current plan of the most modern submarines. A

Between patrols, this group of *Seawolf* submariners lets off steam while on R&R. Second from left is Duke Wall and Chief Mike Wiegenstein is fifth from left. Patrick Bergevin is standing in background, left. *Courtesy of Charles Hinman, USS Bowfin Submarine Museum.*

part of the old Mark One TDC was even retained to drive an additional plotting table and to drive a shipboard attack teacher. FC3c Delbert Mar had to relearn the new TDC. "You train on one model, and then a different model comes in later," he said, "and I'm looking for the switch to turn it on. I wouldn't know how to repair it. Shit, there were so many electrical parts to it."

Those who had come to know and deeply respect skipper Roy Gross found it difficult when it came time for him to depart. Googy Gross had orders to new construction, where he would put *Boarfish* into commission. Before he departed, the crew collected money for special gifts for him. "At the end of the 12th patrol, we knew that Captain Gross and Syverson were leaving," said Hank Thomson. "Quietly, money was solicited from the crew for a gift. We even got money from Syverson that he thought was for Gross' gift." Red Syverson had orders to report to *Sea Fox* as her new Exec.

In a simple ceremony, Cdr. Gross was officially relieved by Lieutenant Commander Richard Barr Lynch, from the Academy class of 1935. Known as "Ozzie" to his contemporaries, Lynch came from service as Exec of the old *Nautilus*. He had been instru-

(Left) This view of *Seawolf* was taken on 9 May 1944 off San Francisco following a Navy yard overhaul at Hunter's Point. *(Right)* Lt. Cdr. Richard Barr "Ozzie" Lynch became the *Wolf*'s third wartime skipper prior to the ship's thirteenth war patrol. *National Archives.*

mental in *Nautilus*' reconnaissance mission of the beaches of Tarawa in preparation for Marine landings there. An amateur shutterbug, Lynch had employed his own German-made Primaflex camera to take periscope pictures of the beaches.[2]

Lynch had been awarded the Silver Star Medal for his actions in helping to save *Nautilus* on her seventh patrol. Attacked by a friendly cruiser and destroyer, *Nautilus* was holed by a dud shell which caused serious flooding. During his five runs on *Nautilus*, he had also been awarded another Silver Star for commanding a rescue mission, the Navy and Marine Corps Medal for serving as officer-in-charge of a rescue mission in which his boat evacuated 29 men from Bougainville Island, and two Bronze Star Medals.

With the departure of both Roy Gross and Red Syverson, *Seawolf* would now be ably commanded by Ozzie Lynch and a new Executive Officer, 30-year-old Lt. Cdr. John Borden Hess, a 1937 Naval Academy graduate from Portland, Oregon. Hess had served on the cruisers *Indianapolis* and *Quincy*, and the survey ship *Hannibal*, before attending Submarine School in New London in 1940. He had made his first two war patrols on the old *S-36*, surviving her sinking but losing all of his personal belongings. He next served as first lieutenant of *Sculpin* for two patrols, aboard which he was considered something of a chess and acey-deucy enthusiast.[3]

John Hess went on to serve as *Searaven*'s engineering officer for one patrol and as the Exec of *Swordfish* for four war patrols. On *Swordfish*'s

This is perhaps one of the final photos of *Seawolf*, taken 9 May 1944, as she departs California with her newly modified superstructure and a forward-mounted 4-inch deck gun. *Official U.S. Navy photo.*

tenth run, she unsuccessfully attacked the Japanese carrier *Shokaku* and her four escorts. Prior to firing, one of the enemy destroyers was only 2,000 yards away and closing fast. "To say that I was scared to death at this point is a pretty fair statement," Hess later wrote.[4]

In addition to the departure of the old skipper and Exec, *Seawolf* had lost Duke Robinson and George Leffingwell. Two other faces joined the wardroom, one new and one old. Ensign John Van Andel from Modesto, California, became the new "George," and took on the commissary officer duties. Plank owner John Bilkey joined the mustang club, being promoted to warrant electrician and taking on the duties of assistant first lieutenant.

Lt. Clary John remained as engineering and diving officer, while Lt. Bob Cox fleeted up to fourth officer. Ed Szendrey and communications officer Marion Asa also remained and Bill Smith would continue on in his role as gunnery officer.

The famous *Seawolf* headed back west toward Hawaii again on 16 May, her new skipper eager to make his mark in the war zone. Ozzie Lynch's crew underwent a brief training period before she was pronounced ready for sea again. *Seawolf* got underway at 1330 on 4 June 1944 for her thirteenth patrol. She departed Pearl in company with *Plaice* and *PC-485* en route to Midway Island. The *Wolf* made a quick trim dive in Mauai Channel, found everything in order for the diving officer, and then proceeded. *PC-485* departed at 2200, leaving *Seawolf* and her surface ship *Plaice* to make the voyage to Gooneyville. Lt. Cdr. Lynch kept *Plaice* within signal distance during the trip and used this ship to conduct all types of exercises, in addition to the usual training dives.

Lt. Bob Cox conned *Seawolf* through Midway's tricky lagoon entrance on 8 June and docked her at 0900. She stayed just long enough to make minor voyage repairs and to refuel. By 1630, she was on her way to carry out Admiral Lockwood's Operation Order No. 194–44 of 3 June. Ozzie Lynch had been assigned the task of photographing Peleliu Island in the Palaus in preparation for the forthcoming attack on that Japanese stronghold. His camera experience and reconnaissance work aboard *Nautilus* made him a perfect candidate for Admiral Lockwood's latest special mission.

Two hours after departing Midway, diving officer Clary John made a deep dive and found conditions satisfactory. En route to Palau on 10 June, *Seawolf*'s lookouts made two different contacts with friendly Catalina patrol planes in the morning hours. On 13 June, *Seawolf* sighted and exchanged recognition signals with *Bowfin* at 1430. One hour later, Lt.(jg) Marion Asa had orders to conduct a training dive. Cupping his hands, he called, "Clear the bridge!" and pulled the diving alarm. The lookouts scrambled down the hatch and the quartermaster dogged it as the *Wolf* was sliding under the waves. Skipper Lynch was pleased with the time until a call came from below: "Flooding in the maneuvering room!"

Electricians and motormacs were doused as sea water sprayed heavily from a busted pipe. Warrant officer John Bilkey and the chief electricians—James Irvin and Mason Poole—sprang into action with their damage control crews. The source was quickly traced to a galvanized iron pipe nipple in the copper pipe system for cooling the main motors—which had given way and caused moderate flooding of the maneuvering room. It was replaced with a copper pipe nipple and, according to Lynch, "a search started to locate any other iron components in the salt water systems."

The next day, Lt.(jg) Asa was troubled with a persistent Japanese Betty bomber. It was first sighted inbound at 1444 and he quickly called for a crash dive. After periscope observations showed all clear, he brought the *Wolf* back to the surface. At 1530, the Betty was spotted and Asa again called for a crash dive. *Seawolf* found the same bomber still patrolling overhead an hour later. Lt. Cdr. Lynch prudently kept his boat down until well after sunset. The APR-1 detected several radars but no planes were spotted.

As *Seawolf* made her way toward her special photography mission at Palau, one of the great naval engagements of history was brewing in the Marianas. U.S. carrier task forces began pounding enemy resistance ashore even as intelligence brought word of a large Japanese fleet approaching. ComSubPac dispatched orders to the *Wolf* on 16 June to change her patrol area temporarily in anticipation of this major engagement.

The ensuing Marianas Turkey Shoot was a carrier battle of epic preportions. During the next few days, *Seawolf* had several more aircraft encounters. It was evident that both American and Japanese forces were in heavy concentration in the area. Lieutenant John's engineers struggled with a broken valve in the main hydraulic plant on 17 June that forced them to resort to hand-operating diving until the damage could be repaired. *Seawolf* ran on the surface during the next two days in search of the Japanese fleet, diving only when bogeys approached on the radar screen.

As opposing fighter pilots engaged in countless dogfights on 19 June, the action was monitored over the radio. Two plane sightings drove the *Wolf* down, but she otherwise remained surfaced and ready for action. At 0720 on 20 June, the radio gang picked up a dispatch directing *Seawolf* to proceed on toward Palau. She was en route on the surface at 1025 when the radar picked up another airplane contact. Bob Cox dived ship as the Japanese plane—believed to be a Kate or possibly a Betty bomber—zoomed down toward the *Wolf*.

As she reached periscope depth, her crew heard numerous explosions, likely from smaller ordnance dumped by the Japanese pilot. Through the periscope, the enemy aircraft was seen to fly within a few hundred feet above the spot where the *Wolf* had submerged. Two different radar signals had been detected prior to the plane's attack, indicating that two planes may have been triangulating *Seawolf* via radar for attack.

Lt. Cox brought her back to the surface an hour later and proceeded on toward Palau. Shortly after Marion Asa had assumed the afternoon OOD watch, he was forced to dive the boat again when the radar detector indicated another plane very close. Another

Japanese bomber was spotted at 1415, flying up and down *Seawolf*'s own course line as if he knew an American sub was lurking below. "Patrolled at periscope depth until he got tired and went away," Lynch logged.

Over the radar detector, Chief Rogers was able to pick up a voice on VHF of an American flyboy in the late afternoon. From the conversations that ensued, *Seawolf*'s crew could hear U.S. aviators discussing what damage they had inflicted upon the Japanese fleet. At 1947, the SD picked up five planes closing at 13–16 miles distance. IFF showed them to be friendlies. They were then picked up on the SJ radar and tracked to be on an easterly course as they winged back toward the American carriers.

Seawolf went deep twice in the next hour to avoid individual planes which closed her position. Ozzie Lynch could not interfere with the American carrier operations, so he stayed down until 2159 before surfacing and continuing on.

Although voices and contacts indicated the American aviators had struck the Japanese, *Seawolf*'s crew could only imagine what had transpired in the late evening of 20 June 1944. They could not guess at how desperate the situation was for the young carrier aviators who were searching out their own task forces long after dark.

The ominous prospects of returning home to his flight deck long after dark vanished from Jack Bramer's mind as his Helldiver nosed over into a steep dive. Strapped in to the rear seat, he faced his twin .30-caliber machine guns aft and searched the skies for Zeros. His pilot, Lt. (jg) Albert Tavel Walraven, kept his aim on the twisting Japanese ship below as his aircraft plunged through the angry black puffs on anti-aircraft smoke.

The Japanese gunners were off their mark as Walraven released his 1,000-pound bomb and pulled out of his dive. His aim was only slightly better. His bomb exploded violently close aboard the Japanese tanker—ripping hull plates with a damaging near-miss.

Al Walraven and his gunner, Aviation Radioman Third Class John Conrad Bramer Jr., were flying with Bombing Squadron Fourteen (VB-14) from the fleet carrier *Wasp* from Task Force 58. Three task groups had begun launching 240 aircraft at 1624 on 20

June. After a handful of aborts, 226 American carrier planes had headed out against a Japanese fleet that was 300 miles from their launch position.[5]

In VB-14's ready room that afternoon, Lt. Cdr. Jack Blitch's pilots and gunners had anxiously listened to the contact reports coming in concerning the Japanese carriers. "The skipper decided that we were probably not going to go after them because they were too far away," recalled Bramer. The next instant, the speakers boomed with the order, "Man your planes."

Once en route, and with the enemy's position clarified as being even further away than originally reported, the situation became intense. "We took off with pretty full knowledge that if any of us got back it would be pure luck because our fuel would have run out," Bramer said.

Wasp's contribution to the late afternoon strike was 12 Helldivers, seven Avenger torpedo bombers, and 16 Hellcat fighters as escort. The other Navy planes were launched from *Belleau Wood, Enterprise, Lexington, Yorktown, San Jacinto, Monterey, Cabot, Bunker Hill, Hornet,* and *Bataan*. Against the Japanese invasion fleet, these aviators managed to damage to varying degrees the carriers *Zuikaku, Hiyo, Junyo,* and *Ryuho*, plus the aircraft tender *Chiyoda*, battleship *Haruna*, and cruiser *Maya*.

Jack Blitch's *Wasp* strike group was unable to find the Japanese carriers in the setting sun. With time and fuel an issue, he lead his pilots against the next prime target—a fleet of oilers and their escorting destroyers. "Our squadron decided to attack the fleet of oilers because if they couldn't refuel their ships, they couldn't make it back to Japan," Bramer explained. *Wasp*'s air group sank two tankers, *Seiyo Maru* and *Genyo Maru*, and damage a third, *Hayasui*.[6]

Recovering from his dive, Lt.(jg) Walraven took on his next challenges. "Anti-aircraft fire was thick," said Bramer, "and Japanese Zeros were all over the place." In the approaching darkness and enemy gunfire, Walraven joined up on two other SB2Cs, two TBFs, and a couple of Hellcats for the homebound leg. "We were kind of a weird group," Bramer admitted, "but we made a formation to head back toward the *Wasp*."

Japanese Zero pilots, angry over the damage to their own flight decks below, made runs on the retiring American planes. In his VB-14 diary, Al Walraven recorded their counterattack.

Bramer called me and said that three Zekes were coming in at six o'clock to tag us for the next dance. He evidently thought he was along just for the ride as I told him to get the guns ready and keep an eye on the Japs...He kept informing me of their positions and then in an excited voice two octaves higher, he said, "Zeke coming in at eight o'clock below."

After a few minutes of silence I asked him, "Well, what happened?" Then his answer, in nearly an inaudible, low, disgusted voice..."Aw, an F-6 came in and shot him down."

Tough—the poor kid had yet to shoot his guns at a Jap fighter.[7]

Wasp's Helldivers and Avengers survived the Zeros, although one Hellcat pilot from VF-14 was lost over the Japanese fleet. The remaining 34 *Wasp* pilots flew in darkness toward TF-58. Three of VT-14's Avengers ran out of fuel and were forced to ditch in the ocean, although all nine men were rescued. Bombing Fourteen's losses were appalling: eleven of 12 bombers made water landings during the night of 20 June 1944, with the loss of nine pilots and aircrewmen. Of the 13 Bombing 14 aviators who survived the sinking of their Helldivers, eight were picked up by Task Force 58 destroyers during the night. Skipper Jack Blitch and two other aviators were picked up by patrol planes two days later on 22 June.[8]

Eighty-six of the 226 planes from TF-58 were lost in the mission after darkness over the Philippine Sea, while even more were destroyed in the night carrier landings. During the next two days, 138 of the missing fliers were rescued—80 percent of those who went into the ocean. Destroyers picked up 85 percent of the downed aviators, amphibious aircraft scooped up seven others, and the balance were picked up by other vessels, including two by submarine.[9]

Lt.(jg) Walraven and ARM3c Bramer were forced to make a water entry when their big Helldiver exhausted its fuel short of the U.S. fleet. "We noticed that our engine was running rough," recalled Bramer. "Our fuel was low and we think our oil line had gotten hit. We finally made the decision to make a crash landing, which we did about 800 miles from the nearest land, which was Luzon."

The pair had not lost a plane in all of their previous missions, but their "Baby Carriage" did not survive 20 June. Their SB2C was so nicknamed by 20-year-old Walraven and 18-year-old Bramer's youthful appearances. Walraven eased up on his airspeed and set his Helldiver into the ocean with a jarring impact. "My face was pushed into the radio gear and I had a broken nose," said Bramer.

With blood streaming from his face, the radioman immediately scrambled out onto his aircraft's wing with their rubber life raft. "I noticed Al wasn't out at all," he said. Walraven was still seated in the cockpit, stunned from the crash landing. "I scurried up there to yank him out," Bramer related. "I'm not a very big person, but I guess I got all of the strength I needed at that time."

Bramer held onto his pilot as "Baby Carriage" sank out from under them. The *Wasp* aviators inflated their yellow life raft and scrambled aboard. The night was dark and cool as they bobbed in their lonely raft in the Philippine Sea. "It was very, very cold," said Bramer. "We just held each other as tight as we could to get warm. The next day, when the sun came up, we were warmer than we wanted to be."

Sunrise brought new hopes to the two men after a lonely vigil in the ocean. "Our first mutual thoughts were to give thanks to God that we were alive and I know that several times during the night I found myself praying that we would be rescued sooner or later," wrote Walraven.[10]

It was hours after the morning sun began baking the fliers that they had their first glimmer of hope. Al Walraven wrote that while he and Bramer "were both trying to absorb all the sun possible, I spotted a plane. It was a patrol bomber that had been sent out to search for us." Soon thereafter, two Hellcats appeared for a while, but they also departed—leaving the men cooking under the hot morning sun and praying for a miracle.[11]

Marion Asa made quick time in decoding the latest dispatch received from Admiral Lockwood at 0620 on 21 June. *Seawolf,* like other Allied subs in the Marianas area, was directed to search for downed U.S. aviators from the previous evening's strike.

"Since we were obviously on the trolley line for returning aviators yesterday evening," Lynch decided to proceed toward that point and search toward the given focal point. He put three engines on the line and ordered his lookouts to keep a sharp eye for liferafts.

Ben Rogers announced a radar contact at 0833 that did not trigger the IFF, indicating an enemy aircraft. Lt. Cox's lookouts sighted the Japanese plane ten minutes later and he made a quick dive. Upon surfacing again, American voices were once again picked up on the VHF radio. Chief Bekke's radio gang then picked up a new dispatch from ComSubPac which directed the *Wolf* to cease her search for aviators and to proceed on her assigned mission to Palau. Apparently, the other subs and surface vessels in the area were deemed sufficient to search for the surviving flyboys.

Ten minutes after receiving these orders, Ozzie Lynch was faced with a moral dilemma. Bekke's radiomen picked up an intelligible voice radio call about a life raft that had been spotted. Using her IFF, *Seawolf* was able to get friendly responses at 17, 18 and 20 miles out. On the bridge, Lynch, Exec John Hess and OOD Bob Cox discussed their chances of breaking off their mission and making a quick search for the life raft that the carrier pilots had spotted.

"After some debate, decided to disobey orders," wrote Lt. Cdr. Lynch. "We could reach Palau no sooner, and felt that we would be of definite rescue service. Reversed course and tried to contact the friendly planes."

Chief Bekke had radiomen James Call and James Johnson try to open up communication with the carrier pilots, who had located survivors. *Seawolf* ran back toward the position during the next 90 minutes until sighting two Hellcats and a Kingfisher at 1100. The planes circled for identification as quartermaster Alfred Kuehn flashed the proper signals from his gun. "Tried to make them understand that we could hear their voice radio but to no avail," wrote Ozzie Lynch.

The Hellcats moved on with their Kingfisher and were heard to locate four survivors. "A Kingfisher must have rescued them as none of the planes returned to get us," wrote Lynch.

At 1225, the radio suddenly crackled with the voice of another pilot saying that he had another life raft in sight and would go get the submarine. The pilot soon appeared on *Seawolf*'s port bow and offered to direct her toward the downed fliers. With an hour's run to the reported location of the raft, Ozzie Lynch ordered all four engines on the line and *Seawolf* was off to the races.

The deadly *Seawolf* was now on a new mission as an angel of mercy for men at peril on the sea.

Jack Bramer assumed the vessel he and Lt.(jg) Walraven had spotted on the horizon must be Japanese. As the ship moved in their direction, "we dove out of the life raft and swam away from it." Minutes later, however, the Helldiver crew saw a heart-warming symbol of their own nationality.

"We saw the American flag and swam back to our life raft," Bramer said. Waving enthusiastically, they soon recognized friendly faces lining the bridge and deck of an approaching U.S. submarine. *Seawolf* cut her engines and coasted up alongside the sunburned aviators at 1315.

On *Seawolf*'s bridge, Ozzie Lynch and John Hess monitored the rescue efforts as *Seawolf* bore steadily down upon the bright yellow life raft. Their lookouts kept a vigilant watch for trouble in spite of the presence of two U.S. Hellcats overhead who were guarding the rescue of two of their own. Chief of the boat Bill Reiland oversaw a deck team who stood ready to haul the fliers aboard.

As the life raft bobbed up alongside the *Wolf*'s starboard side, Lt. Cdr. Lynch jestingly hollered down to the fliers, "We're on a war patrol for about a month. You can come with us or wait for something else to come along."

"We'll come along with you!" Bramer and Walraven replied.

The gunners' mates were topside to assist with the rescue, since *Seawolf* had no bosun's mate aboard. "The gunners' mates kinda took on that role," said Rex Mickey. "We just grabbed the aviators' hands and helped them up out of their rubber raft. We took their firearms and locked them down below in our ordnance room."

Arthur Pierce, the Navy photographer assigned to *Seawolf* for her thirteenth patrol, snapped pictures as the fliers were thrown a line and hauled alongside the *Wolf*. Another shipboard camera appeared to help document the mission of mercy. Ozzie Lynch and John Hess went down to personally greet the fliers and welcome them aboard.

"They were questioned and said the Jap force encountered was about 200 miles away," wrote Lynch. Walraven "could give no further information on [the] whereabouts of any others." Lynch had his deck crew deflate the life raft and stow it below. The aviators were helped into dry clothing and escorted to the pharmacist's mate for a thorough checking over.

The two Hellcats which had stood by the rescue were relieved by two others and they then started a larger search of the area. *Seawolf*'s radio gang sent a dispatch to Comsubpac at 1420, telling of the rescue of the two *Wasp* aviators.

Pharmacist's mate Bill Hadley tended to his new flyboy patients. "He patched up my broken nose as best he could, but there was not much he could do with it," said Jack Bramer. Their exposed skin—particularly their faces and hands—were badly sunburned. Radioman Bramer was assigned a bunk in the forward torpedo room, while Lt.(jg) Walraven joined the officers in the wardroom for the balance of the patrol. Both managed to do a little trading with the *Seawolf* crew.

"They wanted my .38 revolver and Al's, and they bargained for our life raft," said Bramer. "So, we did some swapping. I got some new clothes out of the trading."

With his two new passengers safely below, Lt. Cdr. Lynch wrote in his log: "Decided it was about time to do as we were told and headed for Palau, hoping the violation of orders was justified."

For *Wasp*'s Air Group Fourteen, the rescue was certainly justified. It took more than a week for news of the rescue to reach *Wasp*. In his diary on the night of 30 June, VB-14's Lt.(jg) Ray Heiden scribbled, "Tonight we had the best piece of news in a long time. Al Walraven and his gunner, Bramer, were picked up by the submarine *Seawolf*. It certainly is good news."[12]

The *Wolf* moved into the Palau area on 22 June. "In that ComSubPac's verbal instructions were to accomplish the mission quickly," wrote Ozzie Lynch, "decided to enter first the *Albacore*'s area as she was busy with lifeguard duty."

Seawolf began her special photo reconnaissance of the Palaus on the morning of 23 June. She slid in at periscope depth and documented the two main ship channel entrances. At 0840, OOD Bob Cox sighted a *Chidori* torpedo boat heading into the ship channel. *Seawolf* went deep and ran silently as the *Chidori* passed directly overhead, entering the channel. *Seawolf* sighted aircraft on nearly every observation that day. "Picture taking was disappointing in that the reef kept us at too great a distance from the land," Lynch wrote.

LIFEGUARD LEAGUE: 13TH PATROL

Lt. Cdr. Lynch ignored orders and instead helped search for Navy pilots lost during the Battle of the Philippine Sea in June 1944. *Seawolf* rescued Helldiver fliers ARM3c Jack Bramer and Lt.(jg) Albert Walraven from the carrier *Wasp*. Photos courtesy of John C. Bramer Jr.

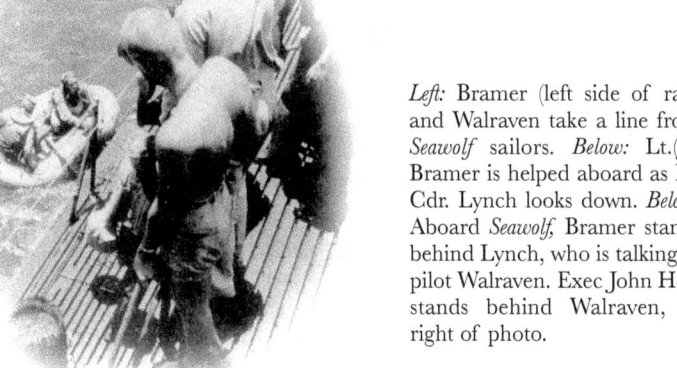

Left: Bramer (left side of raft) and Walraven take a line from *Seawolf* sailors. *Below:* Lt.(jg) Bramer is helped aboard as Lt. Cdr. Lynch looks down. *Below:* Aboard *Seawolf,* Bramer stands behind Lynch, who is talking to pilot Walraven. Exec John Hess stands behind Walraven, to right of photo.

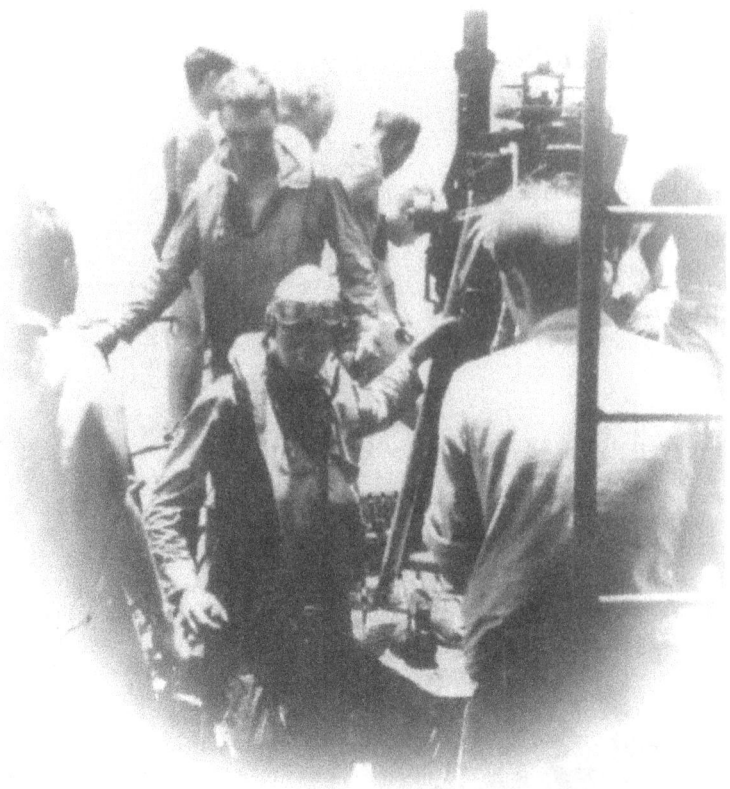

Jack Bramer *(center)*, is followed by Al Walraven as they head below. U.S. Navy photographer Arthur Pierce has his back to the ladder taking additional photos. *Photo courtesy of John C. Bramer Jr.*

The APR-1 detected radar units during the night, two of which were definitely shore-based at Palau. During the early morning hours of 24 June, the *Wolf* shifted to the east coast of Peleliu, where the light conditions were unfavorable for good photography. Lieutenant Cox sighted another *Chidori* torpedo boat at 0809. He remained an hour, but did not echo range.

The airfield at Peleliu started operations at dawn and a continual stream of aircraft was observed to take off and land. Betty bombers were seen to be heading in the direction of Saipan. The bad light prevented any photos during the morning, so *Seawolf* went through the slot between Peleliu and Angaur islands during the noon hour.

With better afternoon light, *Seawolf* was able to photograph the west side of Peleliu and the remaining areas on the west side of Palau.

At 2315, an aircraft was sighted which flew nearly overhead on a westerly course while John Hess and Alfred Kuehn were shooting the stars for their navigational fix. The SJ picked up a radar contact at 7,000 yards, which was believed to be a patrol boat or submarine due to the short range. This was avoided, but an aircraft flew over and dropped a flare near the *Wolf* 15 minutes later. "The answer is not apparent unless his aim for us was bad," Lynch recorded.

Seawolf worked the east coast of Peleliu on 25 June, and enjoyed more favorable light conditions. Ozzie Lynch found this day to be the "most nerve-racking so far," as aircraft circled right over his ship as they prepared to land ashore. Three inter-island sampans were avoided during the day. More photos were taken in the afternoon of the west coast of Angaur Island. Close examination of the prints revealed what appeared to be a concealed radar unit ashore.

The *Wolf* worked the area from Denges Passage to Augulpelu Reef the following day, but the presence of a reef prevented her from getting close-up pictures. An *Otori* gunboat was sighted at 1045 and *Seawolf* went deep as this patrol craft churned by overhead and kept going.

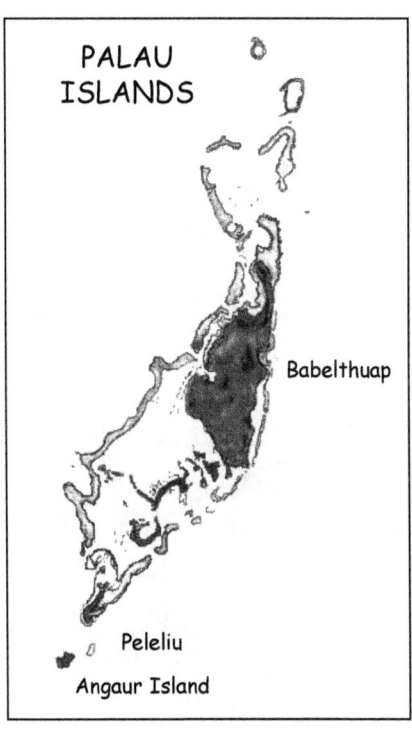

During the day on 27–28 June, the *Wolf* photographed Babelthuap, avoiding several sampans that were sighted inside the reef. She continued photographing the remainder of her assigned area on 29 June and then retired eastward. At 0610 on 30 June, her radiomen sent a dispatch indicating the completion of her mission. The Japanese were on to her presence, however, as Lynch recorded:

"It was decoyed out of us by Japs who very promptly gave an R. Sent it again to CTF 72. This time both the Jap and CTF 72 gave an R."

OOD John Van Andel sighted an aircraft at 1003 and dived for the remainder of the morning. The Navy photographer, Arthur Pierce, used this time to finish making prints of his negatives, which

(Facing pages) Chief Ben Rogers drew a series of cartoons to depict *Seawolf*'s adventures during her thirteenth patrol. Assigned to reconnoiter the Palau Islands, Lt. Cdr. Lynch brought his submarine dangerously close to enemy harbors to take periscope photos. Rogers' cartoon depict *Seawolf*'s rescue of two downed aviators, film being developed in the wardroom, and Japanese planes and patrol boats passing casually by the exposed periscope. *Courtesy of Charles Hinman, USS Bowfin Submarine Museum.*

were able to dry while submerged. *Seawolf* surfaced at 1435 and received the welcome orders an hour later to proceed to Majuro at best speed for voyage repairs. Lynch put on three engines and continued clearing the area.

Jack Bramer and Al Walraven took to life aboard a submarine. Bramer passed time helping out in the radio shack and found that submarine life came with some "interesting experiences" each day.

> They taught both me and Al every aspect of the submarine controls. We stood there in the control room every once in a while and practiced what we learned. We got to try out the bow and stern planes. The diving officer [Clary John] gave us a shot at that.
>
> We encountered all kinds of things. In one instance, they announced the command to dive and there was an inexperienced guy on the flood planes and we headed straight for the bottom. Somebody got to the controls and reversed 'em. We went straight up and broached, which means half of the submarine broke water while the Japanese were flying around trying to sink us.
>
> At night, we had one movie aboard ship. I forget what movie it was, but Gene Tierney was in it. We saw it over and over again. They divided the crew in half and showed the movie in the forward torpedo room. The other half of the men would move aft to keep us in balance. When the movie was over, then we'd trade places.

The next aircraft contact came on 2 July, when two different planes sent the *Wolf* diving for cover. She had one more aircraft contact the following day, but enjoyed a return free from enemy attack. While routining torpedoes on 4 July, a high pressure leak was discovered in one of her Mark 14s, which torpedo officer Bob Cox marked for repair upon docking in Majuro.

During the patrol, Lynch had a sailor with a tooth problem, as he related to Admiral Lockwood later. In making his rounds, he learned from Doc Hadley that a red-haired youngster named "Red" from the forward torpedo room was in misery with an abscessed tooth.[13]

Hadley decided the tooth was poisoning the torpedoman's system and needed to come out. Red's face was swollen abnormally. The problem was that Hadley had no true extractors. From the auxiliary gang, he managed to acquire a pair of long-nosed plumber's pliers. The torpedoman protested about the surgical tool, but as the day progressed his pain increased.

Before the day was out, Hadley had performed his extraction of the tooth. The torpedoman was given some medicinal brandy as a pain-killer and the tooth was yanked out. "Red" opted to keep the tooth as a souvenir, hoping to tape it to the next warhead that *Seawolf* fired.[14]

Seawolf rendezvoused with her escort vessel at 1245 on 6 July while approaching Majuro in the Marshall Islands chain. Another vessel approached and used a throwing gun to send over a dispatch pouch containing essential navigational information on Majuro. Waiting until after dawn on 7 July, the *Wolf* arrived at Majuro and completed her thirteenth war patrol.

She had taken 150 rolls of film of the Palau islands. She also brought back key data on weather, navigation, currents, surf conditions, radio towers, and enemy radar installations. *Seawolf* had moved in close to photograph all beaches and even had photos to show which beaches were covered with barbed wire on Peleliu.

Although *Seawolf* had not conducted a torpedo patrol, her photo reconnaissance mission was equally important to Allied efforts. Admiral Lockwood felt that Lt. Cdr. Lynch's "mission was completed efficiently and successfully." He designated this war patrol as successful for the Combat Insignia Award, particularly in light of the two aviators that *Seawolf* had rescued from the Philippine Sea.

Ozzie Lynch was awarded the Legion of Merit for "maneuvering his ship close to the enemy-held shoreline to obtain photographs of Japanese installations" and for rescuing the two airmen. Lynch in turned recommended three of his men—Exec John Hess, Lt. Bill Smith, and CPhM Bill Hadley—for Bronze Stars.

Seawolf's two special passengers were eager to get back to their carrier. "They left our boat and never looked back," laughed Rex Mickey. "I mean they lit out. I think these poor guys were claustrophobic."

In reality, Al Walraven and Jack Bramer were eager to get back to the carrier *Wasp* and their part of the war. When *Seawolf* docked, both were wearing clothes acquired from the *Wolf*'s crew. Their flight suits and pistols had long since been traded off.

"We decided that we had had enough of the submarine service," Bramer later related. Ironically, soon after returning to *Wasp*, one of their first bombing missions was against Palau. They had a new SB2C Helldiver, nicknamed "Baby Carriage" in honor of their previous aircraft. "We'd been there in a sub, and then we returned to Palau in a plane."

Bramer and his pilot decided they much preferred seeing Palau from the air. "We used to argue with the *Seawolf* crew," Bramer said. "They couldn't imagine how anybody in their right minds would go up and fly and get shot at. We couldn't figure out how anybody in their right minds would get on a submarine and get depth-charged!"

USS *Seawolf* Thirteenth Patrol Summary

Departure From:	Pearl Harbor
Patrol Area:	Philippine Sea/Palau
Time Period:	4 June—7 July 1944
Number of Men Aboard:	87: 78 enlisted and 9 officers
Total Days on Patrol:	32
Number of Torpedoes Fired:	0
Special Missions Accomplished:	Intelligence photography of Palau Islands; rescue of two *Wasp* aviators
Return To:	Majuro

16

Guerrilla Warfare

Fourteenth Patrol   *1–23 August 1944*

Seawolf was an old boat by World War II standards, and even the recent California overhaul did not prevent her engineers from fighting numerous problems during patrol. Once secured at Majuro on 7 July 1944, she would undergo voyage repairs for issues encountered during her thirteenth run.

Submarine Division 141 replacement crewmen and specialists from the tender *Bushnell* came aboard to affect repairs. The engineers replaced and repaired a lube oil pump on the auxiliary engine, overhauled the No. 3 and No. 4 main engine governors, and repaired freon leaks in the refrigeration spaces. *Seawolf* also spent one day in a floating drydock at Majuro to replace a section of negative tank flood valve linkage which was found to be faulty.

The crew received a little time ashore while the repairs were being made and they made the most of it, as usual. Rex Mickey joined some of the torpedo gang in enjoying their gilley juice, torpedo alcohol that was cut down to a less toxic level. "We didn't have a still," Mickey said. "We just poured it through a piece of bread and then drank it."

Although kept highly classified, this special booze was never partaken of while on patrol. "We saved it up until we got in from patrol," Mickey related. The sailors found that is was also highly flammable. "I watched this one guy drinking it. He would heave it back up at his cigarette and blow a fireball up."

Lt. Cdr. Lynch took *Seawolf* out for one day of training to insure the seaworthiness of her repairs. Lynch then received a dispatch on

13 July that his boat was expecting new orders. During the period in Majuro, several transfers took place. Lt. Clary John went on to *Bowfin*, where he would soon serve as her Exec. Lt. Cdr. John Hess would take command of *Pomfret* for her third patrol. In their places, the *Wolf* received Lt.(jg) Ralph Van Dorne Miller (USNR) of Seattle and Lt.(jg) Paul Doane, USN, of Waterford, Connecticut.

Seawolf also received a PCO, Lieutenant Commander Albert Marion Bontier, a native of Missouri. A 1935 Academy man, Al Bontier had a wife back in White Plains, New Jersey. One of his contemporaries described him as "slim and boyish, quick witted and talkative, with thinning, reddish hair that was exposing more and more cranium."[1]

During the early days of the war, Al Bontier was awarded the Silver Star for helping *Spearfish* to sink two ships on her third patrol and for participating in a mission on 3 May 1942 to rescue the last Navy personnel to escape from Corregidor before its surrender to Japanese forces. After making Exec on *Spearfish*, Bontier also later served as Exec of *Sculpin* for one patrol in May–June 1943. He went on to new construction and put *Razorback* (SS-394) into commission on 3 April 1944.

Lt. Cdr. Bontier had the poor luck of running his *Razorback* aground off New London, Connecticut, during his training period. After an official investigation, he was relieved of command and placed back in the prospective commanding officer pool. When he joined *Seawolf* in July 1944, he was assigned as Ozzie Lynch's PCO and also took on the role of Exec with the departure of John Hess.

While at Majuro, several men received promotional notices, including Donald Naze, from TM1c to chief torpedoman. He wrote a letter to his parents back home in Rhode Island on 10 July: "I am now a chief torpedoman, as of June 15th, so you all have a treat on me and send me the bill. Ha! Ha!"[2]

Four enlisted men were transferred, including yeoman John Fox to *Bushnell*. In his place, *Seawolf* received Y2c Allen Bingham from Omaha, Nebraska, on 9 July. "I caught the *Seawolf* in the Marshalls when I came aboard as a relief yeoman from the sub tender *Bushnell*," he said. "One of *Seawolf*'s yeoman had gotten sick." As the new guy, Bingham joined the watch schedule in addition to handling his normal duties of keeping up the ship's paperwork, typing patrol reports, and handling transfers. He considered the *Wolf*'s senior officers to be

Two of the three new officer *Seawolf* received prior to her fourteenth patrol were Lt.(jg) Paul Doane (left) and Lt. Cdr. Albert Marion Bontier (right). Al Bontier soon took command of the *Wolf* and served as her final skipper. *Images courtesy of Charles Hinman, USS Bowfin Museum.*

solid. "Captain Lynch was very good and Bontier was the Executive Officer when I reported aboard."

In addition to new men, commissary officer John Van Andel managed to acquire an ice cream freezer for *Seawolf*'s galley. Fresh foods had been somewhat lacking from *Bushnell*, but the ice cream would be thoroughly enjoyed on the upcoming patrol.

On 16 July, *Seawolf* received dispatch orders directing her to report to Darwin, Australia, for duty with the Seventh Fleet. She departed the Marshalls on 17 July and sailed for Darwin, proceeding on the surface except when dodging aircraft contacts. Ozzie Lynch conducted trim and training dives en route, as well as fire control drills.

As would be expected in any managerial relationship, some of the enlisted men found some officers tougher than others to deal with. For Rex Mickey and his pals, it was Lieutenant Smith.

> He was kinda smart alecky, and we didn't care for him. When he had the deck, we would work on him some. He would holler, "Dive!" and we'd take her down a little too deep. When he ordered the boat to be run up, we'd let her run up a little too high. If he called for, "Up 10 degrees," we'd take her up 15 degrees. Then, pretty soon, the captain

would come down and chew him out for poor depth control. Poor guy!

En route, *Seawolf* also crossed the equator and held her second polliwog initiation of the war. Chief of the boat Bill Reiland's crew of pirates took over the ship from Lt. Cdr. Lynch and dived while the men were inducted in the mystic and ancient orders of the deep.

Several of the crew even tried to escape the hazing. Dallas Malone, Rex Mickey, and John Sadler blocked off the forward torpedo room

En route to Australia before the start of her 14th patrol, *Seawolf* held her second polliwog initiation while crossing the equator. Below is a sample of one of the summons handed out to the polliwogs. *Courtesy of Rex Mickey.*

```
TO WHOM MAY COME THESE PRESENT:    GREETINGS AND BEWARE

WHEREAS:  THE GOOD SHIP SEAWOLF IS ABOUT TO ENTER OUR
DOMAIN CARRYING A SLIMY AND ABOMINABLE CARGO OF DRAFT
DODGERS, FOUR F'ERS, PLOW DESERTERS, MARKET STREET COMM-
ANDOES, CHIT SIGNERS, ASPHALT ARABS, SQUAW MEN AND THE
LIKE LANDLUBBERS FALSELY MASQUERADING AS SEAMEN, OF WHICH
LOW SCUM YOU ARE A MEMBER HAVING NEVER APPEARED BEFORE
US; AND

BE IT KNOWN THAT WE HEREBY SUMMON AND COMMAND YOU
     Mickey G M   USN     TO APPEAR BEFORE THE ROYAL
HIGH COURT ON   July     AT SUCH TIME AS SUITS OUR
PLEASURE AND TO ACCEPT WITH GOOD GRACE THE RIGHTEOUS
PENALTIES TO BE INFLICTED ON YOUR VILE CARCAS.

CHARGES:  Failure to show Mr. Smith how
to fire the 50 cal. machine gun.


DISOBEY THIS SUMMONS UNDER PAIN OF OUR SWIFT AND TERRIBLE
DISPLEASURE.
GIVEN UNDER OUR HAND AND SEAL
ATTENDANT FOR THE KING               NEPTUNUS REX
     DAVY JONES
```

hatch and refused to open it. With *Seawolf* submerged, the shellbacks had no way to get at the unwilling polliwogs. Similar incidents had occurred on other boats during the war. When *Bowfin* crossed the line in 1943, motormac Thomas Malley and other polliwogs had barricaded themselves in the forward torpedo room. *Bowfin* was running surfaced, however, and the skipper let the shellbacks enter through the topside hatch.[3]

Malone, Mickey, and Sadler finally had to give up their hideout on *Seawolf*, as Mickey related. Lt. Cdr. Lynch finally convinced them that they "had to come out sometime, so you might as well give up so we can go on about our business."

Brought forward, Mickey recalled that "they worked Sadler and I over pretty good." He was summarily brought before Chief Reiland—dressed to the part of Davy Jones—and was forced to bow down to kiss his feet. "I bit into his big toe, and they let me have it! They had shockers and socks filled with stuff they beat me with."

"They go through all kinds of antics to cure the polliwog," said yeoman Allen Bingham, as he took the hassling in stride. "For me, the worst was some kind of meal they put in the back of my mouth and then spurted water in it." The initiation was all in good fun, and by the late afternoon all of the no-good polliwogs had been transformed into true shellbacks.

Seawolf arrived at Darwin at 1158 on 30 July 1944, with her SJ radar and an auxiliary engine out of commission. Those men topside found the expanse of the harbor to be stunning. Rex Mickey recalled:

> When we pulled into Darwin, it was amazing to see all the ships in that harbor. You could still see the damage from where the Japanese had bombed the place early in the war. I think they had more ships in that harbor than I ever saw at Pearl Harbor. The Australians were good to us submariners because we helped keep off the Jap ships.

Shortly after docking, there was a brief ceremony at 1330, during which Lieutenant Commander Bontier relieved Commander Lynch as commanding officer of *Seawolf*. Ozzie Lynch had new orders to fly back to Pearl Harbor to take command of *Skate* (SS-305), which had just completed her fifth patrol. Al Bontier would now have his second chance at submarine command. Third officer Bob Cox moved up to executive officer and Bill Smith became engineering officer.

The following two days at Darwin, *Seawolf* underwent repairs, including the installation of a new transmitter for her SJ radar. Due to the lack of a spare lubricating oil pump, her auxiliary engine remained out of commission. Having received word of a new special mission, Al Bontier had his torpedomen unload 12 torpedoes in skids. The skids themselves and other excess torpedo equipment was also offloaded from *Seawolf*.

This photo series shows *Seawolf* making a test dive on 8 May 1944 off California following her Hunter's Point refit. Her new deck gun is just beginning to submerge in the third and fourth images. *Official U.S. Navy photos.*

The crew began loading special cargo on 31 July. Jerry Bekke's radio gang also oversaw the installation of a special model 38Z radio transmitter and receiver for experimental purposes. She was able to successfully communicate with a station ashore during practice. The radio transmitter and receiver used the SJ radar's antenna and it was found to have a limited useful range of no more than 100 miles. Within this, radio stations could be successfully picked up. It was found to be a good device for communicating while submerged and for communicating with a station ashore while engaged in special missions.

The following day, yeoman Allen Bingham removed two names from *Seawolf*'s sailing list. Motormac Bob Devitt was sent ashore to the Naval hospital for medical attention on August 1, and would not be released in time to make the *Wolf*'s departure. Officer Bill Smith was also transferred on this day, with no mention by the skipper in his patrol report. At least one crewman later related that a junior officer had a confrontation topside with Smith, with all crew being ordered to stay below. Not impressed with his attitude, Lt. Cdr. Bontier transferred Smith—who was on the losing end of this conflict—ashore.

Although an offensive that could end a man's career, it was not uncommon for submariners to settle their own scores while the boat was between patrols. Fighting was absolutely prohibited while at sea, but occasionally one had to look the other way to allow a situation to settle itself. Prior to *Gudgeon*'s sixth patrol, her Exec had challenged one of his crewmen to take him on in boxing topside if he wanted to avoid punishment for the whole crew. One young fireman obliged and pounded the XO. Such episodes went unreported in the Silent Service. Any resulting injuries would be attributed to a fall or bump in clearing the hatches. In the case of *Seawolf*, Bill Smith—twice decorated for his service—moved on to help commission *Carp* and he would serve with distinction as her Exec.[4]

The wardroom spirits would be much improved, but *Seawolf* was unable to find another available officer on short notice following this confrontation. She would go out with eight officers instead of her allotted nine. Lieutenant Paul Doane stepped up to handle the duties of engineering and diving officer for the fourteenth patrol.

Seawolf would be departing on a top-secret mission once again. The detailing memorandum and secret operations order would not be disclosed to her crew until well underway, but the men could certainly guess. On the afternoon of 1 August, she received aboard one Filipino officer, one non-commissioned officer, and ten men for transportation to the Philippines. She would also land supplies on a special mission to Tawi Tawi Island of the Sulu Archipelago.

The *Wolf* was underway at 1730 from Darwin, proceeding without incident through the safety lane north from Australia. Paul Doane made a trim dive that evening and she returned to the surface for her run. Ben Rogers used the SD radar intermittently during the night as the *Wolf* zigzagged in the moonlight. Al Bontier conducted training and trim dives en route. A pip appeared at 2031 on the SJ radar on the night of 2 August. It appeared at 9,000 yards and closed rapidly. *Seawolf* submerged and tracked the plane in to 3,000 yards on her port beam as the antenna went under.

Seawolf's guerrillas enjoyed the ice cream freezer while at sea and the movies shown in the torpedo rooms. Bontier found them to be "most cooperative." Allen Bingham recalled that sea life did not agree with all of them. "They were Filipinos and they bunked in the forward torpedo room," he said. "A couple of them got seasick, but they cleaned up their own messes."

Seawolf continued on the surface through the next two days, diving only for aircraft contacts. Just after noon on 5 August, a lookout spotted something on the horizon that initially "appeared to be everything from a sampan to a submarine," as Lt. Cdr. Bontier put it. In the end, it turned out to be a floating tree stump. Calling gunner's mates Francis Ballard, Rex Mickey, and their teams to the deck, he opted to have some target practice. Beginning with a few rounds from the new 4-inch deck gun, Bontier then allowed his gunners to fire their 20mms, .50-calibers, and tommy guns on the stump.

The *Wolf* cleared Sibutu Passage while submerged during the day on 6 August, and then surfaced to head for the point of her first special mission. Plans had been approved to send a reinforcing party to Captain Frank Young on Tawi Tawi Island and also a party to northern Palawan Island.

Captain Young, an America mestizo, had been operating on Tawi Tawi with Lt. Col. Alejandro Suarez. Frank Young had brought the first information of large guerrilla groups existing in central Luzon

to attention in a conference in Brisbane in December 1942. He had been stationed under Colonel Claude Thorp in Pampanga. He left Thorp in July 1942 and traveled via Bicols, Samar, Leyte, Cebu, and Negros to Panay. He there joined German civilian Albert Klestadt and came to Australia via Zamboanga, arriving in Darwin on 12 December 1942. They brought important information of enemy and guerrilla activity in the areas through which they had passed.[5]

On 27 May 1943, Captain Jordan A. Hamner and Lt. Young were dispatched to Mindanao on a special mission. They landed at Tukuran, Zamboanga, established a radio near Zamboanga City, and headed for Tawi Tawi. On that island, 47-year-old Sulu guerrilla commander Alejandro Suarez was contacted. Lt. Col. Suarez had arrived at Tawi Tawi in February 1943 and began working with local guerrillas from Bato Bato. A radio was established there and some supplies were brought to the local guerrillas.[6]

Captain Hamner was evacuated in early 1944, but Young, promoted to captain, was to receive more supplies and reinforcements from *Seawolf*.

The *Wolf* dove at 0433 on 7 August and spent the day reconnoitering Tawi Tawi Island. At noon, a Japanese landing craft was spotted hugging the coastline, and photos were taken of it through the periscope. Shortly thereafter, three more small vessels passed through, one appearing to be a tugboat and the other two small inter-island freighters weighing only about 1,000 tons. "Began to wonder how popular our spot actually was to the Japs," mused Bontier in his report.

Nearing the point of her scheduled evening rendezvous, *Seawolf* sighted the proper security signal on the Tawi Tawi beach at 1530. Bontier waited until after dark to surface about two miles off the beach in a rain squall. Silently, the *Wolf* closed her spot on battery power. When radar range indicated one mile to shore, the lookouts could see the hill top previously selected as their reference point.

Shortly after 1900, *Seawolf* was about 500 yards off the beach of Tongehatan Point on Japanese-held Tawi Tawi. The rain squall kept her well secured. After flashing the proper recognition signals, two men on shore began paddling out toward the submarine. As the rain eased up to a drizzle, Marion Asa and his deck hands were able to shout over to the approaching Allies.

Captain Frank Young and his Filipino guerrilla assistant, Alejandro Suarez, came aboard *Seawolf* at 1930. Lt. Cdr. Bontier wrote that Suarez was "armed with about everything he could carry, including some wicked-looking knives. He could barely get down the hatch with all his paraphernalia."

Bontier had arranged with Young that his four boats were to come alongside, one on each side forward and aft. The *Seawolf* crew would then handle loading each boat so the local guerrillas aboard could paddle them ashore for offloading. The first boat moved alongside at 1938 and the crew commenced loading it with the supplies that had been brought to supply the spies. After a short period, the *Wolf*'s crew found that they could load one boat in ten minutes or all four boats simultaneously in about 20 minutes. The longest part of the mission was waiting for the guerrillas to paddle the 500–700 yards to the beach, unload, and paddle back to the *Wolf*.

Cooks Michael Jurnic, Cal Jancik, and Wayne Cotton had prepared extra sandwiches and snacks for the hungry special forces men while the loading proceeded. "Captain Young meanwhile was below filling the Diving Officer [Paul Doane] with tall tales and filling himself with a large plate of sandwiches," wrote Bontier. "He looked half-starved."

Lt. Doane kept himself informed of the cargo unloading progress. As each ton of supplies was offloaded, he compensated the water in his tanks to keep *Seawolf* neutrally buoyant. He kept her close in trim in case she should be surprised and have to make a quick dive.

The moon rose at 2112 but remained obscured behind the clouds, much to the satisfaction of all. A submarine crew sitting still on the surface just hundreds of yards off enemy soil was not a comfortable lot. At 2150, the weather suddenly cleared, leaving *Seawolf* exposed in a brilliant moonlight. Lt. Cdr. Bontier opted to start an engine to charge his batteries, "knowing that we could now be seen much farther than we could be heard."

By 2205, the crew had completed unloading nine tons of cargo, which had required seven boat trips. Captain Young and Suarez thanked the *Seawolf* crew and rejoined their local friends to row back ashore to continue their all-important recon work. The reinforcing party to Tawi Tawi was led by Lt. Konglan Teo, Sergeant Marcuano R. Daelto, and four other guerrillas. Teo's men landed successfully with weather observation equipment and supplies to run a coast-watching station for the guerrillas.[7]

Bontier then backed his *Seawolf* away from the beach and got underway for the spot of his second special mission. Paul Doane made a trim dive to check his buoyancy after compensating for the nine tons of ballast that *Seawolf* had lost. Doane finally breathed a little easier as his submarine moved into deeper waters—at least temporarily.

"Masts on the horizon!"

Lt.(jg) Ed Szendrey quickly swept the horizon in the indicated direction and had the helmsman swing to the proper course. En route, *Seawolf* moved through an area containing many floating oil drums, likely the flotsam from a maru which had gone to Davy Jones' locker. The masts on the horizon soon proved to be more of the floating debris.

The *Wolf* dived for a radar contact 20 minutes later, but was quickly back on the surface en route to Palawan. Lookouts sighted what Al Bontier deemed to be "a peculiarly shaped and gaudily painted drifting buoy" at 1105. Once again, he called gunners Ballard and Mickey to the bridge to sink it with their .50-caliber machine guns.

Seawolf was forced down two more times by radar contacts that closed her position. After dark, she completed her run in toward the beach on the northern end of Palawan Island. She remained submerged throughout the day on 9 August, studying the area of her special mission. Navigators Bob Cox and Alfred Kuehn found that their available charts did not well cover the conditions within the 10-fathom curve. Try as they might, they could not find a better approach area on their older charts than the spot which the *Wolf* was studying. So, the landings would be made at Pirata Head, Palawan.

Seawolf's second mission involved landing Master Sergeant Eutiquio B. Cabais and five of his guerrillas—Sergeant V. C. Goloyugo, Sergeant J. Cuteran, T/4 T. E. Vergara, T/4 L. Marquina, and T/5 Dagandan. Cabais' Palawan party was to set up position on Dumaran Island for the purpose of establishing communications with North Palawan and the Cuyo Islands and to obtain intelligence on enemy activity in adjacent areas. They were to particularly monitor Puerto Princesa, the Malampaya Sound anchorage, and the sea passage to the Cuyo group.[8]

Lieutenant Doane brought the ship to the surface at 1922 that night and closed the shore on battery power. He kept her trimmed down, with the decks barely awash to minimize her view to watchers ashore. Master Sergeant Cabais and his guerrillas brought their two seven-man rubber boats topside and inflated them. Their thwarts were removed to allow more room for cargo. Lt.(jg) Marion Asa's men also inflated *Seawolf*'s own two-man boat. Lt. Cdr. Bontier had the men hold until he could get a favorable report from the beach before breaking out the cargo.

At 1945, *Seawolf* anchored in 5 fathoms of water (30 feet) with her bow a mere 450 yards from the shore. The tide was flooding at about

Guerrilla Warfare

1 knot, tending to swing the *Wolf*'s stern toward the beach. Bontier put Lt.(jg) Asa and S1c Jack Kenney in charge of the first boat. They boarded with four of the Filipino soldiers with 100 fathoms of 21-thread Jute, waterproofed line, in the first boat, intending to establish a hauling line so that the boats would not have to be paddled. "The line was a dismal failure," Bontier wrote.

Marion Asa's men found that the waterproofed Jute simply would not float as manila line would. The line sank as it was payed out, fouling on coral, because the rubber boat simply could not take the line to the beach. Bontier recorded:

> Considerable time was wasted attempting to utilize the line, because I was convinced that not only would it be a faster method but it would be safer than to try to paddle the rubber boats against the current. After a second boat also failed to run the line, I was forced to abandon the idea and decided to paddle the cargo ashore. Actually, this method worked very well, and no difficulties were experienced in controlling the boats in the current, which, however, never exceeded about one knot.

While Marion Asa's first boat was attempting to run the line ashore, several lights appeared some distance down the beach to the north. "These lights considerably dampened the enthusiasm of all the Filipinos in the boat about this landing operation," Bontier noted. Asa and Jack Kenney finally convinced themselves and the soldiers that the lights were actually caused by friendly natives. They managed to reassure the Filipinos that if these natives did arrive, they could be used to help haul cargo ashore.

Asa and Kenney's first boat returned to *Seawolf* at 2100 after more than an hour of work. They reported that the low water had uncovered a bad coral beach which made landing operations difficult, though not impossible. Not knowing any better place to land the men and supplies, Al Bontier opted to continue with his mission at hand. He had his men continue moving cargo, wishing earnestly that he had more rubber boats to use. He continued to use his crew to load and man the boats, and put the Filipinos ashore to carry the cargo away.

"We've got company!" one of the lookouts suddenly cried out.

The rubber boats were still in full force of moving supplies when the lookouts spotted a fishing boat about a mile down the beach to the south. Bontier had his men send up a flare to illuminate the target. *Seawolf* appeared to have been caught in the act, as the little vessel began moving toward her slowly. The little fisherman appeared to be friendly, but nerves were put to the test. "In view of the fact that we 'looked like a rock' from the beach, and relying upon the flare to ruin the fisherman's night vision, decided to finish the job as planned," Bontier wrote.

The little fishing boat never gave any indication that he had spotted the anchored submarine and continued on its way. The moon rose at 2249 as the landing efforts continued. By 2330, the last boat of cargo had departed for the beach.

The last rubber boat returned to *Seawolf* at 2358 and her own boat was hauled back aboard and stowed. Bontier immediately backed his ship away from the beach, happy to be completed with this tricky mission. His crew had offloaded five tons of cargo in three hours via rubber raft. After completing the task, *Seawolf* left Sergeant Cabais' exhausted Filipino guerrillas with only a few hours of darkness to continue moving supplies ashore before daybreak.

Diving officer Paul Doane made the necessary compensations for the five less tons of ballast aboard and happily dived ship at 0038 on 10 August for a trim dive.

Allen Bingham, one of the watchstanders, felt relief to be underway again. "We sent ammunition, food, and things they needed so they could more or less go terrorize the Japs."

Cabais' men would send their first message on 27 August 1944. Their net control station was originally set up on Dumaran Island but later moved to Batulan, on northern Palawan, in closer proximity to enemy activity. Cabais's guerrillas later established sub-stations at Coron Bay region, Bascuit Bay region, Cuyo Islands region, and a weather post was operated at the net control station.[9]

In September, the party took on the ranks of Lt. Col. Cabais, Maj. Goloyugo, and Captains Cuteran, Vergara, Marquina and Dagandan in order to obtain cooperation of the local guerrillas. The guerrillas then sent all lost airmen and captured Japanese to Cabais, who in turn made arrangements to have them evacuated. In all, one U.S. POW from Puerto Princesa, 13 fliers, and three Japanese—who were previously interrogated by Cabais—were evacuated.

Guerrilla Warfare 385

The Cabais rebels landed by *Seawolf* continued to work in the area into 1945. On 28 February 1945, landings were made on Puerto Princesa, and on that date the Cabais radio net and all guerrilla forces in Palawan passed to the control of the Commanding General, Eighth Army. The Cabais mission was completed soon thereafter.[10]

For some of the old *Wolf* veterans like Bill Reiland, Bud McCoy, and Edward Chapman, these guerrilla landings were especially rewarding. They had been aboard *Seawolf* in Manila when the war broke out in 1941. Now, they had played a part in bringing special agents back to help disrupt the Japanese war efforts in the Philippines.

Seawolf ran on the surface on 10 August, now free of her special cargo and special passengers. At long last, she could hope to use her ten torpedoes against an enemy ship. Her watch was perhaps too enthusiastic. Al Bontier found it to be a "busy day, chasing 'smoke' which turned out to be clouds, and 'masts' which always developed into floating debris." Cruising through the Sulu Sea, *Seawolf*'s chief radioman, Jerry Bekke, sent a detailed report to command on completion of the mission and Bontier's intended patrol plans unless otherwise instructed.

The *Wolf* made an uneventful submerged passage through Sibutu during the early morning hours of 11 August. After crossing the Celebes Sea, she set course to pass south of the Sarangani Islands, located on the tip of Mindanao.

At 0016 on 13 August, she was still 20 miles southeast of the Saranganis when the SJ radar got a pip at 7,000 yards. It was a dark night and she was in a rain squall, with about 500 yards visibility. OOD Ed Szendrey initially thought this contact was on a small ship. The next report from Ben Rogers gave the range at 5,000 yards. Szendrey presumed this must a correction on the range from the first report. Rogers then reported the range to be closing rapidly still and it was immediately understood that this "ship" was an aircraft. Lookout Allen Bingham recalled, "It was at night and we couldn't see him. We picked up this plane on radar and went into a dive."

"Clear the bridge!" Lt.(jg) Szendrey shouted. As the lookouts leaped for the hatch, the plane could be heard coming in on *Seawolf*'s port beam. The Japanese pilot was making a glide bombing run on the *Wolf*. This plane obviously had radar or at least a radar detector which he was using to home in on *Seawolf*'s SJ beam.

The Japanese pilot first released a flare, which he dropped directly over the submarine's after battery just before the conning tower hatch was closed. The pilot then circled and returned to drop a couple of light bombs. Al Bontier found it fortunate that his range was bad "because his deflection was on." The explosions were not loud nor was there the characteristic "click" that preceded a depth charge. This led Bontier to believe that the bombs were light, set very close, and set to explode on contact. Had they been dropped instead of the flare on the first pass, it might have been big trouble.

"We were fortunate that he dropped his bombs a little bit in the rear of us," recalled yeoman Bingham. Other than rattled nerves, there was no damage. After staying down about a half hour, Bontier decided this fellow had no more bombs. He felt it would be wise to clear the area before dawn before more aircraft could be called in to attack. *Seawolf* returned to the surface at 0044 and hauled clear of the Saranganis. He had Chief Bekke radio CTF 72 that *Seawolf* was approaching the Australian safety lane.

The *Wolf* had two more aircraft contacts during the morning before entering the safety lane at 1500. At 1524, another submarine was spotted through the high periscope in the safety lane. After exchanging recognition signals, this was found to be *Flying Fish*. During the next 20 minutes, *Seawolf* and *Flying Fish* passed close aboard, allowing skipper Bontier to obtain a little information on Brisbane from Lt. Cdr. Bob Risser, *Seawolf*'s former Exec. Risser was making his fourth patrol in command of *Flying Fish*.

Seawolf remained cautious in the safety area during the next two days as she approached Brisbane. One SD contact approached to 8 miles, forcing a dive. Lookouts sighted a plane on the morning of 15 August which looked like it was a friendly Liberator, but it would not trigger the IFF.

Seawolf met her escort at 0520 on 17 August for the approach to the Australian port on the west coast. During the day, her crew sighted more than one hundred planes and several ships. She detached her escort at sunset. After daybreak, she made contact with *Geelong* for the transit of Vitiaz Strait between New Guinea and New Britain. "Saw so many ships during the day that I could well understand how the force to kick the Japs out of this area is brought here," Bontier logged.

The escort ship departed at 1600 and *Seawolf* continued on toward Brisbane, giving her position via radio dispatch that evening. The next few days passed uneventfully as the *Wolf* moved down Australia's western coast and past the Great Barrier Reef. *Seawolf* picked up her pilot in the early morning hours of 23 August for passage into Brisbane. She moored to the wharf that morning and ended her fourteenth war patrol after 24 days at sea.

An overhaul would be needed to fix *Seawolf*'s auxiliary engine and the governors on the No. 2 main engine were in bad need of overhaul and adjustments. Her pithometer and high pressure air compressors were also in need of service.

Bontier felt that his men had performed admirably during their special missions. He singled out radar technicians Ben Rogers and Norman Coon, who had gone above and beyond in keeping the finicky SJ radar in operation. "The ship is lucky in having a higher than average percentage of old hands in the crew," Bontier noted. "The officers, however, have experienced approximately a fifty percent turnover, in a period of two months. It is hoped that this high rate is not destined to continue."

Reviewing *Seawolf*'s patrol report, Commander Task Force 72 John Meade Haines wrote that new skipper Bontier had handled two special missions into hazardous areas of the Philippines, both of which were "expeditiously and intelligently accomplished." He therefore designated the patrol as successful for the Combat Insignia, even though no torpedo attacks had been made.

USS Seawolf Fourteenth Patrol Summary	
Departure From:	Fremantle, Australia
Patrol Area:	Philippines
Time Period:	1–23 August 1944
Number of Men Aboard:	78: 70 enlisted and 8 officers
Total Days on Patrol:	24
Total Miles Steamed:	11,164
Total Fuel Consumed:	135,337 gallons
Total Days Submerged:	4
Number of Torpedoes Fired:	0
Special Missions Accomplished:	Delivery of supplies and ten special forces men to two areas in Philippines
Return To:	Brisbane, Australia

17

On Eternal Patrol

Fifteenth Patrol   *October 1944*

Seawolf being lost on patrol was something that gunner's mate Rex Mickey had dreams about. His close friend in the forward torpedo room, John Sadler, confided that he had also woken up in a cold sweat after terrible dreams about the *Wolf*'s demise. Earlier in the year, Sadler and Mickey had actually gone to their executive officer, Doug Syverson, to request transfers. Submariners by nature were a superstitious lot and such dreams were enough to give them bad feelings about staying on board.

Lieutenant Syverson had laughed off their stories with, "You're in the Navy. You do what we tell you to do."

Sadler and Mickey never brought up their troubling dreams again. "We didn't have any chance of being transferred," admitted Mickey. *Seawolf* was in Darwin for almost a month before she departed on her fifteenth war patrol. Before she set sail, Rex Mickey would be among those transferred.

Relief Crew No. 5 from *Nautilus* filled in for *Seawolf*'s crew during their R&R period. A number of these men would come aboard as replacements for the transferees. Nine men were transferred prior to *Seawolf*'s departure from Darwin and eleven were received.

Among these new arrivals on 6 September was Fireman 1/c William B. Beck from New York. He and his buddy Fred Christianson had been working in the relief crew for the past six weeks changing broken crankshafts on Fairbanks engines. "On the sixth week, our chief says, 'You or Beck are up next. I'll flip a coin to see who goes,'" recalled Christianson. "Beck called and lost—lost big time; the boat was *Seawolf*. *Seawolf* went down with all hands...so, my buddy Beck, rest your oars."¹

Another of the men reporting aboard for duty was motormac Bob Devitt, who had been hospitalized prior to the *Wolf*'s fourteenth patrol. Yeoman Allen Bingham took Devitt's place in the hospital on 17 September, as *Seawolf* was getting prepared for her next run. Aboard for only one patrol, he had started feeling ill as *Seawolf* returned to Australia. "I came up with tuberculosis in my right lung and ended up in the hospital. They entered my chest and then I spent the next eighteen months in hospitals before I was discharged," said Bingham. "I was one of only a few guys transferred off after the fourteenth patrol. I guess you could say I was lucky."

Y1c Chester Copas came aboard from *Nautilus* as a replacement for Bingham. Short one officer on her fourteenth patrol, *Seawolf* gained two for her fifteenth. One was Lt. Edward Francis O'Brien Jr., from the Naval Academy class of 1940. A 27-year-old from Massachusetts, O'Brien had previously been awarded both the Silver Star and the Navy and Marine Corps Medal for heroism.

The other new face in the wardroom was Ensign Bill Reiland, the chief torpedoman who had been with the *Wolf* since the start of the war. Effective 1 September, he joined a long list of *Seawolf* CPOs who had become mustangs. The next most senior enlisted man to fill Reiland's former role as chief of the boat was CMoMM Bud McCoy, who had also been aboard since the start of the war. McCoy had a six-month edge in Navy tenure over his fellow chief motormac, Mike Wiegenstein.

Captain Bontier's *Seawolf* departed Brisbane on 21 September 1944 to begin her fifteenth war patrol. Gunner Rex Mickey stopped to watch the *Wolf* as she headed out. "I sat on the dock along that river in Brisbane," he later recalled. "I could see my buddies waving at me as she sailed away. That was the last I ever saw of her."

On 29 September, the *Wolf* reached Manus in the Admiralty Islands. Manus was the largest of the two major islands of the Admiralties, and its Seeadlor Harbor was one of the finest deep-water anchorages—more than six miles wide and 20 miles long—of the Southwest Pacific.²

Seawolf took on fuel, supplies and 17 Army personnel. The Army men were destined for Samar in the Philippines. Just as she had conducted special landings on her fourteenth run, the *Wolf* would support special ops again for her fifteenth. The Army men she took aboard were part of an elite special reconnaissance group known as the Alamo Scouts. Organized by Lt. General Walter Krueger, the Sixth United States Army was activated at Fort Sam Houston in San Antonio, Texas, in January 1943. Later that year, Krueger began developing an Army special intelligence unit, which he dubbed the "Alamo Scouts," due to his close association with the town that was the home of the Texas' famous Alamo mission.[3]

Krueger's first class of Alamo Scouts recruits began training on Fergusson Island, New Guinea, in January 1944. Their advanced training courses included rubber raft handling, knife fighting, first aid, morse code, scouting, patrolling, stalking, river crossing, day and night jungle navigation, sniper training, snares, booby traps, explosives, swimming, mortars, grenades, blinker lights, and knowing which wild plants and bugs were edible for survival. The first class of 24 Alamo Scouts graduated on 5 February 1944.[4]

The Alamo scouts initially participated in the capture of the Admiralty Islands, which was the final step in the Allied efforts to isolate the Japanese stronghold of Rabaul. By mid-July 1944, eight Alamo Scouts teams were operating in the New Guinea area, most being inserted into advanced Japanese areas by PT boats. One team had departed Manus Island's Seeadlor Harbor aboard the submarine *S-47* and inserted onto Waigeo Island on 23 June 1944. One week later, *S-47* picked up the special forces men safely. In their first eight months, the Alamo Scouts performed 36 known missions in the Bismarck Archipelago and New Guinea campaigns. They managed to kill at least 84 enemy soldiers, take 24 prisoners, and rescue approximately 550 civilians without a single loss of their own.[5]

In late September 1944, General MacArthur directed Krueger's special forces teams to begin planning for the invasion of Mindanao, the second largest island in the Philippines. During October, several Alamo Scouts teams were to be inserted onto Leyte and Samar, most boarding PT tenders, which would later dispatch them aboard the smaller PTs.[6]

Seawolf was selected to take one special special unit under Captain Howell Kopp in to Samar to prepare the way for landings on Leyte.

Captain Kopp was a member of the third Alamo Scouts training class, which had graduated from its New Guinea training area on 22 June 1944.

Seawolf was in Manus only long enough to take on fuel and the 17 Alamo Scouts for her special mission. She departed the same day, with 100 men aboard. One man had gotten off the ship at Manus on 29 September. He was warrant electrician John Bilkey, who had orders to report to another ship. Bilkey was the last plankowner aboard and he was the last man to leave her. "I had put her into commission in 1939 and had been aboard her longer than anyone," he said. "I got off on Manus Island, straight east of the northern end of Australia." Bilkey—who had earned a Bronze Star on the *Wolf*—was on hand to bid goodbye to his shipmates as *Seawolf* backed clear of her Manus dock.

John Bilkey went back to Honolulu for a new assignment, only to soon learn that *Seawolf* was "overdue and presumed lost." Distraught over the loss of his close friends and shipmates, he stated many years later, "I consider myself very lucky."

Four days after departing Manus Island, *Seawolf* fell victim to a terrible mistake.

At 0756 on 3 October, *Seawolf* exchanged radar recognition signals with *Narwhal* (SS-167) off Morotai Island, just northwest of New Guinea. Later that day, *Shelton* (DE-407) was torpedoed and sunk by the Japanese submarine *RO-41*, which had fired her last four torpedoes in an attempt to hit an American carrier. The destroyer escort *Richard M. Rowell* stood by the sinking *Shelton*, which would later capsize and sink while under tow. Both destroyers were part of a 7th Fleet task group which included the escort carriers *Midway* (CVE-63)—which would be renamed *USS St. Lo* on October 10—and *Fanshaw Bay* (CVE-70). *Rowell* began searching for the enemy submarine, assisted by two Grumman TBM Avenger torpedo bombers from *Midway*.[7]

Four friendly submarines were known to be in the area. They were directed to give their positions, which three of them did. *Seawolf* was not heard from. On 4 October, *Seawolf* was again directed to report her position and again there was no answer.[8]

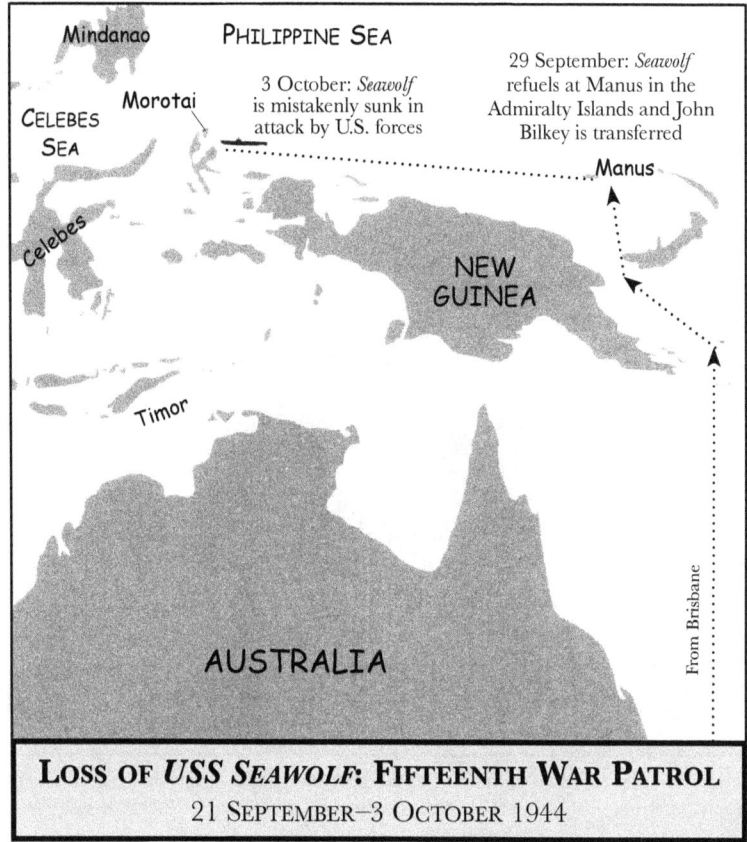

LOSS OF *USS SEAWOLF*: FIFTEENTH WAR PATROL
21 SEPTEMBER–3 OCTOBER 1944

Why Chief Bekke's radio gang did not receive or acknowledge the position request transmissions is unknown. Perhaps there was trouble with the radio gear. In any event, the *Wolf* was sighted and attacked by friendly forces that day. Ensign William C. Brooks Jr., pilot of a VC-65 Avenger, was launched from *Midway* as part of a hunter-killer search. His carrier had dodged the torpedoes fired by the Japanese submarine which hit the destroyer *Shelton*. "They launched me and some other pilots minutes later to go out on anti-submarine patrol to cover certain sectors," he later recalled.

Brooks spotted a suspicious contact close to the enemy sub's expected location and radioman Ray Travers blinkered that day's Morse code signal to the submarine. The submarine below was running on the surface with her decks awash and appeared to Brooks to be charging her batteries. The sub's conning tower was clearly visible

Seawolf was lost with all hands on her fifteenth war patrol when she was mistaken for a Japanese submarine and depth charged by the destroyer escort *Rowell* (DE-403). *Official U.S. Navy photo, courtesy of Mike McCoy.*

but she did not respond to Travers' signals at all. Having exercised his best efforts to determine the submarine's friendly status, Brooks then circled his Avenger back around to make a run with his depth charges. By this time, the sub had submerged. He proceeded to drop two depth charges some 50 to 75 yards ahead of where he guessed the sub had submerged and then released a cannister of fluorescent green dye to pinpoint her last known location. Brooks reported his attack and remained over the location until the destroyer escort *Rowell* quickly arrived on the scene and began making her attack.[9]

Rowell established sound contact on the submarine at 1310 on 3 October. As Captain Harry Barnard prepared to attack, his soundmen picked up a series of long dashes and dots which bore no resemblance to the existing U.S. recognition signals. *Rowell*'s crew assumed that the Japanese submarine which had sunk *Shelton* was attempting to jam their sound equipment. *Rowell* made six attacks on the submarine, including five with forward-thrown "hedgehog" depth charges.

Barnard's log recorded for 1316: "Fired 24-charge hedgehog pattern at contact. Three explosions heard, two large boils [bubbles] observed off port beam, debris observed in the boils." Still circling above, Ensign Brooks noted the explosions. "It sounded like a sheet of plate glass exploding on the ground when the depth charges went off," he remembered. It was later established that the Japanese sub, *RO-41*, made it back to Japan without any U.S. counterattack—leaving little doubt that *Rowell*'s victim was *Seawolf*. Bill Brooks and *Rowell's* skipper were both brought before an inquiry board to describe what they had witnessed during their attacks. Three days later, Brooks was told that a submarine had been sunk and that it was believed to have been *Seawolf*.[10]

Caught by friendly forces, *Seawolf*'s crew had apparently tried in vain to signal the destroyer above that she was attacking a friendly submarine. The deadly hedgehogs exploded with enough force to rupture her old hull and send her on her final dive for the bottom—some 9,600 feet and nearly two miles below. In a sad twist of fate, well-intentioned American forces had managed to do what the most persistent of Japanese anti-submarine efforts had failed to do in fourteen prior war patrols—fatally hole the proud *USS Seawolf*.

Seawolf's loss was announced to the public in late December 1944, and newspaper clippings such as this one from the Spokane *Spokesman-Review* ran throughout the country.
Courtesy of Claude Robinson.

U. S. Sub Sunk.
WASHINGTON, Dec. 28. (AP)— The United States submarine Seawolf is overdue from patrol and presumed lost.

The navy's announcement today said the vessel was under command of Lt. Comdr. Albert M. Bontier of White Plains, N. Y., who is listed as missing.

The 1500-ton submersible was of a class which normally carries a complement of 62 men. No announcement was made of the number lost, but the navy said next of kin of all casualties had been informed.

The Seawolf brought to 34 the total of United States submarines lost during the war—four sunk, 28 overdue and presumed lost, and two destroyed to prevent capture. It was the 239th naval vessel of all types lost since the beginning of the war.

When the time had come and gone for her to have returned from patrol, *Seawolf* was officially presumed to be gone. On 28 December 1944, the Navy Department's Office of Public Information announced in communique No. 564, "The Submarine USS SEAWOLF is overdue from patrol and presumed lost. Next of kin of casualties have been informed."[11]

Seawolf was officially removed from the Navy list on 20 January 1946. She had become one of the 52 United States submarines lost in World War II to all causes. Going on eternal patrol with *Seawolf* were 100 men from all walks of life, some brand-new to the *Wolf* and some seasoned veterans.

There was Al Bontier—having blemished his career by grounding *Razorback*—who had proven himself on the *Wolf*'s previous run. Now, friendly forces had snuffed out his chances of continuing his fine record for *Seawolf*. There was Arnold Bargenquast, who had survived his wounds at Pearl Harbor on 7 December 1941, but did not survive his fourth war patrol. Bargenquast was looking forward to leave in the States after this patrol. His wife Louise was more than seven months pregnant with their first son. There was Lloyd George, the youngest of nine kids from Pennsylvania. There was torpedoman Dallas Malone from Detroit, an ace poker player in his off hours. His best friend in the forward torpedo room was John Colby Sadler, who had dreamed months before of being lost aboard *Seawolf*.

There was CMoMM Edward Chapman, the Indian-blooded Oklahoman nicknamed "Chief" by his buddies. He and his friend Bill Reiland, the newest mustang, had both been aboard since the beginning of the war. The only other man still on *Seawolf* from the start of the war was Bud McCoy, who had kept a private diary for some of the *Wolf*'s patrols. Previously in charge of the forward engine room, McCoy's final role was that of chief of the boat.

Seawolf's chief radarman, Ben Rogers, would long be remembered for his skills as a cartoonist. His illustrations had long decorated the crew's mess and the ship's newsletter put out by Chief Gerald Bekke's radio gang. Finally, there was Captain Howell Kopp and his 16 special forces men. Trained in every aspect of how to survive in the jungle in enemy territory, Kopp's men did not survive their first submarine voyage.

Each man aboard *Seawolf* served an important role in the Silent Service in his own way. Some had been aboard her for the majority of the war, dating back to Lt. Cdr. Freddie Warder. Others were relatively new to the *Wolf*. Quartermaster John Louis Ewing from Springfield, Illinois, had reported aboard on 21 September, the day that the *Wolf* departed from Australia.

October 1944 was a black month for the United States submarine service. In addition to the famous *Seawolf*, four other boats were lost: *Darter* on 24 October due to accidental grounding; *Shark II* on 24 October, due to enemy destroyer attack; *Tang* on 24 October to her own circling torpedo; and the brand-new *Escolar*, lost in October on her first patrol without a word.

These 100 men and the dozens of others who had served on *Seawolf* before them had conducted themselves in fine tradition and had aggressively waged war against Japan. By the time of her loss, *Seawolf* had earned 13 battle stars and was ranked among the top ten scoring United States submarines of the war. Only faulty torpedoes during the first two years of war prevented *Seawolf* from ranking higher on the kill list.

She had carried out special ops missions, rescued aviators, attacked enemy warships, and boldly entered enemy harbors in her efforts to shorten the war. She had received two Navy Unit Commendations for outstanding heroism in action. *Seawolf* was one of only a few boats which was publicly named and praised in the press during the early part of the war.

Her men took it in stride but certainly held their chins a little higher knowing that pictures of their sinkings had been printed in newspapers and *Life* magazine. *Seawolf*'s crew had all entered the submarine service voluntarily, knowing full well the danger that came with the extra pay. They were the men who had helped carry out the war of the *Wolf*.

ON ETERNAL PATROL WITH *SEAWOLF*

Lt. Edward Francis O'Brien Jr. *(left)*, Lt.(jg) Edward John Szendrey *(center)*, and GM2c Francis Arden Ballard *(right)* are among those lost with *Seawolf*. Images courtesy of Charles Hinman, USS Bowfin Museum.

CTM Donald Joseph Naze *(left)*, S1c Patrick Kenneth Bergevin *(center)*, and EM1c Kenneth Judd Flynn *(right)*. Courtesy Charles Hinman, USS Bowfin Museum.

MoMM2c Robert Floyd Devitt *(left)* made two runs on the *Wolf* but went ashore ill for her fourteenth run. He returned to his beloved ship just prior to her ill-fated fifteenth patrol. *Courtesy of Charles Hinman, USS Bowfin Museum.*

Captain Howell Kopp, U.S. Army *(above)*, was the leader of a special forces unit called the Alamo Scouts aboard during *Seawolf*'s final run. *Courtesy of Steven Kopp.*

ON ETERNAL PATROL WITH *SEAWOLF*

S1c Robert Jordan Bennett *(left)* was making his third patrol on *Seawolf.* RM2c James Everard Johnson *(center)* was making his seventh patrol. SC1c Michael Jurnic *(right)* was making his fourth run on the *Wolf.* Images on this page courtesy of Charles Hinman, www.oneternalpatrol.com website.

QM1c Alfred Eric Kuehn *(left)* was making his sixth run. TM2c Dallas Leroy Malone *(center)* was making his seventh patrol. TM2c Robert Thomas Miller *(right)* was making his eighth run on *Seawolf.*

S1c Joseph Albert Morris *(left)* was making his sixth run. EM2c Albert Francis Page *(center)* was making his eighth *Seawolf* patrol. EM3c Wasil Polityo *(right)* joined the *Wolf* just before her eleventh parol.

ON ETERNAL PATROL WITH *SEAWOLF*

MoMM1c James William Saint *(left)* was making his seventh patrol on *Seawolf.* MoMM2c William Hopkins Underhill *(center)* was making his fourth war patrol on the *Wolf.* Ens. John Van Andel *(right)* was making his third patrol. *Images on this page courtesy of Charles Hinman, www.oneternalpatrol.com website.*

S2c Richard Lawrence Miller *(left)* had newly reported aboard *Seawolf.* Lt.(jg) Ralph Van Dorn Miller *(center)* was making his second war patrol on the *Wolf.* F1c Gerald Andrew Steinbecker *(right)* was making his third patrol.

Epilogue

The loss of *Seawolf* was a tough blow for those who had survived war patrols aboard her. Paul Zimmerman was in the Philadelphia Navy Yard when he got the news. "It was a staggering blow," he said. "We got word that she was overdue and presumed missing in action. The story came filtering in there first that she had been sunk by one of the Australian corvettes. Finally, the truth came out that she had been sunk by our own planes and ships. You have to be a submariner to actually feel what it's like to lose another sub or your home sub."

Former electrician's mate Olin Fogle said, "I went on a two-day drunk when I found out about the *Seawolf*. It was tough because you got to know people when you lived with 'em."

Fireman Bill Harlow had transferred to new construction aboard *Tench* (SS-417) after *Seawolf*'s twelfth run. *Tench* was commissioned on 6 October 1944, just days after *Seawolf*'s loss. "I knew nothing of the *Seawolf* being missing at first," Harlow later admitted. "I asked all the fellows, 'What do you hear from the *Seawolf*?' They all looked at each other and then I knew. I didn't sleep or eat for three days, I'll tell you."

Gunner's mate Rex Mickey was one of the fortunate ones to be transferred just before the fifteenth patrol. Shaken by narrowly missing being lost with *Seawolf*, he went on to become "a faithful church member who tried to lead a good life." He named his son Dallas Colby Mickey—who would serve a dozen years in subs himself—after his two forward torpedo room buddies, Dallas Malone and John Colby Sadler, who are on eternal patrol with the *Wolf*.

Aside from those lost with the *Seawolf*, others who served on her were also lost in submarine service. Dave Butler made ensign and was transferred to *Tullibee*, which fell victim to a circular run by one of her own torpedoes in March 1944. Only one man from *Tullibee* was rescued and survived life as a prisoner of war. Lt.(jg) Hubert Gluski transferred off after the *Wolf*'s eighth patrol at Midway. He joined *Runner* and immediately departed with her, only to be lost weeks later.

Officers' cook Basilio Galvan served on *Seawolf* briefly between her second and third patrols but was transferred. Galvan, who had survived the sinking of *Squalus*, was also lost with *Runner*. Two other pre-war *Seawolf* sailors, CTM William S. MacDowell and torpedoman Joseph Perry were both killed during the war at Corregidor. CTM Clarence Weade, a veteran of two runs on the *Wolf*, went down with his ship when *Sculpin* was severely damaged by depth charges on 18 November 1943 and lost with many of her crew.

Clarence Kibbons, a veteran of *Seawolf*'s first nine patrols, joined the new boat *Shark II* as chief torpedoman and her chief of the boat. Outbound on her first patrol in May 1944, *Shark* stopped at Midway, where Kibbons ran into former shipmate Hank Thomson. "He informed me that his brother was killed in Europe in the Army," said Thomson. Sadly, Kibbons and *Shark II* were both lost to a depth charge attack on her third patrol on 24 October in Luzon Strait.

Seaman Bob Lents, a torpedoman striker on *Seawolf* during her pre-war Philippine days, was transferred on to the submarine *Perch*. His boat was lost in March 1942 when she was so badly damaged by Japanese counter attacks that she had to be scuttled. "We all got up on the deck," said Lents. "We opened up all the valves and hatches and kissed her goodbye."

The *Perch* crew took to the ocean and swam until a Japanese destroyer decided to haul the Americans aboard as POWs. "Everybody got off the *Perch*," Lents said, "but not every one of us got out of prison camp." Six *Perch* crewmen died during the war from the lack of food, abuse, and poor living conditions the men suffered. Lents was held on an island in the Celebes, east of Borneo, for most of the war before being shipped to Java just before the war ended. Bob Lents' family was stunned after the war to learn that he was still alive, more than three and a half years after *Perch* was lost. "The Japanese never reported us captured."

Epilogue 403

Statistically, almost one in four men who served in the submarine service in World War II did not survive. By war's end, 52 boats and 3,505 men had been lost. This casualty rate was approximately 22 percent, the highest of all the services. The submariners' loss rate was six times greater than that of the surface Navy. Against such terrible odds, the Silent Service's accomplishments were extraordinary. Although its numbers eventually increased to 288 submarines, it comprised less than 2% of the total U.S. naval strength. U.S. submarines destroyed more than half of Japan's sea losses during the war, in spite of struggling with faulty torpedoes for the first half of the war.

The Japanese merchant fleet lost 134 vessels of more than 1,000 tons during 1942. As American torpedo performance began to improve, this number doubled in 1943 and more than tripled in 1944. The final score might even have been higher if the Japanese had kept better records. Sinking claims were scrutinized at war's end by the U.S. Joint Army–Navy Assessment Committee (JANAC), who dismissed many sinkings because of poorly-kept Japanese records, even in some cases where a ship was witnessed to sink.

The Imperial Japanese Navy also lost many capital warships to U.S. submarines, including a battleship, nine aircraft carriers, 12 cruisers, 43 destroyers, and 23 submarines. *Seawolf*, of course, was among this list, having sunk a destroyer and damaged a cruiser.[1]

In honor of the famous *Seawolf* of World War II, the Navy has taken strides to keep her name alive with more modern submarines. The second submarine with the *Seawolf* name (SSN-575) was nuclear-powered and was launched in 1958. Former crewmen on hand for her commissioning were Rear Admiral Freddie Warder, former motormac Paul Zimmerman, and motormac Orval Cross, among others. The nukie *Seawolf* served with distinction until 1986, earning five Navy Unit Commendations, a Meritorious Unit Commendation and three battle efficiency awards. *Seawolf* took President Dwight D. Eisenhower for a submerged run in 1957 and, in 1958, set an endurance record for remaining submerged for 60 days. SSN-575 was decommissioned on 30 March 1987.

The third new *Seawolf* (SSN-21) was authorized in 1989. She and her sister ships were slated to be the fastest, deepest diving and most heavily armed submarines ever built by the U.S. Navy. She was armed with eight torpedo tubes, 80 inches in diameter, with longer range and a much quieter weapon system. This latest *Seawolf* was launched 24 July 1995 and commissioned on 19 July 1997. Among those on hand for her commissioning was former motormac Charles "Johnny" Johnson, who had made nine runs on the diesel *Wolf*. Johnson would later write:

> I attended the commissioning of SS-197's successor, the SSN-21, in Groton and was treated "royally" by the captain and crew. At the commissioning, I met four ex-*Seawolf* shipmates whom I served with. This helped bring some closure to my range of emotions I felt after World War II. We really had a great ship and a great crew. I know—I was there.[2]

Displacing 9,100 tons and carrying 133 officers and enlisted men, SSN-21 can exceed 800 feet in depth—more than three times the capability of the diesel *Wolf*. In active service as of this writing, the first of the *Seawolf* class of nuclear submarines is based out of the Naval Submarine Base, New London.

Those who had fought on the diesel *Seawolf* in World War II would never forget the sacrifices of their lost shipmates. John Bilkey, who made warrant officer while aboard the *Wolf* and was the last man to leave her before she sailed on her fifteenth patrol, still considers himself lucky. "I put her into commission and I was the last man off her." Bilkey careered the Navy, including 12 years aboard the carrier *Enterprise*. He retired as a captain, and currently lives with his son in Alaska.

Seawolf's three surviving skippers all went on to enjoy long naval careers. Ozzie Lynch was officially credited with sinking five ships while in command of *Skate*. Roy Gross, between his five patrols on *Seawolf* and two in command of *Boarfish*, finished the war as America's eighth best submarine skipper with 14 sinkings to his credit. Gross retired to Connecticut but also spent time in a second home he kept on Saint Maarten in the Virgin Islands.

Freddie Warder—who never accepted the "Fearless Freddie" monicker his crew tried to hang on him—was awarded two Legions of Merit for serving as a wolf pack commander in the Pacific following his *Seawolf* command. He returned to New London in September 1944 to command the Submarine School for two years and a received a third Legion of Merit for his improvements in training submariners. Postwar, he held a number of commands, including that of the *USS Columbus* in 1951–2, and Commander, Cruiser Division Five in 1954. Warder was one of the first World War II submariners to reach flag rank, doing so in 1952. In 1955, Warder was assigned as Assistant Chief of Naval Operations, Undersea Warfare, Navy Department. From 1957 to 1960 he was commander of Submarine Force, U.S. Atlantic Fleet.

While in charge of the Atlantic Fleet submarines, his flagship as of 15 April 1958 became none other than *USS Seawolf*, the world's second nuclear-powered submarine. Warder retired as a Rear Admiral and moved to Ocala, Florida. He lived to age 95, passing away in February 2000.

James "Caddy" Adkins, the first executive officer of *Seawolf* during peacetime, was disqualified from submarines after an unfortunate Atlantic patrol in command of the old *S-21*. Freddie Warder helped him back into a new command in 1944. Aboard *Cod*, Adkins managed to sink four confirmed ships in three patrols.

Seawolf's second Exec, Bill Deragon, attended PCO School at New London in the summer of 1943. He then put *Pipefish* into commission and was her skipper from February 1944 through May 1945. Aboard *Pipefish*, he earned another Bronze Star and two Letters of Commendation. He later commanded the destroyer *Charles C. Roan* and tanker *Salamonte* (AO-26) in addition to a number of administrative billets. Bill Deragon was serving as Administrative Officer of the Portsmouth Naval Shipyard when he retired in July 1960 as a full captain to live in Rhode Island.

Bob Risser married Ruby Archer after the war, widow of Lt. Thomas F. Sharp, who had been lost with *USS Pickerel* during World War II. Risser, *Seawolf*'s exec after Deragon, retired as a captain. He served as ComSubPac's staff ordnance officer for two years and in 1950 became head of underwater ordnance research and development. In addition to other staff positions, Cdr. Risser commanded two submarine squadrons during the 1950s.[3]

Seawolf's fourth Exec, Doug Syverson, served as executive officer of *Sea Fox* from February 1944 until June 1945 on three war patrols, during which time his boat supported the occupations of Iwo Jima and Okinawa. He then assumed command of *Thresher* in June and held that command until December 1945. Syverson then served on the staff of the New London Submarine School into 1948, at which time he assumed command of *USS Sea Robin*. Among numerous other post World War II assignments, Syverson commanded two submarine squadrons and the *USS Hunley* (AS-31), the first submarine tender to be constructed after World War II. Captain Syverson retired from the Navy in November 1967.

Seawolf's fifth executive officer, John Hess, left after her thirteenth run and commanded *Pomfret* for four patrols before the war ended. He managed to rescue seven aviators during lifeguard duties. Hess rose to the rank of captain during his long naval career, during which time he commanded the sub tender *Bushnell* and Submarine Squadron Sixteen before retiring in 1968.

A total of 504 aviators were rescued by submarines during the war, including future President George Herbert Walker Bush. *Seawolf* contributed her two *Wasp* fliers, and as skippers of their own boats, her former officers were instrumental in rescuing many others. Hess' *Pomfret* rescued seven; Deragon's *Pipefish* rescued nine; Lt. Cdr. Dick Holden's *Gato* picked up thirteen aviators; and prewar *Seawolf* officer Bill Kinsella's *Ray* effected 21 rescues.

Dick Holden commanded the training boat *R-14* in Key West after leaving *Seawolf* at Mare Island. He went aboard the new submarine *Blueback* in February 1944 on her maiden run as a PCO. In March 1945, Holden took command of the *Gato*—namesake boat of her class—and commanded her through January 1946. He earned the Bronze Star during *Gato*'s twelfth patrol for rescuing ten U.S aviators in one patrol and even evaded an attack by Japanese aircraft during one of the rescues. Postwar, he would also command *Torsk* and the destroyer *Watts* (DD-567). He retired as a captain in 1957.

Jim Mercer, who took some of *Seawolf*'s famous periscope photos, attended the Navy's postgraduate school at Annapolis from 1945 to 1948. He then commanded the submarine *Burrfish* (SSR-312) from 1949-1951. Bill Whitman went on to *Hammerhead* and later *Charr* after the war, retiring from the Navy as a captain after 30 years' service.

Duke Robinson developed an ulcer after leaving *Seawolf* after her famous twelfth run and he did not make any more war patrols. In the 1950s, he returned to work for Boeing, the company he had been with when the war broke out. Like many others, Robinson was proud to attend the launching of the nuclear *Seawolf*.

Jughead Casler went on to put *Cavalla* into commission as her torpedo and gunnery officer. On her first war patrol, he helped direct four torpedoes into the Japanese aircraft carrier *Shokaku*, one of the six carriers which had brought war to Pearl Harbor in 1941. For helping to sink *Shokaku*, Casler was awarded a Gold Star in lieu of a second Silver Star. Ironically, Casler's *Cavalla* now rests in Galveston, Texas, at Seawolf Park, named for his first wartime submarine.

Casler left *Cavalla* in November 1944 and next commanded the training boat *O-3* through the end of the war. He then had the unique experience of taking command of a German submarine, *U-2513*, in September 1945 after the surrender ceremonies. He even took President Harry Truman aboard his new, high-speed German U-boat off Florida in November 1946 for operations and an extreme deep submergence. Casler commanded *U-2513* for three years before taking a position as ordnance and personnel officer at New London, Connecticut. He retired as captain in 1955.

Pete Lober made ensign at Mare Island while with *Seawolf* and was assigned to *Rasher* as her assistant engineer. Lober earned a Silver Star on *Rasher*, which became another of the war's top-producing submarines. When the war ended, Lober was in Australia and he rode back to the United States aboard *Bluefish*. During his voyage, he was happy to reunite with former *Seawolf* torpedoman Jim Cashero and catch up on old times.

Cashero had been serving on a relief crew in Australia when the war ended. He informed Pete Lober that he had not only escaped *Seawolf*'s loss but that of another submarine as well. He had narrowly avoided being lost with *Escolar*, a new boat he had helped commission that was lost during the same month as *Seawolf*'s loss. Cashero was not moved by his new gunnery officer's plans to conduct battle surfaces and shoot up enemy shipping, something Freddie Warder had avoided in the first year of war. "I told him he'd end up getting us all killed or get us lost on our first run," said Cashero. "He brought me up before the captain on report for bringing down the crew's morale."

In the end, Lt. Cdr. William Millican looked over Jim Cashero's service records and dismissed the whole thing. "War was not as civil as they thought it was going to be," said Cashero. "I decided to get off the boat at Honolulu before *Escolar* left on her first patrol. I had a friend who was the pharmacist's mate. He asked how I was going to get off, and I said, 'Don't worry about it.'"

The psychology of submariners was almost like a sixth sense in some cases. Cashero had such a bad feeling about this upcoming patrol that he downed enough straight 190 proof torpedo alcohol to make himself violently ill. "It tore up my stomach and made me throw up some blood," he related. "The pharmacist said I must have an ulcer and he transferred me to the base on sick leave." The fake ulcer was not sufficient to keep him hospitalized, so Cashero even requested a circumcision to remain off ship even longer.

"I fixed it to where I stayed there long enough to miss her sailing," he said. "Otherwise, I'd be down on the bottom with her." *Escolar* sailed without Jim Cashero. She did conduct a successful battle surface against a small Japanese gunboat, but was lost with all hands while passing through the minefields of Tsushima Straits. Having served on two of the war's lost subs, Jim Cashero returned home to finish high school in Bakersfield, California, and had a long career in carpentry.

Electrician Ed Milas, who had made three patrols on *Seawolf* in 1942, survived a near-fatal depth charge attack on *Seahorse*'s seventh patrol. "We got it for about five hours," he recalled. "They clipped off one of our periscopes. That was the only time I really got scared during a depth charging." *Seahorse* limped back to port. Milas finished out the war with thirteen war patrols under his belt and spent his postwar career working for Western Electric in Chicago.

Paul Zimmerman, who made *Seawolf*'s first nine patrols, served for part of the war as an inspector at Philadelphia's Cramp Ship Building where they were building submarines. When the war ended he was served aboard *Entemedor* (SS-340), which had been launched in the spring of 1945. Due to his years of service, Zimmerman was able to get a quick release. "I had been to refrigeration and air-conditioning school in the Navy, so I fell right into that after the war."

Hank Thomson made eleven of *Seawolf*'s first 12 runs. He was then transferred to new construction where *Tiru* was building at Vallejo, California. "I was transferred to the relief crew awaiting new construction orders and then our crew was transferred to Midway. Every day there, I checked with the squadron for my transfer orders. Finally, they came in September for *Tiru* in Vallejo."

Motor machinist Lew Donche *(above left)*, seen in 2006, made three runs on the *Wolf* under skipper Roy Gross. *Courtesy of Mark Donche.* GM2c Rex Mickey *(above right)* is wearing his *Seawolf* cap in 2005. Mickey was one of the last men to leave *Seawolf* following her 14th patrol. *Courtesy of Rex Mickey.*

Thomson returned to California and worked with the prospective *Tiru* crew until construction stopped before Christmas. He was then transferred to the squadron staff in Saipan. Thomson retired from the Navy in 1956 after 20 years of service. He worked in title insurance for another 32 years before retiring in California.

Charles Woodard, the Corregidor evacuee who later came back aboard *Seawolf* as a regular crewman, found her loss to be a very "sad deal." He retired from the Navy in the mid-1950s as a chief radioman and returned to his current hometown of Norwalk, Connecticut, where "I've always been able to get work in Norwalk. I've done everything, but nothing important."

Lew Donche, a motormac and battle lookout for Lt. Cdr. Roy Gross, later spent 18 years working for the U.S. Postal Service. Delbert Mar served on *Sea Fox* and later ran a grocery store in California for about 30 years. Wilson "Red" Mills, a member of the *Wolf*'s deck gun crew, was serving on *Sumner* (AGS-5) decoding messages in her radio room when he found out that the war was over. Mills attended Michigan State and returned to work for Dow, while working as a farmer on the side.

Allen Bingham, the yeoman transferred off *Seawolf* just prior to her final fifteenth run with tuberculosis, recovered but was discharged from the Navy in 1946 due to his health. Bingham went to school in Connecticut and spent his next 30 years as a teacher.

Members of the USS *Seawolf* crew reunite at a 1976 SubVets Convention in Denver. Left to right are Hank Thomson, Dick Holden, John Sullivan, Edward K. Roszel (a plankowner), and "Swede" Hanson. *Courtesy of Bob Hanson.*

(Below) Brothers Henry "Big Swede" Hanson, left, and Bob "Little Swede" Hanson both made nine patrols on *Seawolf* before the Navy decided to prevent siblings from serving on the same ship. *Courtesy of Bob Hanson.*

Electrician Bob Parden, who made the first seven runs, made chief electrician and put *Dace* into commission. He would put in 20 years in the Navy, mostly aboard submarines. He and his brother had served together on the destroyer *Pillsbury* in the pre-war Asiatic Fleet. *Pillsbury* was later blasted in the Java Sea by a Japanese force which included the heavy cruiser *Maya*. Parden later took great pleasure aboard *Dace* when his submarine managed to torpedo and sink *Maya* in 1944.

Gunner Rex Mickey, another who left just before *Seawolf*'s final run, later served under Roy Gross again. "I got shipped over to Perth and Captain Gross came in with a new construction boat, *Boarfish*. I walked across the gangplank and he just left those admirals he was talking to and greeted me." Mickey and Gross talked for about 15 minutes, during which he said, "If you need a gunner's mate, I'm ready to go." With that, Mickey was back under Gross' command and off to war again. Being a farmboy at heart, Mickey returned to working in the sawmills of Oregon after the war.

Carl Enslin, former *Seawolf* chief of the boat, left after her ninth run and served in administrative positions at New London's submarine base and at Mare Island. In June 1945, he joined *Cavalla*—which had sunk the Japanese carrier *Shokaku* in 1944—and was aboard her when the war ended. Enslin retired from the Navy in 1948 after 20 years' service.

Seawolf Park

Cavalla, the submarine that both Jug Casler and Swede Enslin went on to serve aboard, would later become an important part of the *Seawolf* legacy. Organized in the 1950s, the U.S. Submarine Veterans of World War II association went to great lengths to make sure that America's sub vets were not forgotten—particularly those who paid the ultimate price.

Each of the 50 United States was assigned a World War II lost submarine to honor. Since 52 boats were lost in the great war, Washington D.C. was included, and western New York took on *Grenadier* and eastern New York took on *Runner*.

Less than two dozen of the boats launched before the end of World War II can now be viewed and boarded by the public. This

Cavalla's after torpedo room appears to firing a Mark 16 torpedo. She is now landlocked, her bow pointing toward Galveston Harbor, now one of the busiest cruise ship terminals in the United States. *Author's photo.*

list includes: *Batfish* in Muskogee, Oklahoma; *Becuna* in Philadelphia; *Bowfin* at Pearl Harbor; *Drum* at Battleship Memorial Park in Mobile, Alabama; *Clamagore* in Mount Pleasant, South Carolina; *Cobia* in Manitowoc, Wisconsin; *Cod* in Cleveland; *Croaker* in Buffalo, New York; *Ling* in Hackensack, New Jersey; *Lionfish* in Fall River, Massachusetts; *Pampanito* in San Francisco; *Razorback* in North Little Rock, Arkansas; *Requin* in Pittsburgh; *Silversides* in Michigan; *Torsk* in Baltimore; and, finally, *Cavalla* at Seawolf Park in Galveston, Texas.

Local Sub Vets state chapters began organizing memorials to the men and boats lost during the war. The Texas unit had been chartered as the "Submarine *Seawolf* Commission" and in May 1967, they dedicated a block granite marker capped by a torpedo and bronze plaque to commemorate *Seawolf*. It was located near the battleship *Texas*, but the Submarine *Seawolf* Commission still hoped to add a World War II submarine to their efforts.

In 1958, a bridge was built to Pelican Island, a small island that creates the Galveston Ship Channel with Galveston Island. Seawolf Park was established and was opened in 1974 as Texas' tribute to the lost submariners of World War II. Located within sight of a major

Although her conning tower was modified when she was converted to a hunter-killer submarine in the 1950s, *Cavalla*'s interior still provides visitors the feel of how life was aboard a World War II submarine. Many visitors to *Cavalla* in Seawolf Park assume they are touring the *USS Seawolf*. Photo by David H. Hunt.

cruise ship embarkation area, the park now features the destroyer escort *Stewart* (DE-238) and one of the few remaining World War II submarines, *Cavalla* (SS-244).

Originally, the SubVets commission planned to display the submarine *USS Cabrilla* at Seawolf Park but delays in preparing the park site caused *Cabrilla* to rust and deteriorate. In 1970, the Submarine Seawolf Commission petitioned the Navy to exchange *Cabrilla* for *Cavalla*, which had just been decommissioned and sent to the Reserve Fleet in Orange, Texas. *Cavalla* was saved from being scrapped when the Navy transferred possession of her to the SubVets chapter.[4]

Stewart was commissioned on 31 May 1943 and she successfully completed some 30 escort missions through the North Atlantic. The first destroyer *Stewart* (DD-224), was an old World War I tin can which was caught in drydock when the Japanese had invaded Java in March 1942. Retreating U.S. troops tried unsuccessfully to scuttle her, but she was later refitted and commissioned into the service of the Imperial Japanese Navy. At the end of the war, the U.S. Navy recovered the original *Stewart*, towed her to Pearl Harbor, and recommissioned her into service as the *USS Ramp* (Recovered Allied Military Property). The second destroyer *Stewart* which now resides at Seawolf Park had been so commissioned at a time when the first destroyer was believed to have been lost.

Just as historically significant at Seawolf Park is the proud *Cavalla*, famed for sinking one of the Japanese aircraft carriers which had carried out the 1941 Pearl Harbor sneak attack. She earned a Presidential Unit Citation on her first patrol and eventually destroyed more than 34,000 tons of shipping.

Cavalla continued to serve in the U.S. Navy until she was decommissioned on 30 December 1969. The U.S. Submarine Veterans of WWII Association of Texas obtained possession of the ship on 21 January 1971, and had her moved to Seawolf Park in Galveston. Towed to Pelican Island, she was backed into her prepared slip and secured. A cofferdam was built to lift *Cavalla* six feet above mean low tide and bulldozers pushed dirt against her hull to make her a landlocked sub, thereby being less expensive to maintain. *Cavalla* was opened to the public for tours on 11 April 1971—on the 71st anniversary of the U.S. Submarine Service.[5]

Visitors to Seawolf Park can roam the destroyer escort and the submarine at will. Youngsters eagerly clamber down the ladder into *Cavalla*'s after torpedo room and then make their way forward through the open hatches, curiously inspecting and reflecting upon life aboard a World War II submarine. They move from the maneuvering room forward through the two engine rooms, past the 36 bunks in the after battery compartment, and into the crew's mess and galley. Forward in the control room, visitors can see the two giant wheels that controlled the planes, the Christmas Tree hull opening indicator light panel, and the chart table.

Above, in *Cavalla*'s conning tower, kids can see an early TDC, periscopes, radar, and other equipment—while realizing how cramped this important little compartment was during torpedo attacks. Moving forward past officer's country and the skipper's stateroom, visitors exit *Cavalla* through her forward torpedo room. Some of those who tour assume they have explored the famous *Seawolf* of World War II. Once near the battleship *Texas* on the San Jacinto battlegrounds, the relocated torpedo-capped marker memorial now helps to clarify this fact. On it, the names of the 100 men who are still on eternal patrol with SS-197 are inscribed for visitors to view.

Epilogue

While the plaque memorializes the sailors and special forces men who gave their lives on *Seawolf*, the park itself is a tribute to more than 3,400 other men who also remain on eternal patrol. They paid the ultimate price for our freedom, giving their lives as volunteers of the underseas fighting team known as the Silent Service.

This marker at Seawolf Park—across the bay from Galveston, Texas, on Pelican Island—is a tribute to the men who lost their lives on the *Wolf*'s fifteenth patrol. Author's photo.

Appendix A
Top U.S. Submarines of World War II by Tonnage Sunk (JANAC Credit)

Top Submarines By Tonnage of Ships Sunk
(According to Joint Army-Navy Assessment Committee figures.)

	Boat	Skippers	JANAC Tonnage
1.	Flasher	R. T. Whitaker, G. W. Grider	100,231
2.	Rasher	Hutchinson, Laughon, Munson	99,901
3.	Barb	Waterman, Fluckey	96,628
4.	Tang *	O'Kane	93,824
5.	Silversides	Burlingame, Coye, Nichols	90,080
6.	Spadefish	Underwood, Germershausen	88,091
7.	Trigger *	Benson, Dornin, Harlfinger, Connole	86,552
8.	Drum	Rice, McMahon, Williamson, Rindskopf	80,580
9.	Jack	Dykers, Krapf, Fuhrman	76,687
10.	Snook *	Triebel, Browne	75,473
11.	Tautog	Willingham, Sieglaff, Baskett	72,606
12.	Seahorse	Cutter, Wilkins	72,529
13.	Guardfish	Klakring, Ward	72,424
14.	Seawolf *	Warder, Gross	71,609
15.	Gudgeon *	Grenfell, Stovall, Post	71,047
16.	Sealion II	Reich, Putman	68,297
17.	Bowfin	Willingham, Griffith, Corbus, Tyree	67,882
18.	Thresher	Anderson, Millican, Hull, MacMillan, Middleton	66,172
19.	Tinosa	Daspit, Weiss, Latham	64,655
20.	Grayback *	Saunders, Stephan, Moore	63,835

* Boat lost in action during World War II.

Appendix B
Top U.S. Submarines of World War II by Ships Sunk (JANAC Credit)

Top Submarines By Number of Ships Sunk
(According to JANAC figures.)

	Boat	Ships Sunk	No. of Patrols	Total Days on Patrol	JANAC Efficiency (sinking per)
1.	Tautog	26	13	584	22.46 days
2.	Tang *	24	5	203	8.46 days
3.	Silversides	23	13	701	30.48 days
4.	Flasher	21 (tie)	6	330	15.71 days
5.	Spadefish	21 (tie)	5	220	10.48 days
6.	Seahorse	20	8	407	20.35 days
7.	Wahoo *	20	7	253	12.65 days
8.	Guardfish	19	12	662	34.84 days
9.	Rasher	18	8	390	21.67 days
10.	Seawolf *	18	15	522	29.00 days
11.	Trigger *	18	12	552	30.67 days
12.	Barb	17	12	545	32.06 days
13.	Snook *	17	9	415	24.41 days
14.	Thresher	17	15	596	35.06 days
15.	Bowfin	16	9	390	24.38 days
16.	Harder *	16	6	220	13.75 days
17.	Pogy	16	10	490	30.63 days
18.	Sunfish	16	11	540	33.75 days
19.	Tinosa	16	11	443	27.69 days
20.	Drum	15	13	646	43.07 days

* Boat lost in action during World War II.

Appendix C
Summary of USS *Seawolf* Ship Sinkings

SINKINGS
* Indicates credit officially posted by JANAC post-war.

Patrol and Date	Ship Name	Type	Japanese Gross Tonnage
Fifth War Patrol: Lt. Cdr. Warder			
15 June 1942	Nampo Maru	XPG	1,206*
		Patrol total:	1,206 sunk
Sixth War Patrol: Cdr. Warder			
14 August 1942	Hachigen Maru	C-APK	3,113*
25 August 1942	Showa Maru	Freighter	1,349*
		Patrol total:	4,462 sunk
Seventh War Patrol: Cdr. Warder			
2 November 1942	Gifu Maru	Freighter	2,933*
3 November 1942	Sagami Maru	X-APG	7,189*
8 November 1942	Keiko Maru	XPG	2,929*
		Patrol total:	13,051 sunk
Eighth War Patrol: Lt. Cdr. Gross			
15 April 1943	Kaihei Maru	XAP	4,575*
19 April 1943	Banshu Maru No. 5	XAG	389
20 April 1943	(Gun action)	Sampan	100
23 April 1943	P-39 (ex-*Tade*)	Destroyer	935*
23 April 1943	Nisshin Maru # 2	Tanker	17,579
26 April 1943	(Gun action)	Sampan	100
		Patrol total:	23,678 sunk
Ninth War Patrol: Lt. Cdr. Gross			
27 May 1943	(Gun action)	Sampan	100
20 June 1943	Shojin Maru	Freighter	4,739*
		Patrol total:	4,839 sunk

Tenth War Patrol: Cdr. Gross

31 August 1943	Shoto Maru	Freighter	5,253*
31 August 1943	Kokko Maru	Freighter	5,486*
2 September 1943	Fusei Maru	Freighter	2,256*
5 September 1943	(Gun action)	Sampan	100
5 September 1943	(Gun action)	Sampan	100
		Patrol total:	13,195 sunk

Eleventh War Patrol: Cdr. Gross

29 October 1943	Wuhu Maru	Freighter	3,222*
4 November 1943	Kaifuku Maru	Freighter	3,177*
		Patrol total:	6,399 sunk

Twelfth War Patrol: Cdr. Gross

10 January 1944	Asuka Maru	Freighter	7,523*
10 January 1944	Yahiko Maru	Freighter	5,747*
11 January 1944	Getsuyo Maru	Freighter	6,440*
14 January 1944	Yamatsuru Maru	Tanker	3,651*
16 January 1944	Tarushima Maru	Transport	4,865 (1/2 credit)
		Patrol total:	25,793 sunk

DAMAGE INFLICTED

Patrol and Date	Ship Name	Type	Japanese Gross Tonnage
First War Patrol: Lt. Cdr. Warder			
14 December 1941	Sanyo Maru	Armed AV	8,360
		Patrol total:	8,360 damage
Fourth War Patrol: Lt. Cdr. Warder			
1 April 1942	Naka	Light cruiser	5,195
		Patrol total:	5,195 damage
Tenth War Patrol: Cdr. Gross			
1 September 1943	Durban Maru	APK	7,163
		Patrol total:	7,163 damage
Twelfth War Patrol: Cdr. Gross			
9 November 1943	Amatsu Maru	Tanker	10,567
		Patrol total:	10,567 damage

Appendix D
Comparison of USS *Seawolf* Sinkings by Source

Patrol Number	Wartime Credit	JANAC Credit [1]	John Alden 1999 Credit [2]
1: Warder	0/0 tons	0/0 tons	0/0 tons
2: Warder	Special mission: no torpedoes fired		
3: Warder	Special mission: no torpedoes fired		
4: Warder	3/14,000	0/0 tons	0/0 tons
5: Warder	0/0 tons	1/1,206	1/1,206
6: Warder	2/8,100	2/4,462	2/4,462
7: Warder	3/16,800	3/13,051	3/13,051
8: Gross	3/13,100	2/5,395	6/23,678 [3]
9: Gross	1/4,300	1/4,739	2/4,839 [4]
10: Gross	2/15,300	3/12,996	5/13,195 [5]
11: Gross	2/14,000	2/6,399	2/6,399
12: Gross	4/24,000	4.5/25,793	5/28,230 [6]
13: Lynch	Special mission: no torpedoes fired		
14: Bontier	Special mission: no torpedoes fired		
15: Bontier	USS *Seawolf* lost		
Total by source:	20/109,600	18.5/73,854 [2]	26/95,060

[1] JANAC postwar credits. Includes shared sinking of *Tarushima Maru*, split with *Whale*.

[2] Adjusted ship sinking totals from John D. Alden's *United States and Allied Submarine Successes in the Pacific and Far East During World War II*. Second Edition, October, 1999, pp. S8-S19.

[3] Alden includes two 100-tons sampans destroyed by gunfire which were too small for JANAC to count. He also shows that one Japanese source credits tanker *Nisshin Maru* to *Seawolf* on 23 April 1943. *Nisshin Maru* was originally torpedoed by *Trout* on 7 February 1943 off Miri and was being towed for repairs.

[4] Alden includes a 100-ton sampans destroyed by gunfire which was too small for JANAC to count.

[5] Alden includes two 100-tons sampans destroyed by gunfire which were too small for JANAC to count.

[6] Alden equally credits the *Tarushima Maru* sinking to both *Seawolf* and *Whale*, who shared the kill.

Appendix E
Awards Given to *Seawolf* and Crew

Battle Stars

Seawolf sailors received thirteen battle stars on their Asiatic-Pacific Area Service Medal for participating in the following operations:

1 star: Philippine Island Operation;
 8–26 December 1941 and 10–20 February 1942
1 star: Netherlands East Indies Engagements
 Badoeng Strait; 19–20 February 1942
1 star: Submarine War Patrol, Pacific   Fourth Patrol
 15 February–7 April 1942
1 star: Submarine War Patrol, Pacific   Fifth Patrol
 12 May–2 July 1942
1 star: Submarine War Patrol, Pacific   Sixth Patrol
 25 July–15 September 1942
1 star: Submarine War Patrol, Pacific   Seventh Patrol
 7 October–1 December 1942
1 star: Submarine War Patrol, Pacific   Eighth Patrol
 3 April–3 May 1943
1 star: Submarine War Patrol, Pacific   Ninth Patrol
 17 May–12 July 1943
1 star: Submarine War Patrol, Pacific   Tenth Patrol
 11 August–15 September 1943
1 star: Submarine War Patrol, Pacific   Eleventh Patrol
 5 October–27 November 1943
1 star: Submarine War Patrol, Pacific   Twelfth Patrol
 22 December 1943–27 January 1944
1 star: Submarine War Patrol, Pacific   Fourteenth Patrol
 8–23 August 1944
1 star: Marianas Operation   Thirteenth Patrol
 Battle of the Philippine Sea
 19–20 June 1944

First Navy Unit Commendation

Given for *Seawolf*'s 4th war patrol, the citation reads:

"For outstanding heroism in action during the Fourth War Patrol in enemy Japanese-controlled waters of the Southwest Pacific. Operating dangerously during a period when Japanese naval and air power was at its height, the USS SEAWOLF boldly tracked her targets into treacherously confined and shallow waters to launch in rapid succession one smashing attack after another on heavily escorted convoys and vital enemy combatant units. Undeterred by inevitable countermeasures, she struck daringly to sink or damage thousands of tons of shipping including Japanese cruisers and transports, a freighter and a destroyer, returning persistently to the attack despite exceptionally severe depth charging which started numerous leaks, broke her instruments and put vital machinery out of action. Highly vulnerable during prolonged periods of relentless attack in perilously shallow waters, the SEAWOLF escaped destruction only by the skill, courage and superb seamanship of her gallant officers and men which enabled her to surface and retire from the scene of action after twenty-one hours of submergence and nearly two full days at battle stations."

Second Navy Unit Commendation

Given for *Seawolf*'s 7th, 10th, and 12th patrols, the citation reads:

"For outstanding heroism in action against enemy Japanese shipping during her Seventh War Patrol in the Davao Gulf, Palau and Yap areas, from October 7 to December, 1942; her Tenth War Patrol in the East China Sea from August 14 to September 15, 1943; and her Twelfth War Patrol in the East China Sea, December 22, 1943, to January 27, 1944. Persistent and daring, the USS SEAWOLF penetrated the mouths of enemy-held harbors in bold pursuit of hostile shipping. Striking fiercely n the face of severe depth charges, aerial bombs and gun fire, she launched a series of brilliant attacks which destroyed eleven cargo ships and extensively damaged two freighters. After exhausting all torpedoes, the SEAWOLF tenaciously pursued and demolished by gun fire one enemy ship and turned another back into the attack range of a friendly submarine. Her outstanding performance in combat attests to the skill and superb teamwork of her courageous officers and men and reflect the highest credit upon herself and the United States Naval Service."

Individual Commendations

The following awards were also received by officers and crewmen for service aboard *Seawolf* in World War II.

NAVY CROSS:
Gross, Cdr. Royce Lawrence	12th Patrol
Warder, Cdr. Frederick Burdett	4th Patrol

GOLD STAR (in lieu of additional Navy Cross):
Warder, Cdr. Frederick Burdett	7th Patrol

LEGION OF MERIT:
Gross, Cdr. Royce Lawrence	10th Patrol
Lynch, Lt. Cdr. Richard Barr	13th Patrol

SILVER STAR:
Butler, Mach. David	7th Patrol
Capece, CEM Edmund Currie	4th Patrol
Casler, CQM James Burr	4th Patrol
Deragon, Lt. William Nolin	4th Patrol
Eckberg, CRM Joseph Melvin	4th Patrol
Enslin, CMoMM Carl Henry	4th Patrol
Gross, Cdr. Royce Lawrence	8th Patrol
Holden, Lt. Richard	4th Patrol
John, Lt. Clary Leonard	12th Patrol
Mercer, Lt. James	1-7 Patrols
Reiland, CTM William Frederick Jr.	9, 10, 11, 12
Rogers, CRT Benjamin Franklin	9–12 Patrols
Smith, Lt. William Lee	12th Patrol
Syverson, Lt. Douglas Neil	7th Patrol
Whitman, Lt. William Alexander	8-9 Patrols

GOLD STAR (in lieu of additional Silver Star):
Gross, Cdr. Royce Lawrence	11th Patrol
Syverson, Lt. Douglas Neil	12th Patrol

BRONZE STAR:
Asa, Lt.(jg) Marion Lee	12th Patrol
Bennett, CGM John Nelson	10th Patrol
Bilkey, CEM John Howard	10th Patrol
Casler, Lt.(jg) James Burr	8th Patrol

Cox, Lt. Robert Leon	12th Patrol
Deragon, Lt. William Nolin	8th Patrol
Gervais, TM1c Joseph Rudolph Oliver	7th Patrol
Gross, Cdr. Royce Lawrence	9th Patrol
Hadley, CPhM William Thomas	10th Patrol
Hess, Lt. Cdr. John Borden	13th Patrol
Leffingwell, Ens. George Darrell	12th Patrol
McCoy, MoMM1c Walter Glen	7th Patrol
Reiland, CTM William Frederick Jr.	7th Patrol
Rogers, CRT Benjamin Franklin	10th Patrol
Smith, Lt. William Lee	13th Patrol
Souza, CTM Edwin Enos	1-7th patrols
Thomson, CQM Henry Hansford	12th Patrol
Warder, Cdr. Frederick Burdett	6th Patrol

GOLD STAR (in lieu of additional Bronze Star):

Reiland, CTM William Frederick Jr.	8th Patrol

NAVY LETTER OF COMMENDATION RIBBON:

Mar, Delbert	12th Patrol
Mercer, Lt. James	10th Patrol
Robinson, Lt. Dougald "G"	12th Patrol
Syverson, Lt. Cdr. Douglas Neil	11th Patrol
Warder, Lt. Cdr. Frederick Burdett	3rd Patrol
Warder, Lt. Cdr. Frederick Burdett	5th Patrol

PURPLE HEART AWARD:
Presented to the families of all 100 men who were lost in the fatal depth charging of *Seawolf* on 3 October 1944.

Special thanks to Patrick R. Osborn of the National Archives for his assistance in researching awards for *Seawolf* officers and enlisted men. My apologies for any crewman's award which may have been overlooked. The list for Letter of Commendation Ribbon is considered only a partial list.

Chapter Notes

CHAPTER 1
"WE'RE AT WAR"

1. Frank, Gerald, and James D. Horan. *U.S.S. Seawolf. Submarine Raider of the Pacific* (New York: G. P. Putnam's Sons, 1945), 5.
2. Roscoe, Theodore. *United States Submarine Operations in World War II* (Annapolis: Naval Institute Press, 1949), 26. Frank and Horan, *USS Seawolf*, 26–7. LaVO, Carl. *Back From the Deep. The Strange Story of the Sister Subs Squalus and Sculpin* (Annapolis: Naval Institute Press, 1994), 87.
3. Connaughton, Richard. *MacArthur and Defeat in the Philippines* (Woodstock and New York: The Overlook Press, 2001), 160.
4. Astor, Gerald. *Crisis in the Pacific. The Battles for the Philippine Islands by the Men Who Fought Them—An Oral History* (New York: Donald I. Fine Books, 1996), 1–3.
5. Roscoe, *United States Submarine Operations*, 25–6.
6. Frank and Horan, *USS Seawolf*, 26.
7. Blair, Clay, Jr. *Silent Victory. The U.S. Submarine War Against Japan*, 2 vols. (Philadelphia: J. B. Lippincott Company, 1975), I: 105.
8. Frank and Horan, *USS Seawolf*, 25.
9. Ibid, 25–6.
10. Ibid, 26–7.
11. Blair, *Silent Victory*, 107–8.
12. LaVo, *Back From the Deep*, 88.
13. Blair, *Silent Victory*, 108.
14. Ibid, 107.
15. Ibid, 108.
16. Timothy Holden interview, August 9, 2005.
17. Frank and Horan, *USS Seawolf*, 6.
18. Young, Donald J. *First 24 Hours of War in the Pacific* (Shippensburg, PA: Burd Street Press, 1998), 73, 82–3.
19. Frank and Horan, *USS Seawolf*, 6.
20. Ibid, 28.

21. Warder, Rear Admiral Frederick B. Biographical material and correspondence with Clay Blair Jr. for the book *Silent Victory*, circa 1971-72. Clay Blair Papers, Acc. 8295, Box 70, Folder No. 6. American Heritage Center, University of Wyoming. Hereafter cited as Warder–Blair interview.

22. Ibid.

23. "History of the USS *Seawolf* (SS 197)." Office of Naval Records and History, Ship's Histories Section, Navy Department, 1.

24. Stavros, Sam. "Frederick B. Warder: Letters to Home From *USS Seawolf*." *The Submarine Review* (article date unknown), 119–20; contributed by Ms. Ruth Johnson.

25. Warder-Blair interview.

26. Warder-Blair interview. Stavros, "Frederick B. Warder: Letters to Home From *USS Seawolf*," 121.

27. Frank and Horan, *USS Seawolf*, 19–20.

28. Ibid, 22–3.

29. Ibid, 20–1.

30. Blair, *Silent Victory*, I:133.

31. Blair-Warder interview.

32. Blair, *Silent Victory*, I:133.

33. Ibid, I:113.

34. Stavros, "Frederick B. Warder: Letters to Home From *USS Seawolf*," 120–1.

35. Ms. Ruth Johnson to author, September 22, 2006.

36. Frank and Horan, *USS Seawolf*, 28.

37. Ibid, 28–9.

38. "History of the USS *Seawolf*," 1.

Chapter 2
Faulty Fish

1. Blair, *Silent Victory*, I:112.

2. Frank and Horan, USS *Seawolf*, 10.

3. Ibid, 11–2.

4. Ibid, 6.

5. Ibid, 16.

6. Warder-Blair interview.

7. Frank and Horan, USS *Seawolf*, 33.

8. Ibid, 33–4.

9. Warder-Blair interview.

10. Frank and Horan, USS *Seawolf*, 35.

11. Ibid, 8.

12. Warder-Blair interview.

13. Frank and Horan, USS *Seawolf*, 39.

14. Ibid, 41.

15. Warder-Blair interview.

16. Ibid.

17. Ibid.

18. Newpower, Anthony. *Iron Men and Tin Fish: The Race to Build a Better Torpedo During World War II* (Westport, CT: Praeger Security International, 2006), 19, 22–23, 29.
19. Ibid, 24–25.
20. Ibid, 30.
21. Blair, *Silent Victory*, I:115–18; Newpower, *Iron Men and Tin Fish*, 61–4.
22. Newpower, *Iron Men and Tin Fish*, 39–46.
23. Ibid, 49–53.
24. Frank and Horan, *USS Seawolf*, 43.
25. Stavros, "Frederick B. Warder: Letters to Home From *USS Seawolf*," 120–1.
26. Frank and Horan, *USS Seawolf*, 45.
27. Ibid, 47.
28. Ibid, 49. "Information concerning Chief Petty Officer Walter Glen (Bud) McCoy, courtesy of Mike McCoy, March 11, 2006.
29. Frank and Horan, USS *Seawolf*, 51.
30. Ibid, 51–2.
31. Stavros, "Frederick B. Warder: Letters to Home From *USS Seawolf*," 120–1.

CHAPTER 3
SPECIAL MISSIONS

1. Frank and Horan, *USS Seawolf*, 55–6.
2. Roscoe, *Submarine Operations in World War II*, 29–30.
3. Blair, *Silent Victory*, I:130–1.
4. Frank and Horan, *USS Seawolf*, 57.
5. Ibid, 6.
6. LaVo, *Back From the Deep*, 89-91. Warder-Blair interview.
7. Warder-Blair interview.
8. Blair, *Silent Victory*, I: 132.
9. Frank and Horan, *USS Seawolf*, 58–9.
10. Ibid, 62.
11. Ibid, 62.
12. Ibid, 63–4.
13. Ibid, 64.
14. Blair, *Silent Victory*, I:150.
15. Warder-Blair interview.
16. Frank and Horan, *USS Seawolf*, 67.
17. Ibid, 68–9.
18. Ibid, 66.
19. Blair, *Silent Victory*, I:150.
20. Karl Kramer letter of May 11, 2005.
21. Frank and Horan, *USS Seawolf*, 70–2.
22. Ibid, 74.
23. Elphick, Peter. *Far Eastern File. The Intelligence War in the Far East, 1930-1945* (London: Hodder & Stoughton, 1997), 293–4.

24. Bartsch, William H. *Doomed From the Start. American Pursuit Pilots in the Philippines, 1941-1942* (College Station: Texas A&M University Press, 1992), 215, 238.
25. Ibid, 174.
26. Edmonds, Walter D. *They Fought With What They Had* (Boston: Little Brown & Company, 1951), 138.
27. Ibid, 121–2.
28. Ibid, 206–11.
29. Dow, Gerald Wayne. "Unit History of the 27th Bombardment Group. U.S. Army Air Corps. World War II: 1 January 1940–1 September 1942" (Walnut Creek, CA: Privately published, 2006), 276. Accessed http://www.lindadown.net/pdffiles/bombgplx.pdf on 17 January 2007.
30. Frank and Horan, *USS Seawolf*, 75–6.
31. Ibid, 76–7.
32. Dow, "Unit History of the 27th Bombardment Group," 276.
33. Frank and Horan, *USS Seawolf*, 78.
34. Dow, "Unit History of the 27th Bombardment Group," 276.
35. Ibid, 276.
36. Ibid, 277.
37. Ibid, 277.

Chapter 4
"Into a Hornet's Nest"

1. Frank and Horan, *USS Seawolf*, 81.
2. Blair, *Silent Victory*, 170–1.
3. Frank and Horan, *USS Seawolf*, 81.
4. Ibid, 84.
5. Ibid, 85.
6. "*HIJMS Nagara*: Tabular Record of Movement." Research of Bob Hackett and Sander Kingsepp, 1997-2003. Accessed http://www.combinedfleet.com/nagara_t.htm on 12 July 2005. Hereafter cited as "*HIJMS Nagara*."
7. Frank and Horan, *USS Seawolf*, 87–8.
8. Ibid, 88–9.
9. Ibid, 89–90.
10. Ibid, 90.
11. Ibid, 91.
12. Ibid, 92.
13. Ibid, 93.
14. Warder-Blair interview.
15. Johnson, Charles. "The Life and Death of *USS Seawolf*: SS197." *Polaris*, June 1999, Vol. 43: 3, 12.
16. Frank and Horan, *USS Seawolf*, 94.
17. "*HIJMS Nagara*."
18. Warder-Blair interview.
19. Frank and Horan, *USS Seawolf*, 96.
20. Ibid, 98.

21. Ibid, 99–100.

CHAPTER 5
"FEARLESS FREDDIE"

1. Warder-Blair interview.
2. Frank and Horan, *USS Seawolf*, 101.
3. Ibid, 102.
4. Frank and Horan, *USS Seawolf*, 102.
5. Ibid, 108.
6. "*HIJMS Nagara.*"
7. Frank and Horan, *USS Seawolf*, 105.
8. Ibid, 107.
9. "*HIJMS Nagara.*"
10. Frank and Horan, *USS Seawolf*, 107.
11. Mendenhall, Rear Admiral Corwin, USN (Ret.). *Submarine Diary. The Silent Stalking of Japan* (Chapel Hill, North Carolina: Algonquin Books, 1991), 39.
12. Frank and Horan, *USS Seawolf*, 107–8.
13. Ibid, 108–9.
14. "*HIJMS Naka.*"
15. Frank and Horan, *USS Seawolf*, 109–10.
16. Ibid, 112.
17. "*HIJMS Naka.*"
18. Warder-Blair interview.
19. Frank and Horan, *USS Seawolf*, 112.
20. Frank and Horan, *USS Seawolf*, 112–3.
21. Ibid, 113.
22. Ibid, 114–5.
23. Ibid, 115.
24. Karl Kramer email of July 19, 2005.
25. Frank and Horan, *USS Seawolf*, 117–8.
26. "*HIJMS Naka.*"
27. Frank and Horan, *USS Seawolf*, 118–19.
28. Ibid, 119–20.
29. Ibid, 120–1.
30. Mendenhall, *Submarine Diary*, 50–1. Frank and Horan, *USS Seawolf*, 121.
31. Frank and Horan, *USS Seawolf*, 121–2.
32. Warder-Blair interview.

CHAPTER 6
THE JINX

1. Frank and Horan, *USS Seawolf*, 123–4.
2. Ibid, 124.
3. Ibid, 125.
4. Ibid, 125–6.

5. Ibid, 127.
6. Billie Mallough letter, December 3, 2005.
7. Cline, Rick. *Submarine Grayback. The Life & Death of the WWII Sub, USS Grayback* (Placentia, CA: R.A. Cline Publishing, 1999),64–7. Frank and Horan, *USS Seawolf*, 128.
8. Frank and Horan, *USS Seawolf*, 130.
9. Ibid, 131.
10. Ibid, 133.
11. Ibid, 133.
12. Ibid, 135.
13. Ibid, 136.
14. Ibid, 136–7.
15. Ibid, 137.
16. Ibid, 140.
17. Ibid, 144–5.
18. Ibid, 147.
19. Ibid, 147–8.
20. Frank and Horan, *USS Seawolf*, 148.
21. Blair, *Silent Victory*, I:316–7.
22. Warder-Blair interview.
23. Blair, *Silent Victory*, I:251–3.
24. Frank and Horan, *USS Seawolf*, 151.

CHAPTER 7
"A MARVELOUS SPECTACLE"

1. Frank and Horan, *USS Seawolf*, 152–4.
2. Warder-Blair interview.
3. Ibid.
4. Karl Kramer letter of May 11, 2005.
5. Frank and Horan, *USS Seawolf*, 162–3.
6. Blair, *Silent Victory*, I:266.

CHAPTER 8
SAGAMI MARU

1. Frank and Horan, *USS Seawolf*, 152–4.
2. Blair, *Silent Victory*, I:327.
3. Frank and Horan, *USS Seawolf*, 164.
4. Ibid, 167.
5. Ibid, 168.
6. Ibid, 169.
7. Ibid, 171–2.
8. Ibid, 174–5.
9. Ibid, 175–6.
10. Ibid, 177.

CHAPTER 9
"CIRCULAR RUN!"

1. Frank and Horan, *USS Seawolf*, 178.
2. Tully, Antony P. "IJN *Junyo*: Tabular Record of Movement." Accessed http://www.combinedfleet.com/junyo.htm on September 25, 2005. Three other Japanese carriers—*Zuiho*, *Zuikaku* and *Shokaku*—had reached Kure, Japan, by 9 November 1942, the former two damaged from carrier battles in the Solomons with U.S. forces. Only one other carrier was in the vicinity of Micronesia. *Hiyo* was in Truk during November with disabled engines. See also: Lundstrom, John B. *The First Team and the Guadalcanal Campaign. Naval Fighter Combat from August to November 1942* (Annapolis, MD: Naval Institute Press, 1994), 341–2, 454.
3. Frank and Horan, *USS Seawolf*, 182.
4. Ibid, 182–3.
5. Blair, *Silent Victory*, I:334.
6. Warder-Blair interview.
7. Frank and Horan, *USS Seawolf*, 187.
8. Warder-Blair interview.
9. Ibid.

CHAPTER 10
"HOPE THIS OLD MAN KNOWS HIS STUFF"

1. Frank and Horan, *USS Seawolf*, 190.
2. Ibid, 192–3.
3. Ibid, 195.
4. Ibid, 196.
5. Ibid, 197.
6. Warder-Blair interview.
7. Gross to Blair letter, January 12, 1972, from Gross Papers in Clay Blair Papers, Acc. 8295, Box 70, Folder No. 6. American Heritage Center, University of Wyoming.
8. Blair to Gross letter, May 8, 1972, from Gross Papers.
9. Warder-Blair interview.
10. Blair to Gross letter, May 9, 1972, from Gross Papers.
11. Gross to Blair letter, January 12, 1972, from Gross Papers.
12. W. G. McCoy to parents, 3 March 1943. Courtesy of Mike McCoy.
13. W. G. McCoy to parents, 29 March 1943. Courtesy of Mike McCoy.
14. Mike McCoy, nephew of Bud McCoy, to author.
15. 1943 diary of W. G. "Bud" McCoy, courtesy of Jeff and Harold McCoy.
16. Ibid.
17. Ibid.
18. Ibid.
19. Ibid.
20. Ibid.
21. Ibid.

22. Ibid.
23. Blair, *Silent Victory*, I:421-22. Alden, John. *U.S. Submarine Attacks During World War II* (Annapolis, MD: Naval Institute Press, 1989), 40.
24. Gross to Blair letter, January 12, 1972, from Gross Papers.
25. 1943 diary of W. G. "Bud" McCoy, courtesy of Jeff and Harold McCoy.
26. Margarite Mills to author, August 5, 2005.
27. 1943 diary of W. G. "Bud" McCoy, courtesy of Jeff and Harold McCoy.
28. Blair, *Silent Victory*, I:328-9.
29. 1943 diary of W. G. "Bud" McCoy, courtesy of Jeff and Harold McCoy.
30. Ibid.
31. Alden, *U.S. Submarine Attacks During World War II*, 40. See also Alden, Cdr. John D., USN (Ret.). *United States and Allied Submarine Successes in the Pacific and Far East During World War II. Chronological Listing.* Pleasantville, NY: Self-published, second edition, October 1999. In this follow-up book, Cdr. Alden gives credit to *Seawolf* for finishing off *Nisshin Maru*, p. S-8.
32. 1943 diary of W. G. "Bud" McCoy, courtesy of Jeff and Harold McCoy.
33. Ibid.
34. Ibid.
35. Ibid.
36. Newpower, Iron Men and Tin Fish, 143–51.

Chapter 11
"The Next One's Gonna Get Us"

1. 1943 diary of W. G. "Bud" McCoy, courtesy of Jeff and Harold McCoy.
2. Ibid.
3. O'Kane, Richard H., Rear Admiral, USN (Ret.). *Wahoo. The Patrols of America's Most Famous World War II Submarine* (Novato, CA: Presidio Press, 1987), 248–9.
4. *United States Submarine Veterans of World War II. A History of the Veterans of the United States Naval Submarine Fleet* (Dallas: Taylor Publishing Company, 1984-1990), III: 270-71.
5. 1943 diary of W. G. "Bud" McCoy, courtesy of Jeff and Harold McCoy.
6. Risser, Robert D. Biographical material and correspondence with Clay Blair Jr. for the book *Silent Victory*, circa 1971-72. Clay Blair Papers, Acc. 8295, Box 70, Folder No. 1. American Heritage Center, University of Wyoming. Risser to Blair, 16 December 1971, Clay Blair Papers.
7. Risser to Blair, 16 December 1971 and Risser to Blair, 26 April 1972, Clay Blair Papers.
8. Risser to Blair, 16 December 1971, Clay Blair Papers.
9. Risser to Blair, 26 April 1972, Clay Blair Papers.

Chapter Notes 435

10. Ibid.
11. 1943 diary of W. G. "Bud" McCoy, courtesy of Jeff and Harold McCoy.
12. Ibid.
13. Ibid.
14. Ibid.
15. Ibid.
16. Ibid.
17. Ibid.
18. Ibid.
19. Ibid.
20. Ibid.
21. Ibid.
22. Ibid.
23. Ibid.
24. Ibid.
25. Ibid.
26. Ibid.
27. Ibid.
28. Ibid.
29. Ibid.

Chapter 12
"Looks Like A Long Chase"

1. Margarite Mills to author, August 5, 2005.
2. "Biography for Lewis Sigmund Donche," June 2, 2006, courtesy of interviews with Mr. Donche by his son, Mark Donche.
3. Amy Wiegenstein to author, December 19, 2005.
4. 1943 diary of W. G. "Bud" McCoy, courtesy of Jeff and Harold McCoy.
5. Ibid.
6. Ibid.
7. W. G. Somerville research.
8. Ibid.
9. Robinson to Randy Mohr, 2001 audiotape.
10. 1943 diary of W. G. "Bud" McCoy, courtesy of Jeff and Harold McCoy.
11. W. G. Somerville research.
12. D. G. Robinson to J. Murphy, Naval Weapons Stations, courtesy of Claude Robinson.
13. 1943 diary of W. G. "Bud" McCoy, courtesy of Jeff and Harold McCoy.
14. Robinson to Randy Mohr, 2001 audiotape.
15. 1943 diary of W. G. "Bud" McCoy, courtesy of Jeff and Harold McCoy.
16. Robinson to Randy Mohr, 2001 audiotape.

17. Ibid.
18. 1943 diary of W. G. "Bud" McCoy, courtesy of Jeff and Harold McCoy.
19. Robinson to Randy Mohr, 2001 audiotape.
20. 1943 diary of W. G. "Bud" McCoy, courtesy of Jeff and Harold McCoy.
21. Risser to Blair, 16 December 1971, Clay Blair Papers.

CHAPTER 13
ALL TORPEDOES EXPENDED

1. Accessed www.state.sd.us/military/VetAffairs/sdwwiimemorial/SubPages/profiles/Display.asp?P=140 on October 6, 2006.
2. Risser to Blair, 26 April 1972, Clay Blair Papers.
3. Kimmett, Larry and Margaret Regis. *U.S. Submarines in World War II. An Illustrated History* (Seattle: Navigator Publishing, 1996), 139.
4. W. G. Somerville research.
5. Ibid.
6. Blair to Gross letter, May 8, 1972, from Gross Papers.
7. W. G. Somerville research. Roscoe, *United States Submarine Operations in World War II*, 210. Roscoe cites a captured Japanese document translated during wartime which called the ship *Akatsuki Maru* which was hit by dud torpedoes.
8. Ibid. The wartime document only mentions four dud hits, although Somerville's notes from the Japanese records say five.
9. Ibid.

CHAPTER 14
"A MASTERFUL PERFORMANCE"

1. "Lt. Robinson's Pig Boat Rips Nips; Gets Ash Cans." *Texas City Sun* article published in 1944, courtesy of Claude Robinson.
2. Ibid.
3. W. G. Somerville research.
4. Ibid.
5. Ibid.
6. Ibid.
7. Ibid.
8. Gross to Blair letter, May 18, 1972, from Gross Papers. See also Blair, *Silent Victory*, I:502.
9. W. G. Somerville research.
10. Ibid.
11. "Further Commentary About *Seawolf*." Col. Vernon T. Miller, USAF (Ret.). *Polaris*, Vol. 42, No. 6 (Dec. 1998), 27. Miller, a radioman on *Whale* at the time, later joined the Air Force.
12. W. G. Somerville research.
13. Gross to Blair letter, May 18, 1972, from Gross Papers.

Chapter 15
Lifeguard League

1. Kimmett and Regis, *U.S. Submarines in World War II,* 139.
2. Blair, *Silent Victory*, I:491.
3. Mendenhall, *Submarine Diary*, 33.
4. Hess to Blair, May 23, 1972, Blair Papers.
5. Tillman, Barrett. *The True Story of the Marianas Turkey Shoot of World War II* (New York: NAL Caliber, 2005), 213–6.
6. Ibid, 239.
7. Houston, Charles. *Flying With Iron Angels. The Diaries and Memories of Navy Carrier Pilots Fighting the Pacific War in 1944* (Fresno, CA: Privately published, 2001), 130-1.
8. Ibid, 144–6.
9. Tillman, *Clash of the Carriers*, 276.
10. Houston, *Flying With Iron Angels*, 146.
11. Ibid.
12. Houston, *Flying With Iron Angels*, 164.
13. Lockwood, Charles A., Vice Admiral, USN, Ret. *Zoomies, Subs and Zeros. Heroic Rescues in World War II by the Submarine Lifeguard League* (Philadelphia and New York: Chilton Company, 1956), 187–8.
14. Ibid, 188.

Chapter 16
Guerrilla Warfare

1. Galatin, Admiral I. J., USN (Ret.). *Take Her Deep! A Submarine Against Japan in World War II* (New York: Pocket Books, 1988), 35.
2. Kimmett and Regis, *U.S. Submarines in World War II,* 139.
3. Beynon, Dr. Robert P. *The Pearl Harbor Avenger: U.S.S. Bowfin* (Deland, FL: Just Books 1, 2002), 80.
4. Ostlund, Mike. *Find 'Em, Chase 'Em, Sink 'Em. The Mysterious Loss of the WWII Submarine USS Gudgeon* (Guilford, CT: The Lyons Press, 2006), 150.
5. Willoughby, Charles A. *The Guerilla Resistance Movement in the Philippines: 1941-1945* (New York: Vantage Press, 1972), 117–18.
6. Ibid, 134, 545.
7. Ibid, 45, 167.
8. Ibid, 160.
9. Ibid, 160–1.
10. Ibid, 45.

Chapter 17
On Eternal Patrol

1. Fred Christianson recollections accessed at http://www.geocities.com/tritonbase/who/whosC.htm on January 17, 2006.
2. Zedric, Lance Q. *Silent Warriors of World War II. The Alamo Scouts Behind Japanese Lines* (Ventura, CA: Pathfinder Publishing of California, 1995), 96.
3. Ibid, 33–42.

4. Ibid, 62–89.
5. Ibid, 120–45.
6. Ibid, 147–60.
7. Holmes, Harry. *The Last Patrol* (Shrewsbury, England: Airlife Publishing Ltd., 1994), 128–9.
8. "History of the *USS Seawolf*," 5.
9. Hornfischer, James D. *The Last Stand of the Tin Can Sailors: The Extraordinary World War II Story of the U.S. Navy's Finest Hour* (New York: Bantam Books, 2004), 78–9. William Brooks telephone interview with author, January 26, 2007.
10. Kimmett and Regis, *U.S. Submarines in World War II*, 114–5. Brooks interview.
11. "History of the *USS Seawolf*," 5.

Epilogue

1. Holmes, *The Last Patrol*, 8.
2. Johnson, "The Life and Death of *USS Seawolf*: SS197," 12.
3. Risser biographical material, Clay Blair Papers.
4. "*Seawolf*." Booklet honoring *Seawolf* and the submariners lost during World War II, including ship's histories of *Seawolf*, *Cavalla*, and *Stewart*. Corpus Christi, TX: South Coast Publishing, 1983.
5. Ibid.

Bibliography

DOCUMENTS, MANUSCRIPTS AND COLLECTIONS

Gross, Royce L. Biographical Material. Clay Blair Jr. Papers from *Silent Victory*: Skippers of U.S. World War II Pacific Ocean Submarine Patrols folder at the American Heritage Center, University of Wyoming.

Hackett, Bob, Sander Kingsepp, Allan Alsleban and Peter Cundall. "IJN Seaplane Tender *Kiyokawa Maru*: Tabular Record of Movement." See http://www.combinedfleet-.com/Kiyokawa%20Maru_t.htm. Accessed January 29, 2005.

"*HIJMS Naka*: Tabular Record of Movement." Research of Bob Hackett and Sander Kingsepp, 1997-2002. Accessed http://www.combinedfleet.com/naka_t.htm on 12 July 2005.

"*HIJMS Nagara*: Tabular Record of Movement." Research of Bob Hackett and Sander Kingsepp, 1997-2003. Accessed http://www.combinedfleet.com/nagara_t.htm on 12 July 2005.

"History of the *USS Seawolf* (SS 197)." Office of Naval Records and History, Ship's Histories Section, Navy Department.

Naval Historical Center, Washington D.C. Officer biography sheets for Rear Admiral Frederick B. Warder, Captain Richard Holden, Captain William N. Deragon, Captain James Mercer and Captain James B. Casler.

Risser, Robert D. Biographical material and correspondence with Clay Blair Jr. for the book *Silent Victory*, circa 1971-72. Clay Blair Papers, Acc. 8295, Box 70, Folder No. 1. American Heritage Center, University of Wyoming.

Robinson, Dougald G. Papers, photos, and 2001 taped interview with relative Randy Mohr. Courtesy of his son, Claude Robinson.

Somerville, W. G. Independent research conducted of translations of various Japanese convoy reports and naval publications. Courtesy of David Bouslog.
Submarine Force Museum Archives. Groton, Connecticut. Wendy S. Gulley, archivist. Images sent November 9, 2005.
Tully, Antony P. "*IJN Junyo:* Tabular Record of Movement." Accessed http://www.combinedfleet.com/junyo.htm on September 25, 2005.
Warder, Rear Admiral Frederick B. Biographical material and correspondence with Clay Blair Jr. for the book *Silent Victory*, circa 1971-72. Clay Blair Papers, Acc. 8295, Box 70, Folder No. 6. American Heritage Center, University of Wyoming.
Whitman, William Alexander. Service papers and records provided courtesy of his wife, Pauline Whitman, on 14 October 2005.

CORRESPONDENCE, TELEPHONE INTERVIEWS

Bjerke, Marion and Keith Bjerke. Wife and son of the late Keith A. Bjerke. Telephone interview with Marion Bjerke on August 18, 2005.
Bilkey, John H. Telephone interview of May 14, 2005, with assistance from his son Brian Bilkey.
Bingham, Allen B. Telephone interviews of July 3, 2005, and January 28, 2006.
Bramer, John C. Jr. Telephone interview January 10, 2006. Jack also sent photos from his rescue by *Seawolf*.
Brooks, William C. Jr. Telephone interview January 26, 2007.
Cashero, Francis J. "Jim." Telephone interviews of July 3, 2005, August 3, 2005, and January 7, 2007.
Curtin, Robert R. Telephone interviews March 3, 2006, and April 14, 2006.
Cucchi, Ferdinand V. "Fred." Telephone interview July 7, 2005.
Dombroski, Nancy Valenti. Daughter of Louise Quarto Bargenquast Valenti, who was married to Arnold Bargenquast at the time he was lost with *Seawolf*. Email of August 17, 2005.
Donche, Lewis S. Through his son, Mark Donche, Lew provided copies of his *Seawolf* photos and recollections, including "Biography for Lewis Sigmund Donche," June 2, 2006.
Enslin, Carl H. Jr. and Cora (Enslin) Kobesky. Children of the late Carl H. Enslin Sr. Telephone interview July 3, 2005. The family also mailed a number of photos and papers from their father's collection.

Fogle, Olin R. Telephone interview of April 21, 2005.
George, Joan. Wife of the late Lloyd George. Email of September 7, 2005.
Hanson, Robert N. Telephone interviews of May 1, May 6, and May 21, 2005, and correspondence. Bob provided a number of photos and important papers on *Seawolf.*
Harlow, William R. Telephone interviews of June 28, 2005, and February 9, 2006. Bill also sent photos and papers from his *Seawolf* service.
Hill, Earl G. Jr. Nephew of the late Lloyd George. Email correspondence of August 4, 2005.
Holden, Timothy. Son of the late Richard Holden. Telephone interview August 9, 2005.
Houston, Charles. Author of *Flying With Iron Angels*, Mr. Houston put me in touch with fellow Air Group 18 aviator John Bramer, who was rescued by *Seawolf.*
Kisver, Norman. Telephone interview of August 5, 2005.
Kramer, Karl J. Nephew of the late Lee Bob Parden. Email correspondence from May-July 2005.
Leffingwell, George D. Telephone interview September 16, 2005. Additional correspondence with his daughter, Susan Leffingwell.
Lents, Robert W. Telephone interviews of May 14, 2005, and October 1, 2005.
Lober, Peter N. Telephone interview of April 30, 2005. Pete also sent a 1940 photo of the *Seawolf* crew.
Mallough, Bobbie. Telephone interview of December 3, 2005. Ms. Mallough also provided photos and papers from the collection of her late husband, Kenneth G. "Casey" Mallough.
Mar, Delbert. Telephone interview April 14, 2006.
McCoy, Mike. Nephew of Walter Glen "Bud" McCoy, who was lost with *Seawolf.* Mike shared papers, photos, and wartime letters of Bud McCoy, courtesy of McCoy's two surviving brothers—Claude and Harold McCoy. He also put me in contact with Bill Brooks, the Avenger pilot who made an attack on a submarine that might have been *Seawolf.*
Mickey, Rex L. Interviews of June 28, 2005, and February 9, 2006. Rex also provided photos from his *Seawolf* days.
Milas, Edward L. Telephone interviews of July 17, 2005, and followup. Ed also sent a number of his *Seawolf* papers and photos.
Mills, Margarite. Wife of the late Wilson Mills. Telephone interview of August 5, 2005.

Palazolo, Becky. Daughter of the late Harry Steward. Email correspondence of August 9, 2005.
Rajotte, Lucien T. Telephone interviews: July 3 and July 17, 2005.
Strong, Joseph H. Telephone interview of July 3, 2005.
Syverson, Hope. Wife of the late Douglas N. Syverson. Telephone interview of July 2005.
Thomson, Henry H. Telephone interview April 25, 2005. Hank provided a number of important papers of his *Seawolf* service, photos, and answered dozens of email queries from April 2005 through August 2006.
Whitman, Pauline. Wife of the late William Whitman. Telephone interview of October 7, 2005.
Wiegenstein, Amy. Telephone interview and email correspondence, December 19, 2005.
Woodard, Charles C. Telephone interview March 3, 2006.
Zimmerman, Paul. Telephone interviews April 15, 2005 and August 20, 2005.

ARTICLES

"Commander Warder Awarded Navy Cross." *Baltimore Evening Sun*, February 9, 1943.
"Decorated Navy Admiral Warder Dies." *The Charleston Gazette*, February 4, 2000.
"Further Commentary About *Seawolf*." Col. Vernon T. Miller, USAF (Ret.). *Polaris*, Vol. 42, No. 6 (Dec. 1998), 27.
Johnson, Charles. "The Life and Death of *USS Seawolf* SS197." *Polaris*, June 1999, Vol. 43: 3, 12.
"U.S. Subs at Work. Pictures taken through periscopes give glimpse of the destruction done to Jap ships by the 'silent service.'" *Life*, February 8, 1943, pp. 26–7. Although the submarine's name was not listed by the censors, each photo on this two-page spread was taken by *Seawolf*'s Lt. Jim Mercer.

BOOKS

Alden, Cdr. John D., USN (Ret.) *U.S. Submarine Attacks During World War II*. Annapolis: Naval Institute Press, 1989.

———. *United States and Allied Submarine Successes in the Pacific and Far East During World War II. Chronological Listing*. Pleasantville, NY: Self-published, second edition, October 1999.

Astor, Gerald. *Crisis in the Pacific. The Battles for the Philippine Islands by *the Men Who Fought Them—An Oral History.* New York: Donald I. Fine Books, 1996.

Bartsch, William H. *Doomed From the Start. American Pursuit Pilots in the Philippines, 1941-1942.* College Station: Texas A&M University Press, 1992.

Beach, Edward L., Cdr., USN. *Submarine!* New York: Henry Holt and Company, 1946.

Beynon, Dr. Robert P. *The Pearl Harbor Avenger: U.S.S. Bowfin.* Deland, FL: Just Books 1, 2002.

Blair, Clay, Jr. *Silent Victory. The U.S. Submarine War Against Japan.* Philadelphia: J. B. Lippincott Company, 1975. Reprint.

Bouslog, Dave. *Maru Killer. War Patrols of the USS Seahorse.* Placentia, CA: R.A. Cline Publishing, 1996. Second Printing, 2001.

Calvert, James F., Vice Admiral, USN (Ret.). *Silent Running. My Years on a World War II Attack Submarine.* New York: John Wiley & Sons, Inc., 1995.

Cline, Rick. *Final Dive. The Gallant and Tragic Career of the WWII Submarine, USS Snook.* Placentia, CA: R.A. Cline Publishing, 2001.

———. *Submarine Grayback. The Life & Death of the WWII Sub, USS Grayback.* Placentia, CA: R.A. Cline Publishing, 1999.

Connaughton, Richard. *MacArthur and Defeat in the Philippines.* Woodstock and New York: The Overlook Press, 2001.

Connor, Claude C. *Nothing Friendly in the Vicinity...My Patrols on the Submarine USS Guardfish During WWII.* Mason City, IA: Savas Publishing Company, 1999.

Davenport, Rear Admiral Roy M., USN (Ret.). *Clean Sweep.* New York: Vantage Press, 1986.

DeRose, James F. *Unrestricted Warfare. How a New Breed of Officers Led the Submarine Force to Victory in World War II.* New York: John Wiley & Sons, Inc., 2000.

Edmonds, Walter D. *They Fought With What They Had.* Boston: Little Brown & Company, 1951.

Elphick, Peter. *Far Eastern File. The Intelligence War in the Far East, 1930-1945.* London: Hodder & Stoughton, 1997.

Frank, Gerald, and James D. Horan. *U.S.S. Seawolf. Submarine Raider of the Pacific.* New York: G. P. Putnam's Sons, 1945. This wartime account of *Seawolf*'s first seven patrols was related to the authors by CRM Joseph Melvin Eckberg.

Fluckey, Admiral Eugene B. *Thunder Below! The USS Barb Revolutionizes Submarine Warfare in World War II.* Chicago: University of Illinois Press, 1992.

Galatin, Admiral I. J., USN (Ret.). *Take Her Deep! A Submarine Against Japan in World War II.* New York: Pocket Books, 1988.

Grider, George (as told to Lydel Sims). *War Fish.* Boston: Little, Brown & Company, 1958.

Gugliotta, Bobette. *Pigboat 39. An American Sub Goes to War.* Lexington: The University of Kentucky Press, 1984.

Hinkle, David Randall (editor). *United States Submarines.* New York: Barnes and Noble Books, 2002.

Holmes, Harry. *The Last Patrol.* Shrewsbury, England: Airlife Publishing Ltd., 1994.

Hornfischer, James D. *The Last Stand of the Tin Can Sailors: The Extraordinary World War II Story of the U.S. Navy's Finest Hour.* New York: Bantam Books, 2004.

Houston, Charles. *Flying With Iron Angels. The Diaries and Memories of Navy Carrier Pilots Fighting the Pacific War in 1944.* Fresno, CA: Privately published, 2001.

Kimmett, Larry and Margaret Regis. *U.S. Submarines in World War II. An Illustrated History.* Seattle: Navigator Publishing, 1996.

LaVO, Carl. *Back From the Deep. The Strange Story of the Sister Subs Squalus and Sculpin.* Annapolis: Naval Institute Press, 1994.

Lockwood, Charles A., Vice Admiral, USN, Ret., and Hans Christian Adamson, Colonel, USAF, Ret. *Hellcats of the Sea.* 1955. New York: Bantam Books, 1988. Reprint.

———. *Zoomies, Subs and Zeros. Heroic Rescues in World War II by the Submarine Lifeguard League.* Philadelphia and New York: Chilton Company, 1956.

Lockwood, Charles A., Vice Admiral, USN, Ret. *Sink 'Em All. Submarine Warfare in the Pacific.* New York: E. P. Dutton & Co., Inc., 1951.

Lundstrom, John B. *The First Team and the Guadalcanal Campaign. Naval Fighter Combat from August to November 1942.* Annapolis, MD: Naval Institute Press, 1994.

Maas, Peter. *The Terrible Hours. The Man Behind the Greatest Submarine Rescue in History.* New York: Harper Torch, 2000.

Mansfield, John G. Jr. *Cruisers for Breakfast. War Patrols of the U.S.S. Darter and U.S.S. Dace.* Tacoma, Washington: Media Center Publishing, 1997.

McCants, William R. *War Patrols of the USS Flasher.* Chapel Hill, North Carolina: Professional Press, 1994.

Mendenhall, Rear Admiral Corwin, USN (Ret.). *Submarine Diary. The Silent Stalking of Japan.* Chapel Hill, North Carolina: Algonquin Books, 1991.

Michno, Gregory F. *USS Pampanito. Killer-Angel.* Norman: University of Oklahoma Press, 2000.

Moore, Stephen L. with William J. Shinneman and Robert Gruebel. *The Buzzard Brigade: Torpedo Squadron Ten at War.* Missoula, Montana: Pictorial Histories Publishing, 1996.

Moore, Stephen L. *Spadefish: On Patrol with a Top-Scoring World War II Submarine.* Dallas: Atriad Press, 2006.

Morrison, Samuel Eliot. *History of United States Naval Operations in World War II.* Vol. VIII: *New Guinea and the Marianas. March 1944 - August 1944.* Vol. XII: *Leyte. June 1944 - January 1945.* Boston: Little, Brown and Company, 1958. Reprint, 1988.

Newpower, Anthony. *Iron Men and Tin Fish: The Race to Build a Better Torpedo During World War II.* Westport, CT: Praeger Security International, 2006.

O'Kane, Richard H., Rear Admiral, USN (Ret.). *Clear the Bridge! The War Patrols of the U.S.S. Tang.* Rand McNally & Company, 1977.

———. *Wahoo. The Patrols of America's Most Famous World War II Submarine.* Novato, CA: Presidio Press, 1987.

Ostlund, Mike. *Find 'Em, Chase 'Em, Sink 'Em. The Mysterious Loss of the WWII Submarine USS Gudgeon.* Guilford, CT: Lyons Press, 2006.

Parkin, Robert Sinclair. *Blood on the Sea. American Destroyers Lost in World War II.* New York: Sarpedon, 1995.

Roberts, Bruce and Ray Jones. *Steel Ships and Iron Men.* Chester, CT: The Globe Pequot Press, 1991.

Roscoe, Theodore. *United States Submarine Operations in World War II.* Annapolis: Naval Institute Press, 1949.

Ruhe, Capt. William J., USN (Ret.). *War in the Boats. My WWII Submarine Battles.* McLean, Virginia: Brassey's Inc., 1994.

Russell, Dale. *Hell Above, Deep Water Below.* Tillamook, Oregon: Bayocean Enterprises, 1995. A torpedoman's story of life aboard *Flying Fish.*

Sasgen Peter T. *Red Scorpion: The War Patrols of the USS Rasher.* Annapolis, Maryland: Naval Institute Press, 1995.

Schratz, Captain Paul R. USN (Ret.) *Submarine Commander. A Story of World War II and Korea.* Lexington, Kentucky: The University Press of Kentucky, 1988.

"*Seawolf.*" Booklet honoring *Seawolf* and the submariners lost during World War II, including ship's histories of *Seawolf, Cavalla,* and *Stewart.* Corpus Christi, TX: South Coast Publishing, 1983.

Smith, Steven Trent. *Wolf Pack. The American Submarine Strategy That Helped Defeat Japan.* Hoboken, New Jersey: John Wiley & Sons, Inc., 2003.

Sterling, Forest J. *Wake of the Wahoo*. Philadelphia: Chilton Company, 1960.

Stern, Robert C. *U.S. Subs in Action*. Carrollton, Texas: Squadron/Signal Publications, Inc., 1983.

Tillman, Barrett. *The True Story of the Marianas Turkey Shoot of World War II*. New York: NAL Caliber, 2005.

Trumbull, Robert. *Silversides*. Chicago: P. W. Knutson and Company, 1990. Reprint of 1945 original by Henry Holt and Company, Inc.

Tuohy, William. *The Bravest Man. The Story of Richard O'Kane & U.S. Submarines in the Pacific War*. Phoenix Mill: Sutton Publishing, 2001.

United States Submarine Veterans of World War II. A History of the Veterans of the United States Naval Submarine Fleet. Dallas: Taylor Publishing Company, 1984-1990. Four volumes.

Wheeler, Keith. *War Under the Pacific*. Alexandria, Virginia: Time-Life Books, 1980.

Whitcomb, Edgar D. *Escape from Corregidor*. Chicago: Henry Regnery Company, 1958.

Willoughby, Charles A. *The Guerilla Resistance Movement in the Philippines: 1941-1945*. New York: Vantage Press, 1972.

Young, Donald J. *First 24 Hours of War in the Pacific*. Shippensburg, PA: Burd Street Press, 1998.

Zedric, Lance Q. *Silent Warriors of World War II. The Alamo Scouts Behind Japanese Lines*. Ventura, CA: Pathfinder Publishing of California, 1995.

Index

Italics indicates inclusion in a photo or photo caption.

Adams, Glenn Tommins, 2
Adkins, James Alvin (Caddy), 28, 31, 405
Alamo Scouts, xv, 391–92, 398
Albacore (SS-218), 362
Alden, John D., 235, 244, 421
Almero, Emiliano A., 10
Anderson, James Joseph, 2
Anderson, Owen Stanley, 2
Annapolis, Md., 32
Aparri, Luzon, 51–59, 61–62
Apperson, Asa O., *28*
Arizona (BB-39), 324
Armbruster, Chester Leroy, 2
Arpia, Eugene Furtado, 2
Asa, Marion Lee, 8, 324, 336, 341, 347–48, 353–55, 359, 380, 382–83, 424
Aspro (SS-309), 309
Astarita, John Michael, 8
Augusta (CA-31), 20
"Baby" (washing machine), 32, 63–64, 91, 226
Baker, Auston, 2, 22–23, *28*, 41, 45, 65, 72, 74, 90, 92, 134, 141
Balch, Lloyd Richard, 8
Balikpapan, Borneo, 83
Ballard, Francis Arden, 8, 340, *341*, 344, 378, 382, *398*
Balao (SS-285), 330
Bannister, Jack, 8
Barboni, Albert, 2, 226
Bargenquast, Arnold Frank, 8, 323, *324*, 396
Bargenquast, Louise, 324, 396
Barnard, Harry, 394–95
Bataan (CVL-29), 357

Bateman, Roy Walker, 2, *28*, 282, 307–308, 325
Batfish (SS-310), 410
Beatley, John Wesley, 2, 247, 266
Beck, William Barndt, 8, 391–92
Becuna (SS-319), 412
Bekke, Gerald Edgar, 8, 308, 318, 321, 324, 329, 335–37, 343, 360, 377, 385–86, 393, 396
Bell, Don, 21, 44, 50
Belleau Wood, (CVL-24), 357
Bender, Frank Peter, 6, 90, *92*
Benedict, C. P., 6, 90
Bennett, John Nelson (Gunner), 2, 14, *28*, 40–41, 53, 68–69, *76*, 78, 109, 111, 142, 144, 155, 167, 174, 177, 184, 186, 220, 224–25, 236–37, 242, 245, 258–59, 267, 282, 289, 298–303, 309, 424
Bennett, Robert Jordan, 8, *399*
Bergevin, Patrick Kenneth, 8, *351*, *398*
Bilkey, John Howard, xvi, 2, 13–14, 47–48, 60, *64*, 101, 122, 141, *165*, 206, 214–15, 253, 269, 282, 307–308, 353–54, 392, *393*, 404, 424
Bingham, Allen B., xvi, 2, 372–73, 375, 377–78, 384–86, 390, 409
Bjerk (Bjerke), Keith Allen (Pinky), 2, 33, 125, 146, *183*, 219
Bjerke, Keith, xvi
Bjerke, Marion, xvi
Black Hawk (AD-9), 44
Blitch, John D., 357–58
Bluefish (SS-222), 317, 406–407
Boarfish (SS-327), 351, 404, 411
Bolon, Dallas Victor, 8

Bontier, Albert Marion (Al), xiv–xv, 8, 279, 372, *373*, 375–88, 390–96
Borneo, Australia, 15, 166–71
Borowski, A. A., 6, 70
Bowfin (SS-287), 354, 372, 375
Bramer, John Conrad Jr. (Jack), xvii, 6, 356–59, 361–62, *363–64*, 368–70
Brengelman, Henry Bernard (Hank), 2, 47, *75*, 120, 154, 159, 167, 197–98, 220, 253
Brisbane, Australia, 379, 386–88
Brooks, William C. Jr., 393–95
Brown, John Herbert (Babe), 31, 209
Bryant, Eliot Hinman, 6, 70, 79, 81
Bueno, George B., 10
Bugawisan, Mariano, 2, 22, 45, 72, 90, 101, 199, 219
Bunker Hill (CV-17), 357
Burlingame, Creed Cardwell, 218
Burr, Aaron, 48
Burrows, Albert Collins (Acey), 340–42, 348
Burruss, John Martin, 2, 224, 253, 258, 282
Bush, Pres. George Herbert Walker, 406
Bushnell (AS-15), 325, 329, 371–72, 406
Butler, David, 2, 19, *28*, *182*, 210–11, 214, 219–20, 402, 424
Butler, William R., *28*
Cabais, Eutiquio B., 7, 382–85
Cabot (CVL-28), 357
Cabrilla (SS-288), xiv
Caillier, Leonard Joseph, 2, 309, 350
Call, James Burdell, 8, 287, 335, 343, 360
Campbell, Arthur T., *28*
Cani, John, 2, 283, 323
Canopus (AS-9), 11–14, 16, 21–23, 36, 39, 49, 69–70, 87, 138
Capece, Edmund Currie, 2, *28*, 46, 121, *122*, 139–40, 164, 424
Carithers, James Purcell, 8
Carlisle, Lawrence L., 6
Carnegie, Robert Jack, 8
Carney, James V., *28*
Carp (SS-338), 377
Carpender, Arthur Schuyler, 138–40
Carpenter, John W., 6, 88, 90
Carrico, Harold, 6
Cash, Wilfred Leslie, 8
Cashero, Francis James (Jim), xvi, 2, 79, 89, 125, 136, 139, 148, 163, 177–78, 181–85, 198, 209, 211, 221–22, 268, 407–408
Casler, James Burr Jr. (Jughead), 2, 43, *48*, 49, 52, 54–55, 62, 82, 97, 102, 116, 118, 120, 129, 134–35, 139, 145–46, 149–50, 156, 164, 167, 179, 187, 196, 200, 205, 211, 214, 219, 229, 257, 262, 266, 277, 285, 407, 411, 424
Cavalla (SS-244), xiii, 407, 411–15
Cavite, Philippines, 14–16, 21, 35, 40–41, 50–51, 63, 66–69, 94
Cendonia, O. C., 10
Chapman, Edward (Chief), 8, 33–35, *183*, 208, 307, 385, 396
Chappell, Lucius Henry, 135
Chaumont, (AP-5), 74
Christie, Ralph Waldo, 59, 317
Chubbuck, Wilbur Harver, 2, 33, 219–20
Christianson, Fred, 389–90
Christmas Island, 116–29, 133, 135–36
Cincinnati (CL-6), 34
Clamagore (SS-343), 412
Clark, J. W., 6, 90
Clevenger, Raymond Eugene, 2, 309, 324
Cobia (SS-245), 412
Cod (SS-224), 28, 405, 412
Coe, James Wiggins, 161
Colorado (BB-45), 141
Combs, H. V. Jr., 6, 70
Connolly, Joseph Anthony, 6, 70, 81
Connolly, Paul, *28*
Cook, R. F., 6, 90
Coon, Norman "D", 8, 314, 387
Copas, Chester Mayo, 8, 390

Index

Corbus, John, 218
Corpus Christi, Tex., 32
Corregidor (The Rock), xvi, 17, 46, 65–72, 74, 79–81, 85–87, 92–94, 97, 138, 141, 153, 155–58, 372
Cotton, Wayne Houston, 8, 381
Cox, Robert Leon, 8, 324, 329, 334, 338–39, 346, 348, 353–55, 360, 362, 364, 368, 376, 382, 425
Coyne, Edward J., 2
Crane, Lawrence William Jr., 2, *28*, 47, 181, *182*, 211, 214, 219
Croaker (SS-246), 412
Cronk, Kenneth Eric, 2, 79
Cross, Orval Clyde, 2, *28*, 220, 403
Croxton, Warner W., 6, 90
Cucchi, Ferdinand Victor, xvi, 2, 283
Cunnally, James Patrick, 8
Curtin, Robert Roy, xvi, 2, 283, 287, 318–320, 323, 350
Cuteran, J., 7, 382, 384
Daelto, Marcuano R., 7, 381
Dagandan, 7, 382, 384
Darter (SS-227), 397
Darwin, Australia, 69, 71, 78–82, 90, 141, 373–74, 376, 389–90
Davenport, Roy Milton, 218, 309
Dealey, Samuel David, 325
Delnigro, Albert, 3, *28*
Dempsey, James Charles, 108
Denemore, Howard McLeod (Dinky), 3, 323
Deragon, William Nolin, 1, 11–14, 16, 20–21, *28*, 31, 32, 43, 48, 52, 55, 62, *75*, 77–78, 81–83, 91, 97–98, 102, 114, 116, 118, 120, 133, 136–37, 139, 149, 153–54, 156, 159–60, 167, 177, 186, 189, 195, 197, 200, 205, 210, 214, 216, 219, 226, 229, 255, 257, 282, 405, 424–25
Devilfish (SS-292), 350
Devitt, Robert Floyd, 9, 377, 390, *398*
Dishman, Otis Charles, 3, 19, *28*, 31, 45, 73–74, *75*, 107, 148, 155, 213, 220, 223, 253–54
Doane, Paul, 8, 372, *373*, 377–78, 381–82, 384
Dobbel, Carlton Edgar Jr., 3
Dolphin (SS-169), 33–34
Dombroski, Nancy V., xvi
Donche, Lewis Sigmund, xvi, 3, 283, 345, *409*
Dönitz, Karl, 59–60
Doorman, Karel, 97
Doyle, Walter Edward (Red), 16–17, 69, 80
Drum (SS-228), 412
Dvorak, Frank C., *28*
Ebert, Walter Gale, 247–48
Eckberg, Joseph Melvin (Red), 3, 16, 23, *28*, 33, 51, 53, 56–57, 63, 72, 78, 80, 82, 84, 87, 89, 91, 95, 97, 100–102, *104*, 105–106, 109–11, 114–15, 120–21, 123, 128, 130–33, 138–39, 142, 144, 146, 148–49, 151, 154, 158, 164, 181, 186–87, 192, 195, 198, 205–206, 208, 213, 215–16, 220, 424
Edmonds, Leroy, 3, 325, 347
Elliott, William L., 3
English, Robert Henry, 209–12, 218
Enslin, Carl Henry (Swede), 3, 19, *28*, 39, 68, 73, *124*, 139, 155, 213, 219, 226, 233, 282, 411, 424
Enslin, Carl H. Jr., xvi
Enterprise (CV-6), 357, 404
Entire, R. K., 6, 90
Escolar (SS-294), 397, 407–408
Eubank, William E. Jr., 6, 88, 90
Evans, Melton, *28*
Ewing, John Louis, 9, 397
Faber, Lee M., 3
Faciane, Frank L., 3
Fanshaw Bay (CVE-70), 392
Farragut, David, 214
Fenwick, David James, 3, 308
Ferguson, Joe Carlton, 3, 33, 52, 164
Ferreira, John Frutado Jr., 3, 248
Fife, James Jr., 6, 13, 16, 69–70, *71*, 72, 80, 138, 179

Fixler, Robert Nelson, 9
Flying Fish (SS-229), 309, 386
Fogle, Olin Richard, xvi, 3, 260–61, 338–39, 350, 401
Fox, John Joseph, 3, 372
Flying Fish Cove, 116–29
Flynn, Kenneth Judd, 9, *398*
Francisco, Alberto C., 10
Franco, Peter, 9
Franz, Frank Jr., 3, 12–14, 16, *28*, 49, 53, 64–65, 77, 133, 144, 151–52, 154, 156, 220
Fremantle, Australia, 134–41, 160–66, 179–83
Fria, A. B., 10
Fuller, Vernie Marion, 3, 68, 79
Fulton (AS-11), 253
Galvan, Basilio, 3, 400
Galveston, Tex., xiii, 32, 407, 412–16
Gar (SS-206), 147
Garnett, Philip Weaver, 1, 285, 288, 293, 295
Garrett, George Booth, 3, 314
Gato (SS-212), 404
Geelong, 386
George, Joan, xvi
George, Lloyd, 9, 396
German Navy, 59
Gervais, Joseph Rudolph Oliver (Rudy), 3, *28*, 33, 52, 55, 83, 105, 110, 118, 120, 129–30, 134, 147, 156, 167, 169, 187, 191, 193–94, 197, 204, 210, 253, 425
Gibson, John Stafford Jr., 3, 33–34, 57, 62, 79, 110, 128, 141, 143, 146, 167, *182*, 183, 191–92, 223, 229, 233, 239, 253
Glimsdale, Carlson Eldred, 3
Gluski, Hubert Eugene, 2, 219, 223, 255, 402
Goloyugo, V. C., 7, 382, 384
Gordon, Charles Walter, 3
Goudy, David Ernest, 2, 287, 291, 293–94, 297, 299, 308–310, 324
Grampus (SS-207), 176
Grayback, (SS-208), 142, 161

Grayling (SS-209), 347
Greenling (SS-213), 253
Grenadier (SS-210), 411
Grimes, James, 9, 263, 274, 288, 320, 328, 332
Gross, Royce Lawrence (Roy, Googy), xiv–xv, 1, 216, *217*, 218–19, 221–250, *251*, 257–79, 282–305, 307–22, 324–330, 332–49, 351–52, 404, 409, 411, 417, 419–21, 424–25
Gudgeon (SS-211), 377
Guitarro (SS-363), 282
Gunnel (SS-253), 269
Haddock (SS-231), 260, 309
Hadley, William Thomas (Doc), 9, 254, 311, 347, 362, 368–69, 425
Haggard, Hugh A. V., 108
Haines, John Meade, 387
Hammill, Charles H., 10
Hamner, Jordan A., 379–80
Handley, John Jr., 3
Hannibal, 352
Hanson, Henry Howard Jr. (Big Swede), 3, 34, 38, *39*, 40, 68, 90, 163–64, 177, 211, 214, 225, 259, 282, *410*
Hanson, Marilyn, 214
Hanson, Robert Norman (Little Swede), xvi, 3, 24, 28, 33, *39*, 40–41, 45, 54, 68, 74, 80, 87, 89–90, 96, 115–16, 122, 135, 140, 164, 204, 211, 214–15, 251–52, 282, *410*
Harlow, William Robert, xvi, 3, *310*, 325, 335, 350, 401
Harrington, Earl Lee, 3
Harris, Clifford E., *28*
Harris, John Gordon, 9
Hart, Thomas Charles, 12–14, 16–17, 35, 37–38, 51, 68–69, 87, 89, 151
Heiden, Ray F., 362
Herbig, Robert P., 10
Hershey, Albert Earl, 3, 19, *28*, 46, *76*, 82, 114, 148, 223, 246, 253–54
Hess, John Bordon, 1, 352–53, 360–61, *363*, 364, 369, 372, 406, 425

Index

Hickerson, Peter Thomas, 3
Hickman, Glenn K., *28*
Hicks, James Clarence, 3
Hill, Earl G. Jr., xvi
Hinson, Edward Elie, 3, 164–65, 169, 228–30, 261, 268, 273
Hoevet, Dean C., 6, 90
Holden, Justinian, 18
Holden, Richard, 1, 17–20, 23, *28*, 31, 40, 43, 46, 49, 52–55, 60, 68, 72–74, *75*, 77, 81–82, 85, 95, 99, 101–102, 106–108, 110, 120–21, 125, 130–32, 139, 142–44, 146, 150, 155–60, *165*, 166, 168, 176–77, 179, 181–82, 187, 195, 197–98, 203, 205, 208, 214, 219, 406, *410*, 424
Holden, Timothy, xvi
Holeman, Victor Rollo, 6
Holland (AS-3), 13–14, 16–17, 19, 69, 78–80, 140–41, 179, 181
Hornet (CV-12), 357
Houston, Charles, xvii
Howard, Alfred Herman, 9
Huff, Roy Edward, 9
Hughes, Jimmie Ross, 3, 211
Hunter's Point, CA, 349–52
Hutchison, Robert Lester, 3, *28*, 220
Ibea, Artemio I., 10
Indianapolis (CA-35), 49, 352
Irish, Donald W., 6, 70
Irvin, James Willard, 3, 354
Isabel (PY-10), 160
Jacobs, Tyrrell Dwight, 59
Jancik, Calvin George, 3, 381
Japanese ships: *Akebono Maru*, 118; *Aki Maru*, 317–18; *Amatsukaze*, 118; *Amatsu Maru*, 317–20, 322, 420; *Arashio*, 99, 105; *Asashio*, 99, 105; *Asuka Maru*, 327–28, 330, 420; *Awa Maru No. 6*, 317; *Banshu Maru*, 289, 293; *Banshu Maru No. 5*, 233, 235, 249, 419; *Chiyoda*, 357; *Denmark Maru*, 337, 342–43, 345; *Durban Maru*, 289, 295, 420; *Erie Maru*, 337; *Fumitsuki*, 118; *Fusei Maru*, 289, 295–99, 420; *Genyo Maru*, 357; *Getsuyo Maru*, 328, 330, 332–34, 336, 420; *Gifu Maru*, 188–91, 419; *Hachigen Maru*, 173–74, 419; *Haruna*, 357; *Harushima Maru*, 289; *Hatsukaze*, 118; *Hatsushimo*, 99; *Hayasui*, 357; *Hiyo*, 357; *Hokoku Maru*, 330; *Hokuroku Maru*, 317–18; *Ikutagawa Maru 8*, 330; *Junyo*, 204–205, 357; *Kaifuku Maru*, 316–17, 420; *Kaga Maru*, 317; *Kaihei Maru*, 227–32, 419; *Karukaya*, 328–30, 333; *Keiko Maru*, 201–203, 419; *Kimishima Maru*, 118; *Kinrei Maru*, 330; *Kokko Maru*, 289–90, 420; *Kumagawa Maru*, 118; *Kyokuyo Maru*, 238; *Kyokuei Maru*, 317; *Maya*, 357; *Michishio*, 99, 105; *Minazuki*, 118; *Minegumo*, 118; *Minesweeper No. 18*, 337, 342; *Minesweeper No. 27*, 329–30, 333; *Nagara*, 99, 108, 118; *Nagatsuki*, 118; *Naka*, 118–20, 129–30, 133, 135–36, 418; *Nampo Maru*, 156, 417; *Narita Maru*, 337; *Natori*, 118–19, 127–28, 133, 130; *Natsugumo*, 118; *Nenohi*, 99; *Nikki Maru*, 330; *Nisshin Maru No. 2*, 238–39, 241, 243–44, 250, 417, 421; *Nuwajima*, 337, 342; *Oshio*, 99, 105; *Patrol Boat No. 39* (*Tade*), 240–42, 244, 419; *Reiyo Maru*, 270–72; *RO-41*, 392, 395; *Rokko Maru*, 330; *Ryuho*, 357; *Sagami Maru*, 99, 103–105, 191–95, 197, 200, 210, 218, 419; *Sagi*, 289–91, 306; *Sanyo Maru*, 55–58, 66, 420; *Sasago Maru*, 99, 103–105; *Satsuki*, 118; *Seiyo Maru*, 357; *Shojin Maru*, 270–72, 417; *Shokaku*, xiv, 407, 411, 414; *Shoto Maru*, 289–90, 420; *Showa Maru*, 177, 419; *Suzukaze*, 135; *Suzuya*, 99; *Tarushima Maru*, 344–48, 420–21; *Tama Maru*, 337, 342; *Tsushima*, 317; *Wakaba*, 99; *Wuhu Maru*, 313, 420; *Yahiko Maru*, 328–30, 332–34, 420; *Yamatsuru Maru*, 337–40, 420;

Zuikaku, 357
Jenkins, Maurice (Red), 3, 20, *28*, 114, 124, 155, 157, 184, 189, 199, 220
Jennings, Harry Augustus Jr., 3
Jobe, Clinton, 3, *28*, 46, 122, 140–41, 181, 211
Jobe, Jesse, 3, 170
John, Clary Leonard, 1, 255, 295–96, 309, 311, 314, 317, 321, 324, 328, 333, 336, 348, 353–55, 368, 372, 424
Johnson, Charles Alfred (Johnny), 4, 40, 107, 282, 404
Johnson, James Everard, 9, 287, 360, *399*
Johnson, Ruth, xvi
Jones, Scott Weldon, 4
Juneau, (CL-52), 282
Jurnic, Michael, 9, 381, *399*
Kasloski, Robert John, 4
Kellerer, John Randolph, 4
Kelly, D. M., 6, 90
Kelly's Pool Hall, 34, 45, 51–52, 63, 91, 108, 117, 126, 134, 158, 213, 226, 286, 297
Kennedy, Denver Guy Jr., 4
Kennelly, John J., 2, 219, 264, 272, 281, 285
Kenney, Jack Edward, 9, 383
Kennison (DD-138), 20
Kibbons, Clarence Vernon, 4, *28*, 33, 128, 183, 253, 267, 270, 282, 402
Kimmel, Husband E., 13
Kincaid, Wilford Charles, 4
Kinsella, William T., *28*, 31, 35–36, 406
Kisver, Norman, xvi, 4, 49, 54, 74, 123, 156, 220
Klestadt, Albert, 379
Kobesky, Cora Enslin, xvi
Koehler, Robert Henry, 4, 49
Kopp, Howell Stewart, 10, 391–92, 396, *398*
Klimes, Joseph Jr., 4
Kraght, Henry G., 4, 94

Kramer, Karl J., xvi
Krempa, Charles Stanley, 9
Kropp, Dale Henry, 4, 340, 342–43
Krueger, Walter, 391
Kuehn, Alfred Eric, 9, 360, 364, 382, *399*
Kyuji, Kubo, 99, 103
Lamberson, Arthur Earl, 4, *28*, 33, 76, 146, 183, 247–48, 253
Langford, Robert (Squeaky), 4, *28*, 33, 45, 54–57, 62, 111, 124, 140
Langley (AV-3), 14, 43–44, 46
Larson, James Edward (Swede), 4, 68, 94, 177, 259
Laumann, George Arthur, 4
Lawson, Chester Gelean, 9
Lear, Earl William, 4
Leeman, Merlin Hibbard Jr., 9
Leffingwell, George Darrell (Lefty), xvi, 2, *28*, 55–56, 76, 118, 141, 146, 173, 220, 246, 271, 282, 307–308, 313, 324, 327, 340, 348–49, 353, 425
Leffingwell, Susan, xvi
Lents, Robert W., xvi, 34, 36, 402
Lewis, John Miles, 4
Lexington (CV-16), 357
Life magazine, xvi, 218, 235, 282, 397
Likert, Gilbert Roland, 9
Ling (SS-297), 412
Lionfish (SS-298), 412
Lipham, Carleton R., *28*
Litchfield (DD-336), 34, 209
Llanes, A., 6, 70
Loaiza, Frank (Doc), 4, 64, 70, *75*, 82, 90, 112, 114, 131, 152, 179, 199, 220, 253
Lober, Peter Nicholas, xvi, 2, 15, 20, *28*, 67, 139, 141, 147, 184, 188, 199, 211–12, 214, 219–20, 407
Lockwood, Charles Andrews, *124*, 161–62, 165–66, 168, 179–80, 187, 203, 209, 218, 249–50, 259, 272, 279, 288, 305–306, 317, 324, 336–37, 347–48, 354, 359, 368–69
Louisville (CA-28), 12

Index 453

Luzon, 21–22, 43–44, 54, 72, 86, 154
Lynch, Richard Barr (Ozzie), xiv–xv, 1, 351, *352*, 353–56, 359–62, *363*, 365–76, 404, 424
MacArthur, Douglas, 68–69, 87–88, 391
MacDowell, William S., *28*, 402
Magnuson, William H., 4
Majuro, 368–72
Malang, Java, 95–97
Maley, Paul Leroy, 4, 33–34, 50, 53, 56, 72, *76*, 78, 84, 100, 105–106, 109–10, 120, 123, 130–31, 144, 149, 155, 158, 164–65, 168–69, 171, 186, 192, 195, 228–30, 233, 243, 261, 268, 273, 282
Malley, Thomas, 375
Mallory, William Cleaver, 4, 22–23, 45, 65, 72, 90, 134, 141, 158, 173, 199, 223, 226, 282–83
Mallough, Bobbie, xvi
Mallough, Kenneth Geiser (Casey), 4, *94*, 139, 282
Malone, Dallas Leroy, 9, 224, 236, 268, 286, 374–75, 396, *399*, 401
Mangigian, Paul, 6
Manila, xiv, 11–19, 21–22, 35–36, 41, 43–44, 46, 50–51, 59, 63, 65–71, 78–79, 86–89, 141, 149, 151–55, 161, 219, 385
Manus, 390–92
Maples, John Alonzo Jr., 4
Mar, Delbert, xvi, 4, 220, 227, 245–46, 252, 289, 327, 348–49, 351, 409, 425
Mare Island Navy Yard, 20, 27, 212–19, 407
Marianas Turkey Shoot, 355
Marquina, L., 7, 382, 384
Marrocco, William A., 7, 90
Marston, George Franklin, 9
Mathews, Curtis Arden, 4
Maus, Charles Robert, 9
McAfee, James B., 7, 88–91, *92*,
McCoy, Claude, xvii

McCoy, Harold, xvii
McCoy, Mike, xvii
McCoy, Ralph, 65
McCoy, Walter Glen (Bud), xvii, 9, 65, 74, *76*, 77, 124, 210, 222, *223*, 224, *225*, 226–27, 232–33, 235–37, 241, 243–47, 251–52, 255, 257–60, 265–66, 270–71, 274, 276–79, 283, 286–88, 291, 297, 299, 302, 304, 307, 385, 390, 396, 425
McGee, William Timothy, 4
McGregor, Louis Darby (Sandy), 87
McKinney, Eugene Bradley, 113
McMullen, Clarence R., 4
McNabb, John Joseph, 4
McTavish, John Francis, 4
Mehner, Waldemar Richard, 4
Mercer, James, 1, 23, 52, 54–55, 68, 74, 82, 84–85, 100, 102, 108, 115, 118, 120, 126, 144–45, 148, 150–53, 156, 159–60, 166–67, 173, 176, 183–84, 188–89, 192–95, 200–204, *206*, 207, 210, 214, 218–19, 226–27, 229, 237, 240–41, 244, 257, 260–62, 265, 270, 278, 282, 288–89, 292–93, 295, 309, 406, 424–25
Metz, William T., 4, 19–20, *28*, 32, 34, 108, 116, 124, 140
Michael, Forrest Samuel, 9
Mickey, Dallas Colby, 401
Mickey, Rex LaVere, xvi, 4, 224, 259, 271–72, 274, *275*, 276, 281–82, 286, 302, 304, 312, 350, 361, 369, 371, 373–75, 378, 382, 389–90, 401, *409*, 411
Midway, 225–26, 246–47, 251–55, 257, 278–80, 286–87, 303–305, 311, 321, 325–26, 347, 353–54
Milas, Edward Lawrence, xvi, 4, 141, 163, *165*, 170, 408
Miller, Allen Thomas, 4
Miller, George F., 10
Miller, Ralph Van Dorn, 8, 372, *400*
Miller, Richard Lawrence, 9, *400*
Miller, Robert Thomas, 9, 267, *399*

Miller, Vernon T., 340, 342
Millican, William, 407–408
Mills, Lonnie Tolbert Jr., 4
Mills, Margarite, xvi
Mills, Wilson (Red), 9, 220, 227, 236, 259, 281, 300, 409
Milwaukee (CL-5), 24
Mitchell, Harold Edward, 9
Mocarsky, Alexander Peter (Pop), 4, *28*, 46–47, 73, *76*, 205–207, 220
Moffett (DD-362), 224
Monterey (CV-26), 357
Morris, Edward Lyle, 9, 289
Morris, Joseph Albert, 9, *399*
Morton, Dudley Walker (Mush), 225, 281, 325
Moye, George, 4
Mumma, Morton Claire Jr., 6, 70–71
Munger, Alva Arnold, 4
Murray, Stuart S. (Sunshine), 13, 16–17, 69, 138, 172
Muskallunge (SS-262), 282
Myers, Walter Donald, 4
Narwhal (SS-167), 392
Nautilus (SS-168), 351–52, 354, 389–90
Nazay, George Gilbert, 9
Naze, Donald Joseph, 9, 310, 313, 318, 328, 350, 372, *398*
Neches (AO-5), 48
Needham, George Melvin, 9, 339
Neil, John Spence Jr., 4, 33, 267, 293
Nivison, Clinton LeRoy, 9
New London, Ct., 23, 32, 34, 68, 141, 218, 220, 224, 256, 282, 285, 310, 352, 372, 404–407, 411
Newman, Kenneth Eugene, 4, 300
Newport, R.I., 32
Newton, Joseph Lincoln, 5
New York (BB-34), 23
Nimitz, Chester William, *217*, 281
Noble, Jack Howard, 5
O'Brien, Edward Francis Jr., 8, 390, *398*
O'Brien, Roger Allen, 5, 288, 292–93, 323, 349

Oklahoma (BB-37), 65
Ostrander, Alfred George, 5, 253, 267, 293
Ottaway, Samuel Adrain, 5, 287, 297, 318
Otus (AS-20), 14, 69, 79, 81, 160, 164, 166, 179
Page, Albert Francis, 9, *399*
Page, Leonard Alton, 9
Palau, xv, 44, 337, 354–56, 360, 362, 364–65, 369–70
Palazolo, Becky, xvi
Pampanito (SS-383), 410
Parden, Lee Bob, 5, *28*, 47, 86–87, 133, 177, 220, 411
Pargo (SS-264), 291, 299
Patterson, Donald Richard, 5
Patterson, G. W., 256
Payne, Earl D., 6, 90
PC-485, 353–54
PC-596, 223
Pearl Harbor, 13–16, 20, 32–35, 37, 44–45, 57, 59, 66, 182, 195, 205–212, 218, 221–23, 251, 279, 283–86, 305–310, 321–25, 329, 340, 347–49, 353, 375, 396, 414
Pease, Harlan Jr., 7, 88–90
Pecos (AO-6), 43
Pennsylvania (BB-38), 20, 324
Peralta, George E., 10
Perch (SS-176), 402
Percifield, Willis Merritt, 13, 69
Permit (SS-178), 94
Perry, Joseph Jr., *28*, 402
Perry, Victor Irving, 5
Perth, Australia, 137–41, 163–65, 181–82
Peterson, Elmer Norman, 9
Pew, L. A., 6, 90
Philippine Sea, battle of, xv–xvi, 356–63
Pickerel (SS-177), 403
Pierce, Arthur Calhoun, 5, 361, *364*, 366
Pillsbury (DD-227), 47
Pipefish (SS-388, 255, 282, 405–406

Index

Plaice, 353–54
Plunger (SS-179), 33, 282–83, 307
Poe, James Cavanaugh Jr., 5
Pogy (SS-266), 310
Politylo, Wasil, 9, 350, *399*
Pollack (SS-180), 255
Pollock, Thomas F., 6, 89–90
Pomfret (SS-391), 372, 406
Pompano (SS-181), 249, 286
Poole, Mason, 5, *28*, 47, 122, 269, 308, 354
Portland (CA-33), 283–84
Portsmouth, NH, 13–14, 19, 24–25, 27–28, 31–32, 256
Prien, Günther, 59–60
Pugose, Emil L., 10
Pulsifer, Charles Lee, 5
Quarto, Michael J., 324
Quincy (CA-39), 352
R-11, 36, 47
R-14, 219, 406
Rajotte, Lucien Thomas, xvi, 5, 15, 19, *28*, 73, *76*, 140, 220, 223, 225–26, 246, 254, 258, 262, 277, 300, 309
Ramage, Lawson Patterson (Red), 238, 241, 244
Ramos, O. B., 10
Randazzo, Salvatore (Sandy), 5, 19, *28*, 220, 223, 253
Rasher (SS-269), 211, 407
Ray (SS-271), 406
Razorback (SS-394), 372, 412
Reiland, William Frederick Jr., 8, 33, 128, 143, 147–48, 154, *183*, 198, 210, 215, 220, 223, 227–28, 238, 253, 268, 296, 316, 320, 327, 329, 350, 361, 374–75, 385, 390, 396, 424–25
Requin (SS-481), 412
Rhoads, Guy Benjamin, 9
Riggle, Mahlon Richard, 10
Rimando, Juan F., 10
Risser, Ruby Archer, 405
Risser, Robert Dunlap, 1, 255, *256*, 257, 264, 273, 283, 285, 288, 295, 304–306, 309, 386, 405
Roberts, Lawrence Jr., 5, 268
Robinson, Claude, 285
Robinson, Dougald "G" (Duke), 2, *284*, 285–86, 289–91, 293, 295–96, 298, 300, 302–303, 306, 308–309, 312–14, 324, 326–27, 330, 334, 345–48, 353, 406–407, 425
Robinson, Mary Katherine, 285
Rocaya, Saturnino, 10, 347
Rodriguez, Ireneo R., 10
Rogers, Benjamin Franklin, 10, 229, 231, 247–48, 261–63, 268–69, 287, 294, 298, 300, 303, 311–12, 314, 329–30, 334, 338, 341, 344, 348, 356, 360, *367*, 378, 385, 387, 396, 424–25
Romito, Frank Deminic, 5, 288, 290, 318
Rosete, Tomas, 10, 90, 347
Roszel, Edward K., *28*, *410*
Rowell, Richard M. (DE-403), xv, 392–95
Royal Hawaiian Hotel, 281–82, 323
Ruiz, Ruperto R., 10
Runner (SS-275), 249, 255, 307, 402, 411
Russ, Lamar George, 5
Russell, Benjamin Earl, 5
S-14, 49
S-16, 24
S-21, 28, 256, 405
S-29, 19
S-36, 19, 40, 94, 216–17, 352
S-37, 108
S-38, 72
S-39, 44
S-40, 19, 25
S-42, 12, 47
S-45, 47
S-46, 14
Sadler, John Colby, 10, 224, 268, 374–75, 389, 396, 401
Sailfish (SS-192), 27, 70–71, 167, 279
Saint, James William, 10, *400*
Salmon (SS-182), 16, 109, 113

Salt Lake City (CA-25), 18
Sandridge, Lloyd W., 5, 19, *28*, 73, 75, 140
Sands (DD-243), 12
San Jacinto (CVL-30), 357
Sargo (SS-188), 16, 27, 27, 59, 164, 179, 185, 214
Saury (SS-189), 16, 27, 220, 256, 311
Scamp (SS-277), 247–48
Schultz, Paul Joseph, 5, 283
Scott, Theron, 5
Sculpin (SS-191), 27, 43–44, 46, 50, 135, 352, 372, 402
Seadragon (SS-194), 13, 15, 25, 27, 68, 141, 167, 185
Sea Fox (SS-402), 405–406, 409
Seahorse (SS-304), 408
Seal (SS-183), 16, 140, 206, 208, 284
Sealion (SS-195), 13, 15, 25, 27, 51, 68, 94
Sealion II (SS-315), 94
Searaven (SS-196), 11, 13, 25, 27, 49, 140, 167, 205, 352
Seawolf (H-1) 25–27
Seawolf (SS-197): launching of, 24, 26; commissioning of, 24; details of, 27; equator crossing, 73–77, 374–75; gun actions of, 236–37, 242, 245–46, 258–59, 298–304, 340–44; rescues aviators, 357–64; commandos aboard, 378–85, 390–92; first patrol, 43–66; second patrol of, 71–78; third war patrol, 79–93, fourth war patrol, 96–136; fifth war patrol, 141–62; sixth war patrol, 166–80; seventh war patrol, 183–212; eighth war patrol, 224–250; ninth war patrol, 255–80; tenth war patrol, 286–306; eleventh war patrol, 311–22; twelfth war patrol, 325–48; thirteenth war patrol, 353–370; fourteenth war patrol, 378–88; runs aground, 101, 274; ship's newspaper, 287, 312, 324; final patrol and loss of, xv, 390–98; sinking records of, xv–xvi, 403–404, 417–21; Navy Unit Commendations, 423
Seawolf (SSN-575), 403, 405
Seawolf (SSN-21), 403
Seawolf Commission, xiii–xiv
Seawolf Park, xiii–xv, xvii, 407, 411–15
Seward, N. E., 6, 90
Shark (SS-174), 12, 35, 48, 69–70
Shark II (SS-314), 397, 402
Shedd, Morris H., 7, 90
Shelton (DE-407), 392–95
Sherman, John J., 5
Shoji, Nishimura, 117–18, 129
Short, William Edward, 5
Silversides (SS-236), 218, 412
Skate (SS-305), 375, 404
Skipjack (SS-184), 16, 44, 141, 161, 181, 309
Smith, William Lee, 1, 309, 313, 316, 318, 324–25, 327, 337–38, 341, 343–44, 347–48, 353, 369, 373–74, 376–77, 424–25
Snook (SS-279), 291
Snyder, James Earl (Red), 5, 22–23, 37–38, 65, 107, 151
Sofu Gan (Lot's Wife), 300–303
Souza, Edwin Enos, 5, 14, 21, 32, 52–53, 62, 73, 79, 85, 89, 112, 121, 138, 162, 164, 166, 177, 184, 210–12, 216, 219, 425
Spearfish (SS-190), 16, 27, 113, 372
Sperry (AS-12), 47, 279
Squalus (SS-192), 13–14, 24, 31, 70, 256, 402
Stafford, Robert F., 7, 90
Standley, William H., 59
Stark, Robert Merrill, 5
Steinbecker, Gerald Andrew, 10, *400*
Stephan, Edward Clark, 1, *142*, 143, 145, 155–56, 161
Stephenson, Glenwood G., 7, 88–91, 92
Steward, Harry Carlyle, 5
Stewart (DD-224), 97, 413–14
Stewart (DE-238), xiv, 413–14

St. Lo (*Midway*, CVE-63), 392–93
Strausser, Clarence Elias, 10
Street, John Edwin, 5, 19, *28*, 37–38, 51, 116, 213, 220
Strong, Joseph Hale, xvi, 5, 220, 253
Sturgeon (SS-187), 337
Suarez, Alejandro, 378–81
Sullivan brothers, 282
Sullivan, John Edward (Sully), 2, 22, *28*, 32, 34, 56, *64*, 65, 73, 75, 79, 81, 94–95, 114, 134, 138, 140, 164, 166, 177, 211, 214, 219, 282, *410*
Surabaya, Java, 69, 87, 89, 91–97, 112, 136
Swenson, H. R., 6, 90
Swordfish (SS-193), 27, 59, 67, 69, 168, 352–53
Syverson, Douglas Neil (Red), 1, 20–21, 41, 52, 55–56, 60, 62, 68–69, 71, 74, *75*, 82–83, 86–87, 93, 100, 103, 106, 114, 118, 124, 126, 133, 142, 144–46, 148, 153–54, 159, 164–66, 168–70, 173, 178, 183–85, 188, 196, 200, 210, 214, 219, 229, 237, 242, 257, 260, 266, 268, 273–74, 277, 287–90, 292, 294, 307, 309–10, 313, 318, 324, 327, 346–47, 351–52, 389, 405–406, 424–25
Syverson, Hope, xvi
Szendrey, Edward John, 8, 324, 327, 330, 334, 346–47, 353, 381, 385, *398*
Tamayo, Brigido, 5, 22, 45, 72, 90, 101, 199, 282
Tang (SS-306), 397
Tarpon (SS-175), 44, 78
Tawi Tawi Island, 15, 378–81
Taylor, Wendell Broadus, 5
Tench (SS-417), 282, 350, 401
Teo, Konglan, 7, 381
Texas (BB-35), 412
Thomas, Marion K., 6
Thomson, Henry Hanford (Hank), xvi, 5, 15, 23, 34–35, 38, 47–49, 51–52, 55, 61, 64–65, 74, 98, 106, 112, 115, 120–22, 129, 134–35, 138, 140, 147, 151, 154, 167, 176, 191, 194, 198, 227–28, 255, 259, 267, 274, 281–83, 288, 307, 318, *319*, 327–28, 340, 343–46, 348–51, 402, 408–409, 425
Thorp, Claude, 379
Thresher (SS-200), 406
Times, Charles Henry, 5
Tinosa (SS-283), 260–61
Tokyo Rose, 51, 67, 82, 108, 158, 200
Torpedo failures, xiv, 57–60, 66, 110, 121, 135, 145–48, 155, 161–62, 169, 171–73, 179–80, 184, 192–96, 198, 213, 221, 228, 231, 239, 247–49, 264, 268, 295–97, 305–306, 320–22
Torsk (SS-423), 412
Tranquilly, Robert, 5, 233
Travers, Ray, 393–94
Tremblay, George S., *28*
Trinity (AO-13), 43
Trout (SS-202), 238, 244, 249, 421
Truant, HMS, 108
Truman, Pres. Harry, 407
Tullibee (SS-284), 402
Tuna (SS-203), 283
Tunny (SS-282), 248–49
Ultra dispatch, 109, 148, 188, 227, 259, 317, 336–37
Underhill, William Hopkins, 10, *400*
Utter, Harmon T., 6, 89–90
Van Andel, John, 8, 353, 366, *400*
Vance, Reginald F. C., 7, 88, 90–91
Vanzant, Ralston B., 6, 70, 81
Vergara, T. E., 7, 382, 384
Vincennes (CA-44), 18
Vineyard, John Leslie, 5
Vitello, Donato, 5
Voge, Richard G., 27, 71, 202, 317, 336, 340
Wahoo (SS-238), 225, 281
Wainwright, Jonathan Mayhew (Skinny), 18
Wall, Vernon Palmer (Duke), 10, 253, *254*, 259, 287, 298, 300, 336, 340,

351

Walraven, Albert Tavel, 6, 356–59, 361–62, *363–64*, 368–70
Warder, Frederick Burdett (Fearless Freddie), xiv, 1, 13–14, 16–17, 20, 24–25, 27, *28*, 31–32, 35–36, *37*, 38–41, 43–47, *48*, 50–60, 62–75, 77–84, 93–132, 135–39, 141–42, 144–62, 164–80, 181–211, 213–14, 216–18, 227, 236, 247, 250, 257, 397, 403–405, 407, 417, 419–21, 424–25
Warder, Mary, 27, 40, 62, 66
Warrego, 82
Warren, Thomas Wilson, 10, 244, 267
Wasp, xvi, 356–59, 362, 369–70, 406
Watts, Evan E., *28*
Weade, Claiborne Hoyt, 5, 94, 402
Wenchow, 173
Whale (SS-239), 340–48, 421
Whitman, Pauline, xvi
Whitman, William Alexander (Whit), 2, 68–69, 74, 82, 94, 154, 157, 165, 167–68, 176, 178, 214, 219, 223, 229, 231, 236, 238, 246, 249, *252*, 257–58, 263–64, 268–71, 273, 278, 286, 288–89, 295, 298, 300, 312, 318, 324, 406, 424
Wiegenstein, Amy, xvi
Wiegenstein, Gene, 284
Wiegenstein, Michael Paul, 10, 283, *284*, *351*, 390
Wilcoxen, Ray H., *28*
Wilkerson, Robert Goley, 5
Wilkes, John, 13, 16–17, 28, 35, 43, 57–59, 66, 68–69, 71, 79, 81, 95–96, 116, 161
Wilkinson, Gerald, 7, 88, 90–91
Williamson, L. H., 6, 90
Wilton, Henry Peter, 5, 141, 173, 199
Wise, Braynard L., 10
Woodard, Charles Clark (Duke), xvi, 5, 6, 70, 74, 164–66, 409
Wright, Harold Lankford (Gus), 5, 22, 45, 65, 72, 90, 93, 101, 117, 134, 141, 173, 199, 223, 226, 253
Wyatt, David Bernard, 10
Yersick, Paul Anthony, 5
Yorktown (CV-10), 357
Young, Frank, 378–81
Young, Robert Porterfield, 10
Zimmerman, Paul, xvi, 6, 16, 20–22, 31, 34, 39–40, 45, 53, 57, 61, 74, 94–96, 123, 136, 151–52, 174, 209, 215, 219, 237, 257, 276, 282, 401, 403, 408
Zirkel, Frederick Andrew, 6, 34, 82–83, 107, 114, 148, 164
Zuel, Edward Andrew, 10

If you enjoyed *War of the Wolf*, check out this other great World War II submarine book by Stephen L. Moore from Atriad Press!

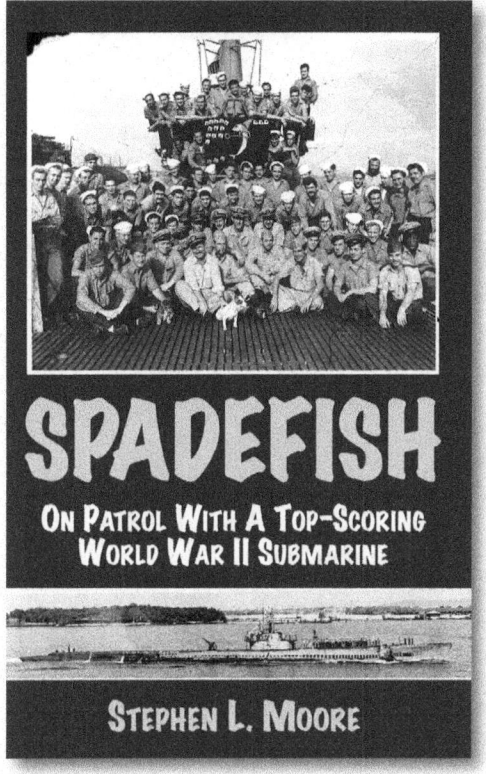

480 pp. Includes crew rosters, sinking records, maps and numerous photos. $24.95

Relive the war patrols of one of America's top submarines of World War II, as recounted by dozens of her officers and enlisted men! *Spadefish* destroyed six Japanese ships on her maiden war patrol and at least four vessels on each succeeding patrol, including the aircraft carrier *Jinyo*.

"The breadth and depth of Moore's spadework for *Spadefish* is evident. The most remarkable features of the *Spadefish* book are the battle narratives and the lavish detail of life aboard the boat.."—*America In WWII Magazine,* October 2007

"This is a GREAT submarine book! Steve Moore has written a very detailed but also very readable book about a top submarine that has not, until now, received the credit it was due. Liberally sprinkled with interesting photographs, it is a fine tribute to one of the leading submarines of WWII."—James F. Calvert, Vice Admiral, USN (Ret.), author of *Silent Running*

To order a copy of *Spadefish,* visit www.atriad.press.com or call 972-671-0002

Written, published and
printed by Texans

Stephen L. Moore, author
Atriad Press, LLC, and
America's Press, League City, Texas
www.americas-press.com

www.ingramcontent.com/pod-product-compliance
Lightning Source LLC
Chambersburg PA
CBHW060447170426
43199CB00011B/1128